SUBUNITS IN BIOLOGICAL SYSTEMS

PART A

BIOLOGICAL MACROMOLECULES

A Series of Monographs

SERIES EDITORS

SERGE N. TIMASHEFF and GERALD D. FASMAN

Graduate Department of Biochemistry
Brandeis University
Waltham, Massachusetts

SUBUNITS IN BIOLOGICAL SYSTEMS

PART A

Edited by

Serge N. Timasheff and Gerald D. Fasman

Graduate Department of Biochemistry
Brandeis University
Waltham, Massachusetts

1971

MARCEL DEKKER, Inc., New York

MARCEL DEKKER, INC.
95 *Madison Avenue, New York, New York* 10016

LIBRARY OF CONGRESS CATALOG CARD NUMBER 70–145883
ISBN 0–8247–1187–4

PRINTED IN THE UNITED STATES OF AMERICA

PREFACE

From its inception, the philosophy of the *Biological Macromolecules* series has been to bring together principles and approaches from various areas of biological research. In previous volumes, unifying themes have been the relation between structural details and stability, fine structure, and the electrostatic properties of large molecules. The theme of this and several volumes to follow is *subunits*.

The choice of subunits as a timely topic to examine in detail stems from the rapidly growing realization that both the structural stability, dynamics and biological function of living organisms or their component parts are to a great extent the results of the assembly and fine interplay of sub-structures, each with its peculiar properties and function. For the purpose of these volumes, we have defined the term "subunit" in its broadest sense. We regard as subunits discrete structural entities within a biological system which have a structural stability essentially independent of that of their neighbors. Within this definition, subunits may be covalently bonded to each other, or they may be physically separable entities. In this manner, the lobes of a t-RNA cloverleaf may be regarded as individual subunits, as may be the component protein molecules of an allosteric enzyme assembly, or the various components of ribosomes or membranes.

It is our hope to present a rather comprehensive coverage of the topic of subunits. The plan for these volumes contains general chapters, describing the principles, both geometric and thermodynamic, of subunit assembly, as well as pointing out some of the principal approaches that may be used in the investigation of such molecular assemblies. The logical evolution then is to specific systems of increasingly higher complexity of assembly. For example, chapters on the assembly of nucleic acids and polynucleotides lead to descriptions of ribosomes and polyribosomes. Similarly, the structure of muscle is discussed in progressing order of assembly. A large section will be devoted to subunit enzyme and other interacting protein systems. Finally, membranes, both of animal and bacterial cells, will be discussed as examples of biologically important, highly complicated assemblies. This approach is exemplified in the contents of the present volume. Three chapters are devoted to simple subunit proteins (hemocyanins, hemerythrins, phycocyanins). These are followed by a more complicated assembly, that of the TMV protein. Three chapters (myosin, actin, muscle fibril) develop the assembly of muscle proteins. Finally, the widely occurring complicated microtubule systems are described. The approaches chosen by the individual authors exemplify the many tools available to us at present, and the finesse with which they have treated their individual systems helps to blend their contributions into a cohesive volume.

Our original wish was to present this material in what we regard as a logical order. However, editors' wishes are not always fulfilled, and it is well known that not all contributors have identical energies of activation. We have decided that freshness of material is more important than logical order of presentation. As a result, each volume in this sub-series will be published as the proper number of contributions become available. We hope that, when all the material is in print, the readers will find a logical sequence that can be assigned to the individual chapters, the several parts giving a well-rounded critical survey of the field.

Waltham, Massachusetts   SERGE N. TIMASHEFF
                                                                          GERALD D. FASMAN

CONTRIBUTORS TO THIS VOLUME

Donald S. Berns, *New York State Department of Health and Albany Medical College, Albany, New York*

Michiki Kasai, *Nagoya University, Nagoya, Japan*

Irving M. Klotz, *Northwestern University, Evanston, Illinois*

Max A. Lauffer, *University of Pittsburgh, Pittsburgh, Pennsylvania*

Susan Lowey, *Harvard Medical School, Boston, Massachusetts*

Fumio Oosawa, *Nagoya University, Nagoya, Japan*

Frank Pepe, *University of Pennsylvania, Philadelphia, Pennsylvania*

R. E. Stephens, *Brandeis University, Waltham, Massachusetts*

E. F. J. Van Brugger, *Laboratorium Voor Structuurchemie Rijksuniversitiet te Groningen, Groningen, The Netherlands*

K. E. Van Holde, *Oregon State University, Corvallis, Oregon*

CONTENTS

SUBUNITS IN BIOLOGICAL SYSTEMS
PART A

CHAPTER I

THE HEMOCYANINS

K. E. van Holde

DEPARTMENT OF BIOCHEMISTRY AND BIOPHYSICS
OREGON STATE UNIVERSITY
CORVALLIS, OREGON

AND

E. F. J. van Bruggen

LABORATORIUM VOOR STRUCTUURCHEMIE
RIJKSUNIVERSITEIT TE GRONINGEN
GRONINGEN, THE NETHERLANDS

1

I. INTRODUCTION

Far less well known than the hemoglobins, the hemocyanins serve as oxygen carriers for a wide variety of invertebrate species. These proteins appear to have little in common with the other respiratory pigments. They contain copper, not iron, which accounts for their typical blue color, and have no porphyrin. One molecule of oxygen is bound per two copper atoms. The hemocyanins exist not in blood cells but as giant molecules dissolved in the hemolymph, with molecular weights ranging to over 9×10^6. Their similarity in function to the other blood pigments (hemoglobins, chlorocruorins, and hemerythrin) coupled with the complete difference in structure provide a remarkable example of convergent evolution on the level of molecular function. Some of the unique aspects of their structure and function have made these proteins the objects of considerable interest and study over the past 50 years.

In this paper we attempt to provide a general view of current knowledge about hemocyanins and to point out those areas in which confusion still exists. This is not intended to be an exhaustive review, in the sense that not *every* piece of published work is covered. Rather, we have attempted to select those data that either give definite answers or pose definitive questions. A number of other reviews of hemocyanins have been published over the years (*1–6*). The reader is referred to these for additional information.

II. THE OCCURRENCE AND PREPARATION OF HEMOCYANINS

The blue color of certain molluscan bloods was noted as early as the seventeenth century by Swammerdam (*7*). In 1866, Blasius (*8*) pointed out that the pigment is a protein which functions in oxygen transport. Leon Frédéricq (*9*), in 1878, in a study of the physiology of the octopus carried out at Naples, recognized hemocyanin as a special chromoprotein and gave it its name.

Hemocyanin occurs freely dissolved in the hemolymph of invertebrates belonging to the phyla Mollusca and Arthropoda. It is found in terrestrial and freshwater species but mainly in marine forms. For the Mollusca most of the hemocyanin studies have been made with species belonging to classes of the Gastropoda (snails), especially the subclasses Prosobranchia and Pulmonata, and the Cephalopoda, especially the orders Decapoda and Octopoda. Hemocyanin has also been reported in the hemolymph of chitons belonging to the subclass Polyplacophora of the class Amphineura.

Thus far no hemocyanin has been found in the classes Monoplacophora, Scaphopoda (tooth shells), and Bivalvia (e.g., clams, oysters), nor in the subclass Opisthobranchia of the Gastropoda. Almost all of the protein present in the hemolymph of Mollusca is hemocyanin: 98% in *Octopus vulgaris*, 95% in *Helix pomatia* (*10*). The concentration of hemocyanin often depends on age and season. In *O. vulgaris* the hemocyanin content varies from 60 to 114 mg/ml and in *H. pomatia* from 15 to 50 mg/ml (*11*).

In the phylum Arthropoda the presence of hemocyanin is clearly shown for the order Decapoda (crabs and lobsters) belonging to the subclass Malacostraca and the class Crustacea. It is also found in the orders Araneida (spiders) and Scorpionida (scorpions), both belonging to the class Arachnida, and in the *Limulus polyphemus* (horseshoe crab) one of four remaining species of the order Xiphosura of the nearly extinct subclass Merostomata. Weaker indications for the presence of hemocyanin are reported for the orders Isopoda, Amphipoda, and Euphausiacea belonging to the Malacostraca, and for the other orders of the Arachnida. No hemocyanin is found in the classes Progoneata (e.g., the Diplopoda multipedes), Chilopoda (e.g., centipedes), Hexapoda (insects), and for the subclass Entomostraca of the Crustacea. The concentration of hemocyanin in the hemolymph of various species varies over the wide range from 0.3 to 100 mg/ml (*5*). The arthropod hemolymph also contains, in many cases, a coagulation protein (about 2 mg/ml) and small quantities of other proteins of low molecular weight (*11–13*).

The hemolymph is generally obtained by introducing a canula or syringe into a vessel or by direct collection of blood running from an opened vessel. For example, snail hemolymph is collected either by heart puncture (e.g., *H. pomatia*), or after a short 2-cm-deep incision in the midline of the foot (e.g., *Haliotis* species) (*11*). Squid hemolymph can be obtained after introducing a canula or syringe into the dorsal vessel (e.g., *Sepia officinalis*). The molluscan hemolymphs do not coagulate; a small precipitate that forms after standing can be removed by low-speed centrifugation. Crab or lobster hemolymph (e.g., *Homarus vulgaris*) is collected by bleeding a living animal after removal of a leg. Spider (e.g., *Avicularia metallica*) hemolymph is obtained with a syringe in the heart region about 3–5 mm under the skin. Scorpion hemolymph is collected by introducing a syringe between the third and fourth dorsal segments at a depth of about one-seventh of the body thickness. All arthropod hemolymphs are stirred until coagulation is complete, whereafter the clot is removed by filtration or centrifugation.

Since the hemolymph is often so rich in hemocyanin, many studies have been carried out with the solutions obtained after dialysis of the hemolymph against the desired buffer. Further purification is possible via ammonium sulfate precipitation or ultracentrifugation. Hemocyanin can be stored at 2°C in solution (after addition of a few drops of toluene) or as ammonium sulfate precipitate. Freezing causes damage to the molecules. Lyophilization is possible for a number of hemocyanins after the addition of sucrose at a 2.75 : 1 weight ratio of sucrose to protein (14).

III. THE CHEMICAL STRUCTURE OF HEMOCYANINS

In this section we discuss two questions of a purely chemical nature: the manner of binding of copper by hemocyanins and the primary structure. The information available on both subjects is meager, especially in contrast to the detailed kinds of physical information described in Sections IV and V.

A. Copper Binding by Hemocyanins

All hemocyanins are copper proteins. No evidence for a porphyrin or other metal-binding moiety has been discovered; although early experiments were successful in the isolation of "copper-rich" degradation products (15,16), no conclusive evidence that they contained other than amino acids has been presented. The copper atoms are evidently bound in pairs, for in every case so far investigated there are two atoms of copper present for every molecule of oxygen bound at saturation (5). Another kind of evidence for the involvement of two copper atoms in the binding site is given later in this section. All molluscan hemocyanins contain one binding site for approximately 50,000 daltons of protein; in almost all arthropod hemocyanins the corresponding weight is about 75,000 daltons (See Table I). The sole apparent exception, the hemocyanin from *Callinectes sapidus*, seems of interest. However, no conformation of this exception has been reported.

The binding of copper is quite strong; values of 10^{17}–10^{19} have been obtained for the binding constant (17,18). The metal cannot be removed from undenatured protein by such strong complexing agents as *o*-phenanthroline or EDTA. Cyanide, however, is effective in removing 85–100% of the copper to produce an *apohemocyanin* which exhibits physical properties rather like those of the native protein (19–21). The process can be reversed to give nearly quantitative yields of active protein provided that *monovalent* copper is used (19,22,23). The reconstitution experiments have been used to confirm the belief that the active sites involve pairs of copper ions (24). As can be seen in Fig. 1, oxygen-binding capacity is proportional to the *square* of the copper content; this would be expected for random addition to paired, nearly identical binding sites.

TABLE I

Copper Content of Hemocyanins

Source of hemocyanin	Copper, wt %	Molecular weight per single binding site	Ref.
Arthropoda:			
Callinectes sapidus	0.132	96,300	*21*
Cancer magister	0.166	76,600	*27*
	0.163	78,100	*10*
Eriphia spinifrons	0.167	76,100	*21*
Homarus vulgaris	0.169	75,200	*21*
Limulus polyphemus	0.170	74,800	*21*
Maia squinado	0.170	74,800	*21*
Palinurus vulgaris	0.170	74,800	*21*
Average	0.168[a]	75,800	
Mollusca:			
Eledone moschata	0.252	50,400	*21*
Helix ligata	0.250	50,800	*21*
Helix pomatia (α)	0.241	52,700	*28*
Helix pomatia (β)	0.261	48,700	*28*
Murex brandaris	0.246	51,700	*21*
Murex trunculus	0.257	49,500	*21*
Octopus macropus	0.254	50,000	*21*
Octopus vulgaris	0.250	50,800	*21*
Octopus vulgaris	0.245	51,900	*20*
Average	0.251	50,600	

[a] Excepting *C. sapidus*.

Almost nothing is known about the ligands holding the copper to the protein. SH groups have been suggested on very indirect evidence (*25*), but this seems to be largely ruled out by two facts: (1) removal of copper releases *at most* one SH per four copper atoms (*26,27,21*), and blocking of SH does not inhibit reconstitution of apohemocyanin (*26*).

Perhaps a stronger case can be made for the involvement of histidine. Three pieces of evidence exist:

(1) Differences in titration curves between hemocyanin and apohemocyanin indicate about one group per copper atom with a pK value approximately that of histidine is freed when the copper is removed [(*26*); see also Konings (*28*)].

(2) Photooxidation of histidine has been shown to correlate with a loss in the characteristic copper spectral absorption (*29,30*).

(3) The circular dichroic spectra of hemocyanins resemble those of certain complexes of copper with histidine-containing peptides (*31*).

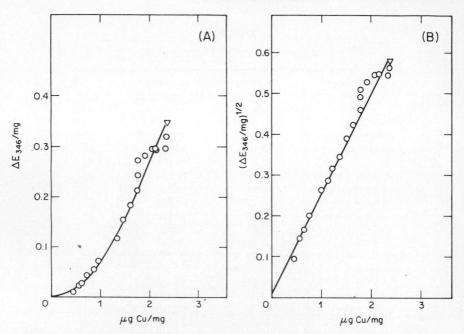

Fig. 1. Oxygen-binding capacity of *H. pomatia* hemocyanin as a function of copper content. Oxygen-binding capacity is measured by the change in extinction coefficient at 346 nm per milligram of protein, when the protein is saturated. The circles represent hemocyanins prepared by regeneration of apohemocyanin to the levels indicated on the absisca; the triangles represent results for the native protein. In (A) ΔE_{346}/mg is plotted vs. the copper content, whereas in (B) the square root is used. The fact that the graph is linear with the square root shows that the binding sites involve pairs of copper atoms. [From data of Konings et al. (*40*).]

None of this evidence is especially convincing. Furthermore, even if one histidine per copper is involved, the nature of the other ligands is still obscure. Copper is normally tetra- to hexa-liganded.

Finally, it should be pointed out that the oxidation state of the copper is still uncertain. While further discussion of this point is deferred to Section VI, to be taken up in conjunction with the oxygen-binding process, it is illustrative of our lack of detailed chemical knowledge about the active site of hemocyanins.

B. Amino Acid Composition

Although the amino acid sequence of no hemocyanin has been determined to date, the amino acid compositions of a number have been measured (*21,32,33*). These results are summarized in Table II. Examination of these results shows that these proteins, as a group, are characterized by relatively large amounts of acidic groups (up to 26%) and a fairly high content of nonpolar amino acids.

TABLE II

Amino Acid Compositions of Hemocyanins[a]

Source of hemocyanin	Lys	His	Arg	Try	Asp	Thr	Ser	Glu	Pro	Gly	Ala	Cys	Val	Met	Ileu	Leu	Tyr	Phe	Ref.
Arthropoda:																			
Callinectes sapidus	4.6	5.8	4.7	1.2	13.4	5.0	4.6	10.9	4.7	6.5	6.8	0.9	6.7	2.4	4.8	7.3	4.0	5.5	21
Eriphia spinifrons	4.1	7.1	4.8	1.6	13.3	5.4	5.6	10.2	4.6	5.9	6.2	0.5	6.7	2.6	4.4	7.4	3.8	6.1	21
Homarus vulgaris	4.8	6.6	4.8	1.2	13.1	6.1	4.3	11.3	4.6	6.0	6.0	0.8	6.6	2.3	5.0	7.5	3.5	5.6	21
Limulus polyphemus	5.9	7.5	5.1	—	11.6	5.1	4.9	11.4	4.2	5.9	4.9	1.6	6.7	2.4	5.4	8.4	3.6	5.2	21
Palinurus vulgaris	4.5	6.7	4.9	1.3	14.9	4.9	4.0	11.1	4.8	6.2	5.6	0.8	6.4	3.0	5.1	7.2	3.3	5.6	21
Mollusca:																			
Cymbium neptuni	4.7	4.7	4.7	3.0	11.2	5.1	5.9	11.0	4.7	6.1	6.6	0.5	5.6	2.3	4.2	8.9	4.4	6.3	33
Eledone moschata	4.6	6.0	3.9	1.7	12.1	5.1	5.5	9.6	5.3	5.8	6.5	2.1	5.8	2.4	5.3	8.9	4.1	5.5	21
Helix pomatia (α)	4.4	5.7	4.4	1.5	11.4	5.4	5.4	9.9	5.4	5.9	6.9	1.5	5.9	1.0	4.9	9.6	4.9	5.9	32
Helix pomatia (β)	4.7	5.7	4.4	1.5	11.9	5.7	5.4	9.9	4.9	5.9	6.7	1.7	5.7	1.2	4.9	8.9	4.9	5.7	32
Murex brandaris	4.2	6.8	4.7	1.6	11.9	5.1	4.7	11.2	5.1	6.1	6.5	1.6	5.6	2.1	4.0	8.7	3.7	6.3	21
Murex trunculus	4.6	6.4	4.6	1.6	11.9	5.0	4.8	11.2	5.0	6.2	6.6	1.4	5.7	2.3	4.1	8.7	3.9	6.2	21
Octopus macropus	4.8	5.5	4.4	1.4	11.2	6.0	5.3	9.8	5.3	5.7	6.4	1.8	5.5	2.3	5.5	9.2	4.4	5.5	21
Octopus vulgaris	4.8	5.2	3.8	1.7	11.7	5.2	5.0	9.5	5.7	5.2	6.7	2.4	6.2	2.6	5.2	9.0	4.3	5.7	21
Pila leopoldvillensis	3.2	4.1	5.3	3.0	11.0	5.0	5.7	11.5	6.2	6.0	7.6	1.1	6.0	1.1	4.4	8.7	3.9	5.5	33
Strophocheilus terrestris	3.9	4.4	4.4	2.8	11.8	5.6	6.0	10.7	5.6	5.8	6.5	0.9	5.6	0.5	4.9	9.0	4.6	5.8	33

[a] Values given in mole percent of amino acids.

The amino acid composition data allow us to consider a problem raised initially by the different copper contents of the arthropod and molluscan hemocyanins. These data (see Table I) indicate a rather large difference in the "single-site subunit" size for the hemocyanin from these two phyla (75,000 vs. 50,000 daltons). End group analysis on a typical molluscan hemocyanin [*H. pomatia* (*34*)] indicates a chain weight of about 24,000 daltons. This corresponds to one copper binding site per chain. If the analogous situation prevails in the case of arthropod hemocyanins, a chain of 37,000 daltons would be expected. It is difficult to see how a chain of 37,000 daltons can bear close evolutionary relationship to one of 24,000 daltons. This, together with differences in the quaternary structure of the two classes (see Sections IV and V) and in the chemical behavior of the copper (see Section VI) have led some investigators to believe that the arthropod and molluscan hemocyanins are in fact unrelated proteins.

Recently, a method has been developed by Teller (*35*) that allows the use of amino acid composition data to test relatedness in proteins. If two proteins have, respectively, mole fractions X_i and Y_i of each amino acid i, the function

$$D = \frac{1}{N} \sum_{i=1}^{N} (X_i - Y_i)^2 \qquad (1)$$

provides a rough measure of homology. Obviously, if the proteins are identical in composition, $D = 0$, while large values of D indicate little relatedness. Considerable question still exists as to the significance of moderately small values.

In Table III are reproduced some calculations by Teller and his associates for a number of hemocyanins and some unrelated proteins. Values of $D \leq 0.05$ are indicated in boldface. It seems likely from these data that some homology exists not only within the classes of molluscan and arthropod hemocyanin, but also *between* these classes. While firm conclusions must await sequence studies, it seems unlikely that the molluscan and arthropod hemocyanin could have evolved independently from unrelated proteins. Therefore we cannot look to such an explanation for the difference in weight per copper atom, or the other differences referred to above. A possible explanation (see Section IV) is that chains of about 25,000 daltons are common to all hemocyanins, but that in the arthropod proteins not all chains carry copper atoms.

C. Hemocyanins as Glycoproteins

At least one hemocyanin, that of *H. pomatia*, is known to be a glycoprotein. Both α- and β-hemocyanin (see Section IV) contain about 9%

TABLE III

Correlation of Amino Acid Compositions of Hemocyanins

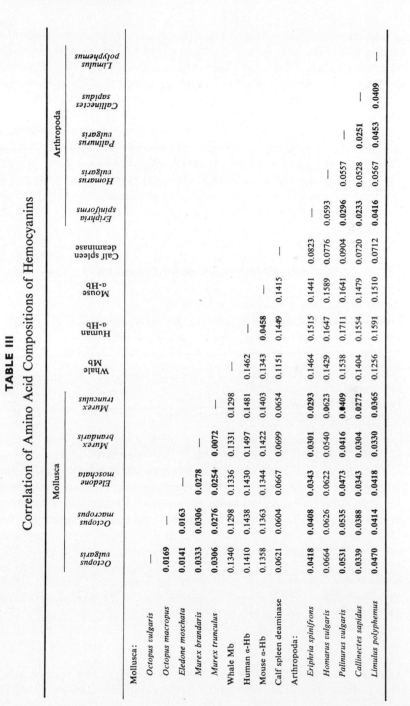

	Mollusca									Arthropoda				
	Octopus vulgaris	*Octopus macropus*	*Eledone moschata*	*Murex brandaris*	*Murex trunculus*	Whale Mb	Human α-Hb	Mouse α-Hb	Calf spleen deaminase	*Eriphria spiniforms*	*Homarus vulgaris*	*Palinurus vulgaris*	*Callinectes sapidus*	*Limulus polyphemus*
Mollusca:														
Octopus vulgaris	—													
Octopus macropus	0.0169	—												
Eledone moschata	0.0141	0.0163	—											
Murex brandaris	0.0333	0.0306	0.0278	—										
Murex trunculus	0.0306	0.0276	0.0254	0.0072	—									
Whale Mb	0.1340	0.1298	0.1336	0.1331	0.1298	—								
Human α-Hb	0.1410	0.1438	0.1430	0.1497	0.1481	0.1462	—							
Mouse α-Hb	0.1358	0.1363	0.1344	0.1422	0.1403	0.1343	0.0458	—						
Calf spleen deaminase	0.0621	0.0604	0.0667	0.0699	0.0654	0.1151	0.1449	0.1415	—					
Arthropoda:														
Eriphria spinifrons	0.0418	0.0408	0.0343	0.0301	0.0293	0.1464	0.1515	0.1441	0.0823	—				
Homarus vulgaris	0.0664	0.0626	0.0622	0.0540	0.0623	0.1429	0.1647	0.1589	0.0776	0.0593	—			
Palinurus vulgaris	0.0531	0.0535	0.0473	0.0416	0.0409	0.1538	0.1711	0.1641	0.0904	0.0296	0.0557	—		
Callinectes sapidus	0.0339	0.0388	0.0343	0.0304	0.0272	0.1404	0.1554	0.1479	0.0720	0.0233	0.0528	0.0251	—	
Limulus polyphemus	0.0470	0.0414	0.0418	0.0330	0.0365	0.1256	0.1591	0.1510	0.0712	0.0416	0.0567	0.0453	0.0409	—

carbohydrate, which includes residues of glucosamine, galactosamine, fucose, galactose, glucose, xylose, mannose, and an as yet unknown sugar. No neuramic acid is present. The single carbohydrate chain is probably heterogeneous and is attached to an asparagine residue in a chain of 24,000 daltons (36). Recent evidence suggests that most hemocyanins may be glycoproteins (37).

If it is proved that hemocyanins are in fact generally glycoproteins, the possibility of a carbohydrate being involved in the copper binding site should at least be considered. It should be noted that early attempts (15,16) to detect a copper-binding prosthetic group produced copper-rich materials of unknown composition. It is possible that they may have contained copper carbohydrate complexes produced by the strongly alkaline solution used.

IV. STATES OF AGGREGATION

The hemocyanins are remarkable in the variety of subunit aggregates that they form. These include some of the largest functional protein structures known, ranging in "molecular" weight up to 10^7 or greater. While most of our knowledge of the structures of such particles comes from electron microscopy, their existence, weight, and homogeneity were first demonstrated by such classic solution techniques as sedimentation and light scattering. We consider the results of such studies first, and turn to electron microscopy in the following section.

A. The Existence of Discrete States of Aggregation

The hemocyanins were among the first proteins to be investigated in the pioneering ultracentrifuge studies by Svedberg and his collaborators (see, for example, References 38 and 39). From these experiments two principles became clear, which have guided much of the subsequent work in this field:

1. *Depending upon the solution conditions, a given hemocyanin can exist in a number of discrete states of aggregation.* The kind of situation frequently encountered is shown in the sedimentation diagram of *H. pomatia* hemocyanin in Fig. 2. Three distinct species are observed to be present under the chosen conditions of pH, ionic strength, and temperature. These components appear to be homogeneous, and other evidence indicates that they are, or very nearly so. Furthermore, these seem to be the *only* components present in appreciable amount. However, if the pH of the solution is changed, the pattern of aggregation states changes; this is illustrated in Fig. 3, drawn from the early data of Eriksson-Quensel and Svedberg (38).

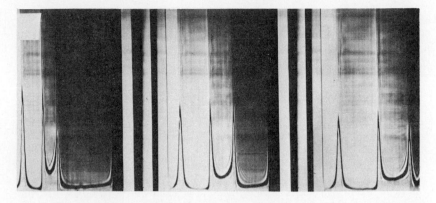

Fig. 2. Sedimentation of *H. pomatia* hemocyanin in 0.1 ionic strength tris–HCl buffer at pH = 7.85. The rotor speed was 37,020 rpm; photographs taken 15, 20, and 25 min after attaining speed. The 20, 60, and 100 S components can be seen. The elevation of the base line between the 60 and 100 S components depends upon rotor speed; it indicates that the dissociation of 100 S hemocyanin is favored by the higher hydrostatic pressures lower in the solution column. [Data of Konings (*28*).]

It is evident that under some conditions only one component is present, while under others more are found; furthermore, the changes with pH are abrupt. This is further brought out in the more recent data of Konings and

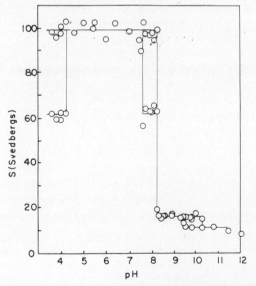

Fig. 3. A pH-stability diagram for *H. pomatia* hemocyanin. The sedimentation coefficients of components observed at various pH values are indicated. [Taken from the data of Eriksson-Quensel and Svedberg (*38*).]

co-workers (*40*) presented in a somewhat different fashion in Fig. 4. We consider later such questions as the reversibility of such changes and their dependence on factors other than pH. For now, let us consider a second fundamental observation.

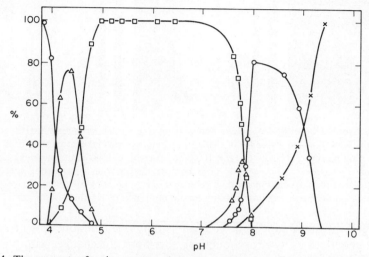

Fig. 4. The amounts of various aggregation states of the α-hemocyanin from *H. pomatia* present at different pH values. Concentrations were determined from sedimentation velocity experiments, corrected for radial dilution. □,100 S; △, 60 S; ○, 20 S; ×, 11 S. [From Konings (*28*).]

2. *Within each of the two classes of hemocyanins, arthropod and molluscan, there is remarkable uniformity in the sedimentation coefficients observed.* This is most clearly brought out in Figs. 5 and 6, in which the results of all the sedimentation velocity studies we have been able to find are summarized. While not all species have been demonstrated to exhibit all of the components common to a given class, it is evident that there exist a limited number of size classes for the hemocyanins of each phylum. We discuss each group briefly.

The *Arthropod hemocyanins* exist primarily as structures with sedimentation coefficients of about 5, 16, and 25 S. In a few cases components with sedimentation coefficients† of about 34 and 60 S are also observed. In most

† Examination of Fig. 5 illustrates a general problem in the nomenclature of such components. While the use of an approximate *s*-value is relatively unambiguous, there is the difficulty that *s* is concentration dependent. Thus the "34 S" component of *Limulus* hemocyanin actually has a sedimentation coefficient at infinite dilution of nearly 40 S. Nevertheless, since most sedimentation velocity experiments are carried out at a concentration of about 3 mg/ml, we use for designation the approximate *s*-value found in this range of concentration.

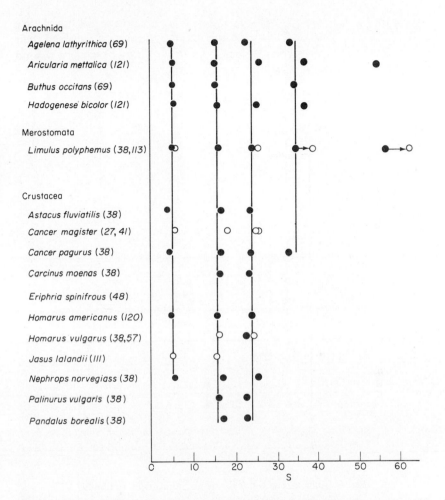

Fig. 5. A diagram illustrating sedimentation coefficients reported for the hemocyanins of arthropod species. The filled circles denote values obtained at finite concentration (usually about 3–5 mg/ml). The open circles are results extrapolated to infinite dilution. All data are corrected to water at 20°C. References are indicated.

cases the 25 and 16 S components are the principal constituents of the hemolymph. The molecular weights of a number of the purified components have been determined; some† are listed in Table IV. That the 16

† We have chosen not to include a number of molecular weight values that were determined prior to 1940. For many of these data, there is no assurance that homogeneous preparations were being used, and effects of solution nonideality and association–dissociation equilibria were generally not considered. Data obtained prior to 1940 are summarized in the book by Svedberg and Pedersen (39).

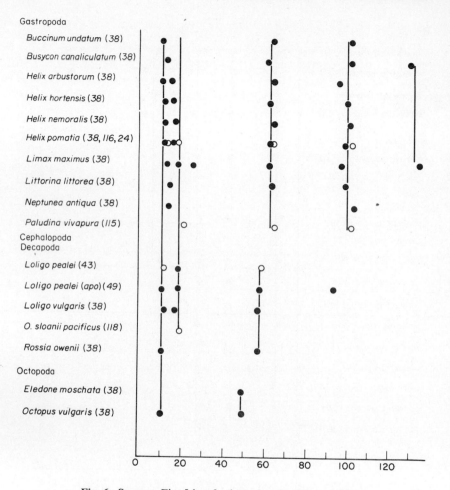

Fig. 6. Same as Fig. 5 but for hemocyanins from molluscs.

and 25 S components are homogeneous is demonstrated by the data for *C. magister* shown in Fig. 7. Evidently, the 25 S particle is formed from two 16 S particles. Further, the data in Table IV indicate that the 34 S component results from another dimerization. The relationships are elegantly confirmed by the electron micrographs shown in Section V. The relationship of the 60 S *Limulus* hemocyanin to the smaller units is not clear from these data; the problem is discussed later.

At high pH, or in the presence of denaturing agents, the 16 S arthropod hemocyanins can be dissociated into smaller units. These "5 S particles" do not generally appear to be homogeneous (see Fig. 7) but correspond to roughly one-sixth of the 16 S unit. These particles have nearly the right weight to contain a single oxygen-binding site (about 75,000 daltons). They appear to be made of still smaller subunits; how and whether their size is related to the single-site unit (50,000 daltons) for the molluscan hemo-cyanin is still not clear. Ellerton et al. *(41)* have suggested that the 5 S arthropod subunit may contain three chains of about 25,000 daltons each, one pair of these forming an oxygen-binding unit. In this way an otherwise curious dilemma concerning the difference in size of the "basic units" of the arthropod and molluscan hemocyanins could be resolved (see Section III).

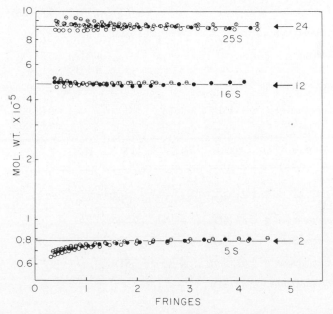

Fig. 7. Results of sedimentation equilibrium molecular weight studies of *C. magister* hemocyanin. Each component has been investigated in a number of experiments, using the Yphantis technique; data from individual experiments are shown by different kind of points. [From Ellerton et al. *(41)*.]

The *molluscan hemocyanins* appear to be built on an entirely different plan of subunit aggregation. The smallest unit reported has a sedimentation coefficient of about 11 S (see Fig. 6) and a molecular weight of about 4×10^5. It evidently dimerizes readily to form a 20 S particle, which then can form pentamers of molecular weight about 4×10^6 and decamers of $M \cong 8 \times 10^6$ (see Table IV). The latter two components predominate in the hemolymph of most molluscs. Still larger aggregates have been observed: a 130 S component is found in at least two species, and there is

TABLE IV

Molecular Weights of Hemocyanin Components[a]

Source of hemocyanin	Molecular weight, $\times 10^{-5}$					
Arthropoda	60 S	34 S	24 S	16 S	5 S	Ref.
Crustacea:						
Cancer magister	—	—	9.4 (SE)	4.7 (SE)	0.78 (SE)	*41*
			9.5 (SD)	—	—	*27*
Eriphia spinifrons			9.5 (LS)	4.5 (LS)	0.8 (LS)	*48*
Homarus americanus	—	—	8.3 (SD)	—	—	*57*
Jasus lalandii	—	—	—	4.6 (A)	0.88 (A)	*111*
				4.5 (LS)	—	*112*
Merostomata:						
Limulus polyphemus	33 (SE)	19 (SE)	—	—	0.65 (SE)	*113*
					0.8 (SD)	*114*

Mollusca	100 S	60 S	20 S	11 S		Ref.
Gastropoda:						
Helix pomatia	89 (SD)	43 (SD)	10 (SD)	—		*115*
Helix pomatia (α)	87 (SD)	—	9.9 (SD)	—		*116*
Helix pomatia (β)	90 (SD)	—	8.9 (SD)	—		*116*
Paludina vivapara	87 (SD)	—	11 (SD)	—		*115*
Pila leopoldvillensis	87 (SD)	—	—	—		*117*
Cephalopoda:						
Loligo pealei	—	38 (SE)	7.7 (SE)	3.9 (SE)		*43*
Omnatostraphes sloanii pacificus	—	—	6.1 (SD)	—		*118*

[a] The expressions in parentheses following each number indicate the method of measurement: SE, sedimentation equilibrium; SD, sedimentation plus diffusion; LS, light scattering; A, Archibald method.

some evidence for 155 and 175 S particles in *Busycon canaliculatum*. These probably represent trimers and tetramers of the 60 S hemocyanins (see Section V for electron micrographs of such particles). In fact, Condie and Langer (*42*) have reported that indefinite polymerization can occur near the isoelectric point.

There appear to be slight differences between the cephalopod hemocyanins and those of gastropods. This is evident in both the sedimentation coefficients (Fig. 6) and in the molecular weight (Table IV). At present, these differences, if real, are hard to understand in view of the evident structural similarities (Section V).

As in the case of the arthropod hemocyanins, the various states of aggregation of the molluscan hemocyanins appear to be molecularly homogeneous [see, for example, van Holde and Cohen (43)]. However, such apparent homogeneity may mark subtle heterogeneity, for it has long been known (44–47) that the hemocyanins of some gastropods contain at least two components, termed α- and β-hemocyanin. These seem to have about the same molecular weight and sedimentation coefficients, but they differ in dissociation behavior. The α-hemocyanin can be dissociated from 100 to 60 S particles in 1 M NaCl at pH 5.7; the β component is stable in the 100 S form under these conditions. As Table V shows, the ratios of these components vary in different gastropod hemocyanins. It may well be that such heterogeneity is more widespread than has been recognized to date.

TABLE V

α-β Composition of Some Gastropod Hemocyanins[a]

Species	Percent α	Percent β
Helix arbustorum	100	0
Helix pomatia	75	25
Helix hortensis	30	70
Strophocheilus terrestris	30	70
Helix aspersa	25	75
Pila leopoldvillensis	0	100
Cymbium neptuni	0	100
Buccinum undatum	0	100

[a] Data from Konings (28).

A fundamental puzzle in the behavior of the molluscan hemocyanins lies in the difficulty of obtaining dissociation beyond the 11–14 S stage. Dissociation of these molecules has never been attained by the use of agents such as urea, guanidine hydrochloride, or sodium dodecyl sulfate. However, it has recently been shown (Dijk et al., in preparation) that 70% formic acid leads to dissociation into subunits with $s^{\circ}_{20,\,w} \cong 3.0$ S, and $M = 50,000$. Furthermore, a small amount of material with $M \cong 24,000$ is present. Since performic acid or reducing agents are ineffective in dissociation of the 14 S *Helix* hemocyanin, it seems that these smaller subunits

are not held together by disulfide bonds. These subunits also display a
marked microheterogeneity, confirming the observation of heterogeneity
in the larger particles.

B. Conditions for Stability and Interconversion of Different Aggregation States

In general, the larger aggregates of both the molluscan and arthropod
hemocyanins are stable near neutral pH; at low or high pH, dissociation
occurs. This is evident in Figs. 3 and 4 for the *H. pomatia* hemocyanin, and
Fig. 8 shows that a similar result obtains for the hemocyanin of the crab
C. magister. A large number of similar "stability diagrams" have been
prepared by Eriksson-Quensel and Svedberg (*38*) and are reproduced in
Svedberg and Pedersen (*39*). The transitions with changing pH are exceed-
ingly abrupt and are not fully understood. For example, while in many
cases such transitions are reversible they do not appear to depend upon
total protein concentration in the expected manner. Table VI shows some
data obtained (*28*, *40*) for the α-component of *H. pomatia* hemocyanin.
The equilibrium between 20 and 60 S particles simply does not behave

TABLE VI

Dependence of the Ratio of α-Hemocyanin Components on the Total
Protein Concentration[a]

Buffer	Hemocyanin concentration mg/ml	Wt % present as:			
		11 S	20 S	60 S	100 S
0.1 *M* Tris–HCl, pH = 7.7	2	11	—	17	72
	1	13	—	25	62
	0.75	16	—	29	55
	0.50	17	—	30	53
0.1 *M* Sodium borate, pH = 8.1	4	—	47	48	5
	2	—	51	45	4
	0.7	—	46	51	3
	0.2	—	44	53	3
0.1 *M* Tris–HCl, pH = 8.9	4	38	62	—	—
	3	40	60	—	—
	2	40	60	—	—
	1	46	54	—	—
	0.5	54	46	—	—

[a] Data from Konings (*28*).

according to the mass action law. DiGiamberardino (*48*) has demonstrated the same behavior in the hemocyanin of a crustacean, *E. spinifrons*. It has been suggested [Konings et al. (*40*); DiGiamberardino (*48*)] that these results may be explained on the basis of microheterogeneity: that there may exist in an apparently homogeneous protein a number of isomers, each with a different and very sharp pH transition region. At any given pH each component would be either fully associated or fully dissociated. There is clearly a need for further investigation of this problem.

The dissociation–association processes are in many cases affected by the presence of divalent cations. Numerous investigators have noted that magnesium ion tends to stabilize larger structures. Others have found that calcium, strontium, and barium have a similar effect. An example of the effect of Mg^{2+} is seen in Fig. 8; the 25 S *Cancer* hemocyanin dissociates

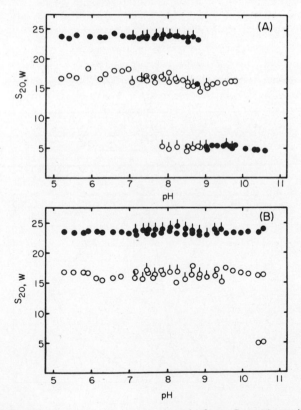

Fig. 8. pH-Stability diagrams for the hemocyanin from *C. magister*. All data were obtained at concentrations of 3.5 mg/ml. At each pH the predominant component is indicated by the filled circles. (A) Results in solutions not containing added divalent ions. (B) Results in solution to which 0.01 M $MgCl_2$ had been added. [From Ellerton et al. (*41*).]

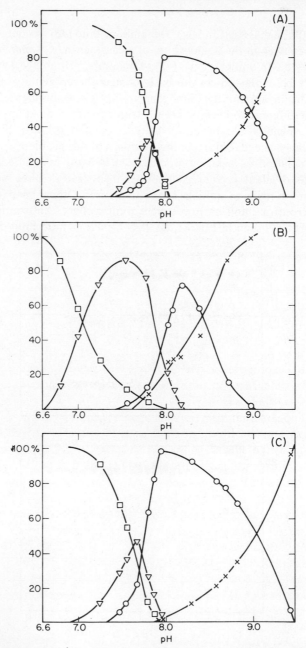

Fig. 9. The effect of copper removal and replacement on the stability of α-hemocyanin from *H. pomatia*. Percentages determined from sedimentation velocity, experiment corrected for radial dilution. (A) α-hemocyanin; (B) apo-α-hemocyanin; (C) reconstituted α-hemocyanin; □, 100 S, ▽, 60 S, ○, 20 S, ×, 11 S. [From Konings et al. (*40*).]

into 16 S units and then into 5 S units at high pH in the absence of Mg^{2+}, but in 0.01 M MgCl$_2$ the 25 S unit is stable to pH $>$ 10. This is not simply an ionic strength effect; an equivalent concentration of NaCl has no effect. Addition of Mg^{2+} to 5 S units at pH 10 causes quantitative reassociation to 16 S particles. Surprisingly, 25 S particles are not produced in this way; the dissociation is not entirely reversible. If the pH of the solution is lowered to 7.0, some 25 S material is formed, but most remains as 16 S. The difficulty in interconversion between these species is further indicated by the fact that while both are observed in the hemolymph they are not in equilibrium. It is possible, for example, to separate homogeneous samples of 25 and 16 S *C. magister* hemocyanin by sucrose gradient centrifugation (*41*). A similar result was noted by DiGiamberardino (*48*) with *Eriphia* hemocyanin.

Very high ionic strengths (ca. 1 M) also appear to affect some of these equilibria. This has already been noted as the basis for the distinction between the α- and β-hemocyanins of the gastropods.

Removal of the copper from hemocyanin, to form the apo protein, seems to have an effect on the stability of the larger aggregates. Cohen and van Holde (*49*) found that the 59 S apohemocyanin from *Loligo pealei* dissociated into 19 S particles at a somewhat lower pH than the native hemocyanin in solutions containing 0.01 M MgCl$_2$. In magnesium-free solution the two proteins behaved identically. Similarly, Konings et al. (*40*) observed the differences shown in Fig. 9 between apo, native, and reconstituted α-hemocyanins from *H. pomatia*.

It now appears that the association–dissociation equilibria of some hemocyanins may be linked to the oxygenation of the protein. In an investigation of the effects of pH and fraction oxygenation on the component distribution in *L. pealei* hemocyanin, DePhillips et al. (*50,51*) found the results shown in Fig. 10. Evidently, either the deoxygenated or wholly oxygenated proteins exist mainly in the 60 S form, whereas partially oxygenated molecules dissociate into subunits. Very recent experiments indicate an effect of oxygenation on the 100–60 S equilibrium of the hemocyanin from *B. canaliculatum* (*52*).

Clearly, the effects of pH, oxygenation, and divalent cations on these equilibria present a complex picture, the details of which are still obscure. Much more research is required before we can hope to understand the whole pattern of behavior.

V. MOLECULAR ARCHITECTURE OF HEMOCYANINS

A. Introduction

In this section we discuss the molecular architecture of representative

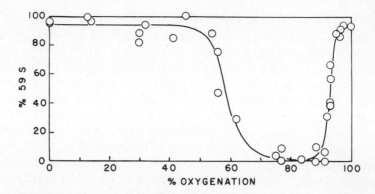

Fig. 10. The effect of oxygenation upon the component equilibria of the hemocyanin from *L. pealei*. The equilibrium involves a 59 S component which dissociates into 19 and 11 S subunits. The percentage of 59 S component is shown as a function of percent oxygenation in the pH range 6.4–9.0. [From DePhillips et al. (*50*).]

hemocyanins from the phyla Arthropoda and Mollusca as revealed by electron microscopy.

It was shown in Section IV that hemocyanin molecules dissociate reversibly into smaller units depending upon the pH, ionic strength, and other factors. Undissociated molecules from different biological origins vary widely in size. This was clearly illustrated by the sedimentation coefficients presented in Figs. 4 and 5. It is possible to group the different components into classes and the following questions then arise:

1. What is the structural relation between the components of each species?

2. What is the structural relation between the components of each phylum?

3. What is the structural relation between arthropod and molluscan hemocyanins?

Electron microscopy is a very suitable technique to study these questions as it gives direct information on the structure of each separate molecule.

B. Specimen Preparation and Electron Microscopy

The early electron microscope studies of hemocyanin were made with unstained or shadowed molecules (*53–57*). It was possible to demonstrate size differences between a component and its dissociation products, but details of the molecular structure remained obscure. The great breakthrough came from the technique of negative staining, first used by Hall (*58*) and Huxley (*59*), and modified to a routine method by Brenner and

Horne (*60*). We describe here how this technique is applied to the study of hemocyanins.

A solution of hemocyanin at a protein concentration of 10–1000 μg per milliliter is mixed with a 0.5–2.0% solution of a heavy metal "stain" (e.g., uranyl acetate, uranyl oxalate, uranyl EDTA, potassium phosphotungstate, and so on). The mixture of protein and stain is transferred to a carbon-coated specimen grid. Thus in the electron microscope the unstained protein molecules are contrasted against a heavy metal stained background. There are many procedures by which this can be done, and the reader is referred to Reference *61* for a full description. Uranyl stains were found very suitable for following the dissociation of the molecules because they seem to have a fixing influence on the structures present in a given situation. However, the tungstate stains tend to dissociate the molecules into smaller components, an effect similar to that caused by a high concentration of NaCl. Therefore the tungstate stains are less suitable for the study of the molluscan hemocyanins than for the study of the arthropod hemocyanins, in which such dissociation is less frequently encountered. In any case the dissociation can be diminished by a prior fixation with formaldehyde or glutardialdehyde.

Electron micrographs are usually taken at magnifications of 20,000–200,000 with high-resolution electron microscopes operating at 60 or 80 kV. Certain variations are possible in the choice of the illuminating system and the condenser and objective apertures. The use of an anticontamination device cooled with liquid nitrogen improves the results considerably.

For an explanation of contrast patterns in electron micrographs, the wave properties of the electrons must be taken into account. These lead to phase-contrast patterns with properties determined by the amount of defocus (*62*). It has been shown by van Dorsten and Premsela (*63*) that for thin, amorphous specimens there is a "structureless" region in or close to focus where these phase contributions are minimal. It is only in this so-called parafocal region that one can expect to detect small amounts of amplitude contrast attributable to heavy metal. This is important because at this time only structural conclusions from amplitude contrast patterns are allowed.

C. Results for Arthropod Hemocyanins

Several authors have published electron micrographs of arthropod hemocyanins (*61,64–70*). Analogous photographs have, however, led them to different models for the molecular structure.

It was shown in Section IV that arthropod hemocyanins can exist as particles with sedimentation coefficients of about 5, 16, 24, 35, and 60 S.

We now correlate these particles with the structures found by electron microscopy.

1. 5 S Structure

Although clearly visible in the electron microscope as very fine material (Fig. 11), this structure has not been carefully studied and characterized until now.

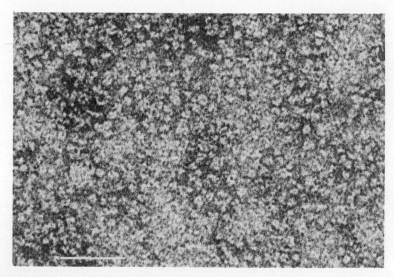

Fig. 11. *Limulus polyphemus* hemocyanin molecules negatively stained at pH 10 with potassium phospнotungstate (scale, 500 Å).

2. 16 S Structure

This molecule is found as main component in the hemolymph of the spiny lobster *Palinurus vulgaris* (Decapoda, Crustacea). Staining with potassium phosphotungstate (Fig. 12) shows the projection of these molecules as squares, rectangles, or hexagons with a maximum dimension of 100–125 Å. This electron micrograph was taken at an underfocus setting of the objective lens of the electron microscope. This favors the appearance of certain course periodicities. We observe a spot on each corner for the squares, two rows of three spots for the rectangles, and a dubious central spot with six spots around it for the hexagons. As pointed out in Section B, it is not necessarily correct at this moment to interpret these phase structures as subunit structures. Nevertheless, this has been done in the past. Different models have been proposed: 12 units in a truncated tetrahedron

[Levin (*66*)], 12 units in a hexagonal prism [di Giamberardino and de Haën (*68*)], 6 units in a trigonal antiprism [Wibo (*69*)] and 8 units in a cube [van Bruggen et al. (*67*)]. It seems better to approach this problem of the detailed structure from electron micrographs taken in the parafocal region, which unfortunately are not available at present. Certainly, the molecule has some kind of a prismatic structure as indicated by the contours of the molecular projections.

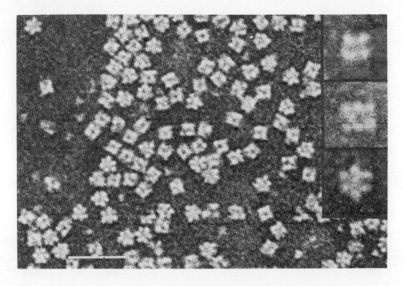

Fig. 12. *Palinurus vulgaris* hemocyanin molecules negatively stained at pH 7 with potassium phosphotungstate (scale, 500 Å).

3. 24 S Structure

Lobsters and crabs have this structure as major components in their hemolymphs. Figure 13 was taken from *H. vulgaris* (Decapoda, Crustacea) hemocyanin negatively stained with potassium phosphotungstate. We clearly observe dimeric structures measuring 235 × 100 Å, consisting of a hexagonal profile connected with a rectangular or square profile. Wibo (*69*) showed for the lobster *H. vulgaris* the existence of a left- and a right-type structure using a superposition technique. It is possible to dissociate these dimers into monomers that are not distinguishable by electron microscopy from the 16 S structures shown in Fig. 12. Therefore models for this 24 S structure always consist of the combination of the 16 S structure in two different orientations, usually with rotation of one unit by

90° with respect to the other. Details concerning the linkage of monomers to form dimers are unclear at this moment. More data can be expected from studies of tilted molecules or hemocyanin crystals, which are being undertaken at the present time.

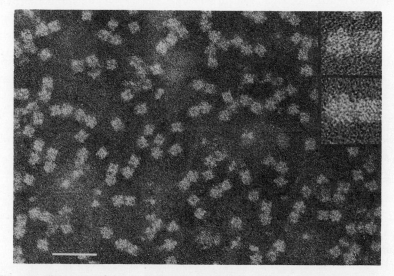

Fig. 13. *Homarus vulgaris* hemocyanin molecules negatively stained at pH 7 with potassium phosphotungstate (scale, 500 Å).

We should also mention here a structure with a slightly lower sedimentation constant (21 S) found by Wibo (*69*) as the main component in the hemolymph of spiders belonging to the Araneomorphaes families Agelenidae, Lycosidae, Dictynidae, Clubiomidae, and Salticidae. Here three types of dimeric structures are found. In most cases a combination of two squares connected along their sides is observed. In addition, the combination of a hexagon with a square and the double hexagon combination can be rarely observed.

4. 34 S Structure

This structure is found as the main component in the hemolymph of scorpions belonging to the order Scorpiones and spiders belonging to the order Araneida, with exception of the families mentioned. The hemocyanin of the scorpion *Hadogenese bicolor* is shown as an example of this structure in Fig. 14 (*70,71*). The molecules are observed as large rectangles measuring 235 × 230 Å. Parallel to the longest side the molecules have a

20-Å-wide gap crossed by two narrow bridges. Thus the molecule is divided into two equal parts. It is possible to dissociate these tetramers into dimers and monomers. The monomers show the square and hexagonal contours of the previously described 16 S structures. Separate dimers have nearly all possible combinations of a hexagonal and a square profile. Most common is the combination of a hexagon and a square in which a side of the square is connected with a side or a corner of the hexagon. We further see two squares connected along their sides or at the corners and two hexagons connected along their sides.

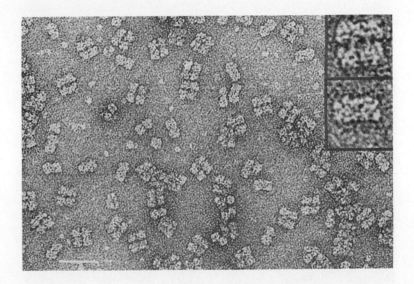

Fig. 14. *Hadogenese bicolor* hemocyanin molecules negatively stained at pH 6.5 with uranyl oxalate (scale, 500 Å).

It is difficult to deduce an exact model for the 34 S particle. The likely conclusion is that this structure results from the combination of two 25 S particles in a probably parallel (but perhaps antiparallel) orientation.

5. 60 S Structure

This structure plays a major role in the hemolymph of the horseshoe crab *L. polyphemus* (Xiphosura, Merostomata) and has also been found as a minor component in the hemolymph of the spider *A. metallica* (Theraphosidae, Mygalomorphes, Araneida) (*71*).

Electron micrographs of *Limulus* hemocyanin (Fig. 15) show a number of different projections:

1. A roughly circular projection with a diameter of about 240 Å.

2. A rectangular projection of 240×210 Å composed of two parallel rows each measuring 240×90 Å. This structure differs slightly from the 24 S structures described earlier. The dimensions are about the same, but they show more contrast (more protein?) and the bridges over the gap are missing.

3. Large squares consisting of four smaller 90×90 Å squares connected at their corners.

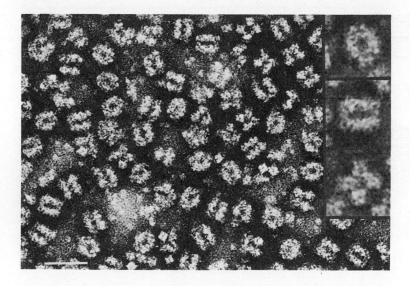

Fig. 15. *Limulus polyphemus* hemocyanin molecules negatively stained at pH 7 with sodium phosphotungstate (scale, 500 Å).

It is possible to dissociate this hemocyanin into dimers and monomers with the same variety of structures as the dissociation products of the 34 S structure.

The value of the sedimentation constant (60 S) and the square appearance of many of the molecular profiles make it plausible that we are dealing with a structure that is an octomer of the 16 S particles. The only recorded measurement of a molecular weight for the 60 S Arthropod hemocyanin (3.3×10^6) is roughly eight times the weight found for the 16 S particle

($\sim 4.5 \times 10^5$) (see Section IV). The precise arrangement of the 16 S mono-mers in the octomer is uncertain. Wibo (69) has interpreted structure 2 as the projection of a tetramer and structure 3 as the projection of an octomer; van Bruggen has given the opposite opinion (70,72).

Fahrenbach (73) recently presented electron micrographs of thin sections of *Limulus* hemocyanin crystals found in the cyanoblast, a cell type circu-lating in the hemocelic fluid. Mutually perpendicular cross sections of these crystals showed hexagonally packed hollow-appearing cylinders with diameters of about 190 Å and with a 260-Å center-to-center spacing. There is a longitudinal stacking of the hemocyanin molecules comprising the cylinders, with a faint 100-Å periodicity. Analysis of these electron micro-graphs with optical diffraction (74) showed for the stacked molecules a periodicity of 200 Å, each period consisting of two layers. Furthermore, the molecules in the stacked cylinders are probably rotated with respect to each other. Indications for an angle of rotation of 90° were obtained.

In summary we can say that the *Limulus* hemocyanin octomer appears to be a two-layered structure with some kind of a polarity. Every layer consists of four 16-S structures. The mutual orientation of these structures is still doubtful.

D. Results for Molluscan Hemocyanins

Depending on biological origin and environment, molluscan hemo-cyanins can occur as particles with sedimentation coefficients of 11, 20, 60, 100, and even 130, 155, and 175 S (see Section IV).

1. 11 S and 20 S Structures

Figure 16 shows the dissociation products (low salt, high pH) of hemo-cyanin from the freshwater snail *Pila leopoldvillensis* (Ampullariidae, Gastropoda) (75). It is difficult to correlate the observed structures with the sedimentation constants of 11 and 20 S found for the same solution. The smallest particles often exhibit a square profile having sides of about 65 Å corresponding to a very roughly estimated molecular weight of about 100,000. A corresponding component is not visible on the sedimentation patterns and therefore probably formed during the specimen preparation procedure. In addition structures of about 200×200 Å, sometimes with a subdivision of three parallel lines, can be seen. Assuming for these struc-tures a height of 65 Å leads to an estimated molecular weight of 900,000; this corresponds to a sedimentation constant of 20 S (see Section IV). The particles with a size between 65 and 200 Å probably represent the 11 S structures; it is not possible to characterize these better.

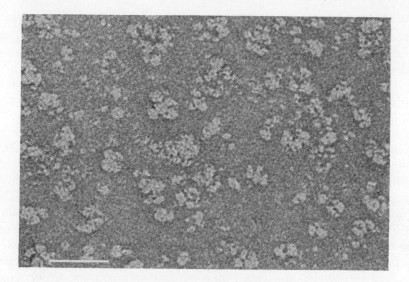

Fig. 16. *Pila leopoldvillensis* hemoycanin molecules negatively stained at pH 8.5 with uranyl oxalate (scale, 500 Å).

2. 60 S Structure

This structure is found as a main component in the hemolymph of squids and octopi (*61,69,76*) and as a stage in the dissociation of gastropod hemocyanin. Figure 17 shows as an example the hemocyanin of *O. vulgaris* (Octopoda, Cephalopoda). The molecules are observed as circles with a diameter of 350 Å or as rectangles of 350 × 170 Å. The circle has an outer ring with no evident substructure and a thickness of about 65 Å and an inner ring which often exhibits a fivefold symmetry. The rectangles show a subdivision into three rows parallel to their longest side. It is possible to dissociate this hemocyanin into structures that show great correspondence with the particles described as the 11 and 20 S structures.

These observations are in agreement with a cylindrical type of molecule. The diameter of the cylinder is 350 Å, the height 170 Å. The cylinder has a fivefold rotation axis and is built from three stacks of subunits perpendicular to the axis. The cylinder wall has a thickness of about 65 Å. The level at which those subunits observed as an inner ring in the circular projections lie has not yet been determined.

3. 100 S Structure

This structure occurs as the main component in the hemolymph of the Gastropoda. The most detailed electron microscope studies have been

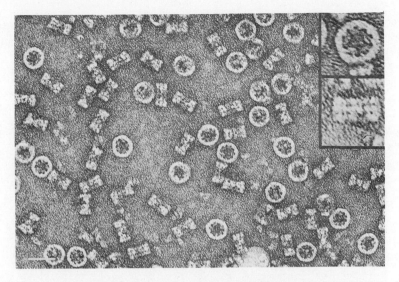

Fig. 17. *Octopus vulgaris* hemocyanin molecules negatively stained at pH 6.5 with uranyl oxalate (scale, 500 Å).

carried out with the hemocyanin of the Roman snail *H. pomatia* (Stylo-matophora, Pulmonata) (*61,70,76*). The molecules are observed as circles with a 10-fold rotational symmetry or as rectangles with a subdivision into six parallel rows (Fig. 18). The circle measures 350 Å across, the side of the

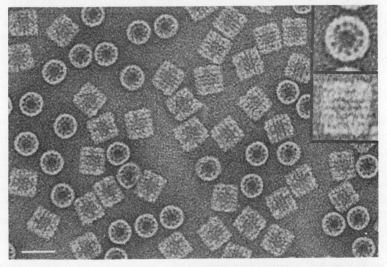

Fig. 18. *Helix pomatia* hemocyanin molecules negatively stained at pH 7.0 with uranyl EDTA (scale, 500 Å).

rectangles perpendicular to the rows is 380 Å, while the other side ranges from 350 to 390 Å. Circles and squares were shown to be different aspects of the same cylindrical molecule (76).

An increase in the pH dissociates the 100 S structure via halves (60 S) into smaller components (20 and 11 S) (Fig. 19). The half-molecules are observed as circles, which show great similarity to the 60 S circular projections of cephalopod hemocyanin described above. They clearly show a fivefold symmetry. The other aspect of the halves is a rectangle of 190 × 350 – 390 Å with a subdivision of three rows parallel to their longest side. These rectangles show an asymmetric staining pattern, thus indicating a

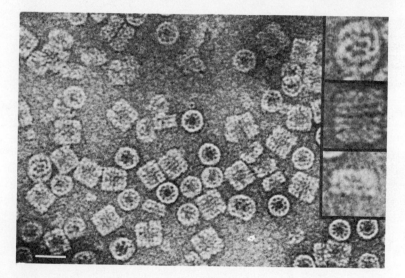

Fig. 19. *Helix pomatia* hemocyanin molecules negatively stained at pH 7.7 with uranyl EDTA (scale, 500 Å).

polar character for the half-molecules. This polarity is not observed in the rectangular views of the 60 S structure from squid and octopi (see Fig. 17). Perhaps this lack of asymmetry explains why these particles do not readily dimerize. The dissociation of 60 S particles into smaller structures occurs sectorwise (Fig. 20). The structure of the dissociation products shows great similarity to the *Pila* hemocyanin dissociation described above (see Fig. 16).

These results indicate the 100 S structure to be a cylindrical molecule with a diameter of 350 Å and a height of 380 Å. This cylinder has a fivefold axis of rotation and is built from six layers of subunits perpendicular to its axis. The first dissociation step occurs perpendicular to the cylinder axis into two three-stacked 190-Å-high cylinders. The further dissociation

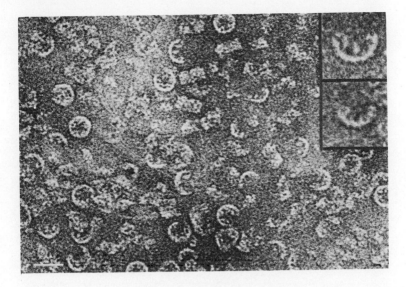

Fig. 20. *Helix pomatia* hemocyanin molecules negatively stained at pH 8.2 with uranyl EDTA (scale, 500 Å).

is a sectorwise splitting of these 190-Å-high cylinders into five sectors, each sector next dissociating into two probably equal parts along an unknown direction.

The polarity of the half-molecules indicates a different situation for the outer and inner stacks of the whole molecule, which may be attributable to the local presence of an inner ring structure. The detailed structure of the cylinder wall is also unsolved. Konings and associates (*28,77*) have proposed a hexagonal array of units based mainly on biochemical data.

4. 130, 155, and 175 S Structures

These structures are found under certain conditions in hemocyanin of the whelk *B. canaliculatum* (Rhachiglossa, Prosobranchia) (Fig. 21). The molecules are observed as circles measuring 350 Å across and as rectangles with 6, 9, 12, and more rows. A closer look shows that it is often possible to divide the striation pattern into one six-row and several three-row structures. It is clear that we are again dealing with stacked cylindrical molecules of 350-Å diameter. The 9-row structure probably corresponds to the 130 S structure, the 12-row structure with 155 S, and the 15-row structure with 175 S, and so on. It has been shown that the sedimentation coefficients correspond nicely to those predicted for such structures (*78,79*).

Indefinite polymerization can occur for any hemocyanin near its isoelectric point (*42*). Crystallization is possible after the addition of ammo-

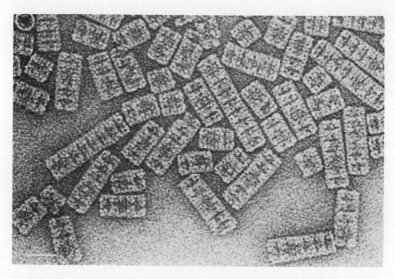

Fig. 21. *Busycon canaliculatum* hemocyanin molecules negatively stained at pH 7 with uranyl EDTA (scale, 500 Å).

nium sulfate (Fig. 22), and at the isoelectric point under salt-free conditions (*80*). Study of these crystals with optical diffraction or, when they grow large enough, with X-ray diffraction, will hopefully give more structural information about hemocyanins in the not-too-distant future.

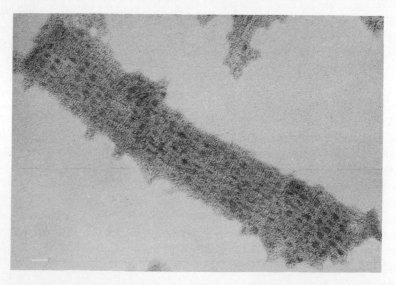

Fig. 22. Crystal of *P. leopoldvillensis* hemocyanin molecules negatively stained with uranyl acetate (scale, 500 Å).

E. Summary

We have shown the 16 S structure to be the building block for the arthropod hemocyanin molecules. Although details of the 16 S structure are not clearly established, there are strong indications for some kind of prismatic structure. The 24 S structure is a dimer of the 16 S structure and is most often seen as a combination of a hexagonal profile with a square profile. The 34 S structure is again a dimer of the 24 S structure in a probably parallel (but perhaps antiparallel) orientation. The 60 S structure is most likely a two-layered octomeric structure with some kind of a polarity. Every layer consists of four 16 S structures.

In contrast, the molluscan hemocyanin molecules exhibit in all cases circular projections with the same diameter and symmetry and rectangular projections with an increasing number of rows. This clearly underlines their cylindrical character. From a structural point of view, *H. pomatia* hemocyanin cannot be considered simply as a dimer of *O. vulgaris* hemocyanin, and *B. canaliculatum* hemocyanin is more complex than a mere polymer of *H. pomatia* hemocyanin. This can be concluded from the staining patterns of rectangular projections of the cylinders.

It is further concluded that there exists no evident structural relation between the hemocyanins from Arthropoda and Mollusca at the quaternary level of organization. This of course does not exclude other types of structural relation.

To avoid complications possible structural difference between α-, β- and apohemocyanin have not been discussed. Thus far, however, we have not been able to show any such structural difference by electron microscopy (*61*).

VI. OXYGEN BINDING BY HEMOCYANINS

All hemocyanins reversibly bind oxygen,† and in most cases appear to play a physiological role analogous to that of hemoglobin in higher animals, that is, the transport of oxygen to tissues. However, the fact that copper rather than iron is the binding site metal, and the high multiplicity of binding sites in some hemocyanin components, give rise to certain unique aspects in their oxygen-binding behavior. We discuss first the spectral changes accompanying oxygen binding, for these changes are commonly utilized in experimental studies of this reaction, and the spectra provide a little information about the binding sites and changes in the protein accompanying binding.

† Hemocyanins are also known to bind carbon monoxide (*19,122*), at the same site as oxygen. There is also evidence for the formation of an NO complex (*80*). However, we shall restrict the discussion here to the much more extensively studied O_2 binding.

A. Spectral Changes in Oxygen Binding

The spectrum of a deoxygenated hemocyanin, while often distorted by the light scattering exhibited by these large particles, is in general a typical protein spectrum (see Fig. 23). Contributions from the copper are not obvious.

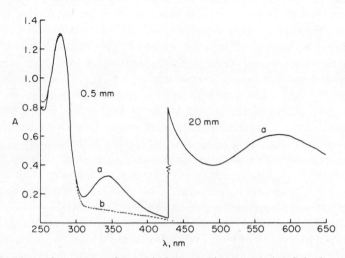

Fig. 23. Absorption spectra of *L. pealei* hemocyanin. a, oxygenated; b, deoxygenated. Note that different path lengths have been used in the different regions of the spectrum.

Upon binding oxygen, hemocyanin exhibits a profound change in absorption spectrum. Strong bands in the neighborhood of 350 and 400–800 nm are observed (Fig. 23). Early measurements by Redfield (*81*)

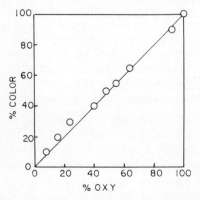

Fig. 24. Absorbance (by colorimetric determination) in the visible region as a function of percent oxygenation (gasometric) for *L. polyphemus* hemocyanin. [From date of Redfield (*81*).]

(Fig. 24) indicate that the absorption in the visible (in excess of scattering) is proportional to the fraction oxygenation, but the relationship has not been checked with the precision available with modern instruments. Since the band at about 350 nm is considerably more intense, the absorption at this wavelength is generally used at the present time as a measure of oxygenation. This band, similar to the one at 600 nm, disappears completely upon deoxygenation (see Fig. 25); it has been assumed that the absorbance at this wavelength, corrected for scattering, is proportional to the fraction

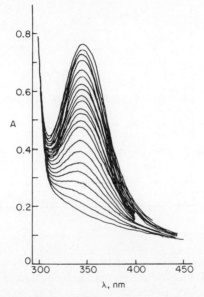

Fig. 25. Spectrum of *B. canaliculatum* hemocyanin as a function of oxygen pressure, in the region about the 350-nm band. [H. A. DePhillips, unpublished.] Oxygen pressures are 0, 0.93, 1.87, 2.78, 3.70, 4.60, 5.51, 6.41, 7.31, 8.20, 9.08, 9.96, 10.83, 11.70, 13.41, 15.11, 17.62, 21.69, 29.49, 52.19, 95.62, and 619 mm.

oxygenation. While the assumption is doubtless *approximately* valid, it is also now known that there are conformational changes accompanying oxygenation (see below). If this is true, it is possible that the extinction coefficient may change over the oxygenation range. This, together with possible changes in degree of association upon oxygenation (with accompanying changes in light scattering), may lead to some error in the spectrophotometric determination of oxygen-binding curves. This point should be kept in mind in considering the curves shown in Section VI. It is possible that a correct oxygen-binding curve for hemocyanin is yet to be determined.

It is now known that the absorption spectra of oxyhemocyanins are not as simple as had been supposed. Circular dichroism (CD) spectra (*82–84*) reveal that in both molluscan and arthropod hemocyanins the broad band in the visible portion of the spectrum is in fact multiple (see Fig. 26). In most cases two or three bands are revealed in the spectral region between 400 and 800 nm. Furthermore, this has been supported in one instance by spectra recorded at liquid nitrogen temperature, which give enhanced resolution (82). However, evaluation of this data involves the assumption that no pronounced conformational changes occur in aqueous glycol solutions at low temperatures.

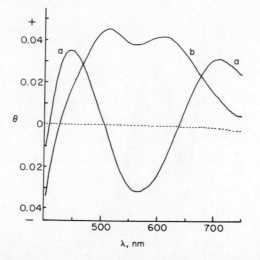

Fig. 26. Circular dichroism spectra of the hemocyanins from *L. pealei* (a) and *C. borealis* (b) in the visible region of the spectrum. Both solutions are fully oxygenated; concentrations are 16.8 mg/ml (a) and 24.9 mg/ml (b). Data are given in degrees of ellipticity. Cell length, 2 cm. [Unpublished data of Nickerson et al. (84).]

While it is difficult to assign exactly either maximal wavelengths or extinction coefficients to the weaker bands, estimates are given in Table VII. Data from a number of other copper proteins are included for comparison. These results are considered again in Section VI.D in connection with the nature of the binding site.

Recent studies [Nickerson et al. (84)] show that the CD spectra for molluscan species are qualitatively different from those of arthropod hemocyanins.

B. Oxygen-Binding Data

Quantitative measurements of the binding of oxygen by hemocyanins extend back over at least 45 years. A large number of careful determinations of oxygen binding were carried out for a variety of hemocyanins in the

TABLE VII

Spectral Properties of Hemocyanins and Some Other Copper Proteins[a]

Protein	$\frac{Cu}{mole}$	$\frac{Cu(II)}{mole}$	EPR	Spectrum							
				Band I		Band II		Band III		Band IV	
				λ, mμ	ε	λ, mμ	ε	λ, mμ	ε	λ, mμ	ε
Oxyhemocyanin from:											
Octopus vulgaris	b	c	—	347	8900	440	<500	570	500	700	<500
Loligo pealei	b	c	—	345	8900	440	d	570	370	700	d
Helix pomatia	b	c	e	346	9000	d	d	580	d	d	d
Other Cu proteins:											
Ceruloplasmin	8	4	+	332	500	459	140	610	1300	794	260
Laccase (fungal)	4	2	+	330	1000	d	d	610	1000	650	≪1000
Plastocyanin	2	2	+	—	—	460	590	597	4900	770	1700
Rhus vernicifera blue protein	1	1	+	—	—	450	970	608	4030	850	700

[a] Most of the data from van Holde (82); original references are given therein.
[b] Hemocyanins contain many coppers per active molecule.
[c] The oxidation state of copper is not known (see Section VI.D).
[d] Not measured.
[e] A very weak band has been detected; see Reference 98.

period before 1940 (see, for example, References *85–90*). However, at this time little was understood about the association–dissociation processes that take place in hemocyanins, and it is difficult in many instances to know the state of aggregation of the materials that were studied. Furthermore, for some of the earlier data conditions of pH and buffer composition are not adequately described. Therefore the early literature in this field must be regarded with some caution.

It was recognized quite early that oxygen binding by hemocyanins is often of a cooperative nature. Many of these data were interpreted in terms of the Hill equation

$$\frac{x}{1-x} = Kp^n \qquad (2)$$

where x is the fraction saturation and p the oxygen pressure (usually given in millimeters). Expressed in logarithmic form Eq. (2) becomes

$$\log \frac{x}{1-x} = \log K + n \log p \qquad (3)$$

which indicates that a graph of $\log x/(1-x)$ vs. $\log p$ should be a straight line. If $n = 1$, the Hill equation describes noncooperative binding; for $n > 1$, positive cooperativity† is indicated. Values of $n < 1$ can result from either negative cooperativity or a heterogeneity in binding sites. Two facts should be kept in mind: (1) The Hill equation, as given by Eq. (2) or (3) with $n > 1$, is never strictly applicable over the entire p-range. Such a situation would correspond to total cooperativity, an "all-or-none" binding situation for each molecule. In real cases the graph is sigmoid, with regions with slope $= 1$ for very low or very high p (see Fig. 27). Wyman (*91*) has pointed out that this shape can be extreme in the case of a molecule with *many* binding sites; there can be a nearly discontinuous change in binding over a very small interval of ligand concentration. (2) The exponent n does not in general have the physical meaning often ascribed to it: the number of binding sites per molecule. This would only be true for all-or-none cooperativity; in real cases n is less than the number of sites per molecule.

In Fig. 27 are shown some Hill plots from a relatively recent study of oxygen binding. Two facts should be noted: (1) the curves are of the form described above, and indicate cooperativity, and (2) there is a pronounced effect of pH upon the oxygen binding. This latter phenomenon has also been observed in oxygen binding by hemoglobins and is termed the *Bohr*

---

† The binding is said to exhibit positive cooperativity if the binding of one ligand facilitates the binding of more ligands. If the binding of a ligand *inhibits* further binding, the system is said to exhibit negative cooperativity.

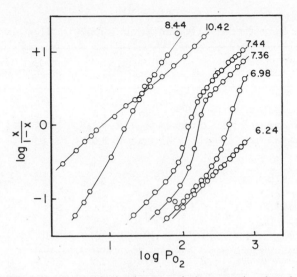

Fig. 27. Graphs of the oxygen binding by *L. pealei* hemocyanin, plotted according to Hill's equation [Eq. (3)]. Data are obtained at the pH values indicated on the graph. Data of DePhillips et al. (*50*).

effect. A positive Bohr effect, defined as an increase in binding strength with increasing pH, is usually thought to be physiologically useful; lower pH values in the tissues promote the unloading of oxygen from the blood. We shall see that a number of hemocyanins have negative Bohr effects— lower pH values strengthen binding. It has been suggested that such behavior facilitates respiration in a carbon dioxide–rich environment.

In Table VIII are listed data on oxygen binding by a number of hemocyanins. The data are mainly chosen from the more recent literature and are intended to be representative rather than comprehensive. Maximum values of n are listed, and a measure of the strength of binding is given by P_{50}, the oxygen pressure for half-saturation. Finally, the sign of the Bohr effect is given for each.

Examination of these data, as well as others in the literature, leads to the following conclusions:

1. The binding is, in many cases, cooperative. This is not unexpected, since near neutral pH the hemocyanins invariably exist as multisubunit structures. The values of n observed appear to be far smaller than the number of binding sites probably present per molecule. This is not unexpected, however, for all-or-none binding is not expected. More surprising is the number of instances in which noncooperative binding is observed in these large molecules. For example, it has been shown that while binding

TABLE VIII

Some Representative Oxygen-Binding Data for Hemocyanins

Source of hemocyanin	pH	Divalent cations?	n_{max}	P_{50}, mm Hg	Bohr effect	Ref.
Arthropoda:						
Limulus polyphemus	7.43	No	1.0	2.1	−	81
Cardisoma guahumi	7.55	Yes	2.64	3.5	+	119
Homarus americanus	7.7	Yes	3.6	2.5	+	120
Mollusca:						
Helix pomatia (α)	7.6	Yes	2.3	11	Small+	24
Helix pomatia (α)	7.6	No	1.1	10	+	24
Helix pomatia (β)	7.6	Yes	4.6	21	−	24
Helix pomatia (β)	7.6	No	1.1	5	−	24
Loligo paelei	7.36	Yes	3.9	150	+	50
Busycon canaliculatum	8.77	No	1.0	2.5	−	88
Busycon canaliculatum	8.2	Yes	2.0	7.9	?	92

is cooperative in both α- and β-hemocyanins from *H. pomatia* when 0.01 M Mg^{2+} and 0.01 M Ca^{2+} are present, noncooperative binding is found when these ions are absent (Table VIII). Similarly, the hemocyanin from *Busycon* appears to exhibit cooperative binding of oxygen only in the presence of divalent ions. Finally, even though the 11 S subunit of *Loligo* hemocyanin contains about eight subunits, its binding curves at pH 10.5 indicate $n = 1$ over most of the range, and $n < 1$ in part of the curve. A similar result has been found by Konings et al. (*24*) for succinylated hemocyanin from *H. pomatia*. This absence of interaction within the multisite subunit raises perplexing questions. In the latter two cases (*Loligo* hemocyanin at high pH and succinylated *Helix* hemocyanin), the material has been dissociated to the 11 S stage. Does this mean that cooperativity only exists *between* subunits? What is the role of the divalent ions? They appear to both stabilize quaternary structures and promote cooperative binding. Yet it has been observed by DePhillips and van Holde (unpublished) that *Loligo* hemocyanin at pH 7.4 still shows cooperative binding in the *absence* of Mg^{2+}. More work is required in this area.

2. The strength of oxygen binding, as measured by the P_{50} at neutral pH, varies enormously over this class of proteins. In general, as might be expected, the P_{50} values are low for bottom-dwelling, inactive species and high for free-swimming, active animals.

3. The Bohr effect can be either positive or negative. However, that the situation is far from simple is shown by the results of Konings et al. on the Bohr effects in α- and β-hemocyanin from *H. pomatia* (see Table VIII).

Apparently, the change in behavior with pH is *opposite* for these two components. Since the question of multiple components has hardly been considered for hemocyanins other than that of *H. pomatia*, we face the appalling prospect that the oxygen-binding behavior observed for any whole hemocyanin may be the sum of behaviors for mixtures of two or more very different components. If this is true, the task of unraveling this problem will become formidable indeed.

C. Evidence for Molecular Changes Accompanying Oxygen Binding

It is becoming quite clear that rather significant changes occur in the hemocyanin molecule when oxygen is bound. It is perhaps simplest to consider, in turn, evidence for or against changes at the levels of secondary, tertiary, and quaternary structures.

1. Secondary Structure

It is generally believed that changes in secondary structure (amount of α-helix, β-conformation, and so on) are manifest in changes in the CD or optical rotatory dispersion (ORD) in the far UV [see, for example, Reference *93*]. On this basis, there is *no* evidence that oxygenation produces any such changes. For example, recent studies show that oxygenation produces no change at all in the far UV CD of the *Busycon* hemocyanin (*52*). A similar result has been found for *L. pealei* hemocyanin by both CD (*50*) and ORD (*49*). In any event, the result is not surprising and may be expected to be general. A change in the secondary structure of a protein is a profound change in conformation, and likely more than should be expected from the binding of a small ligand.

2. Tertiary Structure

Evidence that the environment of certain side-chain chromophores has changed is tentative evidence for changes in tertiary structure (*94*) (although such changes can also result from quaternary structure modification— see below). There is some evidence for such changes from CD studies of oxy- and deoxyhemocyanins (see Fig. 28). However, we are unable at the present time to describe these changes in more detail.

3. Quaternary Structure

As already pointed out in Section IV, the state of aggregation of *L. pealei* hemocyanin is markedly sensitive to oxygenation (*50*), and similar effects have been observed with the protein from *B. canaliculatum* (*52*). Furthermore, small differences in sedimentation coefficient of the oxy-

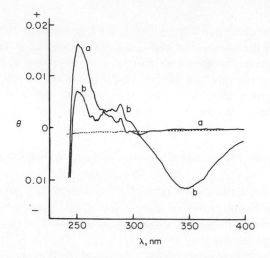

Fig. 28. A comparison of the CD spectra of deoxygenated (a) and oxygenated (b) *Busycon* hemocyanin. Each is at a concentration of about 1 mg/ml in a 0.5-cm cell, and at pH = 7.8. Note the changes in the aromatic region (250–300 nm) of the CD spectrum. Ellipticities are given in degrees. [Nickerson et al. (*84*).]

genated and deoxygenated forms of the 60 S *Loligo* hemocyanin suggest that the arrangement or spacing of subunits may differ in the two states (*50*). General arguments have been proposed (*95*) for a dependence of subunit interaction energy upon binding of small ligands, so the observation of such phenomena is not surprising. However, consideration of the apparent complexities of the association equilibria of hemocyanin components (see Section IV) suggest that it may be some time before these phenomena are understood in any detail. It seems worthwhile to point out that a high-molecular-weight hemocyanin particle, which may contain as many as 200 binding sites, approaches an "extended system" in the sense defined by Wyman (*91*). The changes accompanying partial saturation of such a large set of binding sites may be complex indeed.

In a very recent publication, Brunori (*109*) has shown by temperature-jump measurements that the kinetics of oxygen release from *Octopus* hemocyanin follow a two-stage process. Brunori suggests that the slower stage in this reaction may correspond to a conformational change.

D. Oxygenation and the Oxidation State of Copper

It is surprising, considering all of the studies that have been carried out on hemocyanins, that as fundamental a question as the oxidation state of the copper is still unsettled. As far as deoxyhemocyanin is concerned, a number of lines of evidence strongly suggest that the copper is in the Cu(I) state:

1. The apparent absence of a typical Cu(II) spectrum in deoxyhemocyanins. However, it should be pointed out that this may not be conclusive, for the copper content of most hemocyanin solutions studied has been very low, and the bands might escape detection if very weak.

2. The absence of an ESR signal.

3. The fact that copper is not readily removed by typical strong Cu(II) chelators such as EDTA, whereas CN^- and 2, 2'-biquinoline, which are good chelators for Cu(I), are effective.

4. The fact that copper can be reintroduced into the apoprotein only as Cu(I).

The situation with respect to the oxyhemocyanins is much less clear. An absorption spectrum similar to those of many Cu(II) complexes (but much stronger) and very like those of a number of Cu(II) proteins (see Table VII) is observed. However, it is generally believed that undenatured samples of hemocyanin give no ESR spectrum (96). Heirwegh et al. (97) and Bayer and Fiedler (20) found that while fresh oxyhemocyanin showed no ESR signal, one appeared upon aging. Conversely, Andree et al. (98) observed a very weak signal which does not change with aging of the protein. A weak signal in fresh *Homarus* hemocyanin has been reported by Blumberg (99); a similar result was found for *Jasus lalandii* hemocyanin by Boas et al. (100). As further evidence for the divalent state, it should be noted that the Cu(I) reagent, 2,2'-biquinoline, does not remove all of the copper from oxyhemocyanin, in contrast to its effect on the deoxyprotein (25, 101). Finally, it should be pointed out that it is possible to irreversibly oxidize the copper in hemocyanin to Cu(II), by the use of H_2O_2. The resulting methemocyanin is incapable of oxygen binding. Some groups report that this material exhibits an ESR signal (96), others have not been able to detect any change (98).

To explain the apparently contradictory observations concerning the oxidation state, a number of postulates have been put forward. These may be roughly summarized:

1. The copper is in the Cu(I) state in both oxy- and deoxyhemocyanins. This may explain the absence of strong ESR, but seems to require that the absorption bands, in both the near UV and visible, be charge transfer bands. No really comparable sets of bands are recorded for other charge transfer complexes, and the similarity of the hemocyanin spectrum to those of known Cu(II) proteins seems hard to explain in this way. Finally, the effect of biquinoline on oxyhemocyanin must be explained.

2. The copper is all, or partly, in the Cu(II) state. It has been suggested (25) that electron transfer to oxygen may occur, leading to a set of states

$$Cu(I) - O_2 - Cu(I) \rightleftarrows Cu(I) - O_2^- - Cu(II)$$
$$\rightleftarrows Cu(II) - O_2^= - Cu(II) \qquad (4)$$

If this is true, the absorption bands (at least in the visible) can be identified as arising from d–d transitions; their intensity can be explained, as in other Cu(II) proteins, as resulting from an abnormal binding site geometry (*102*). If, for example, a tetrahedral site appropriate to Cu(I) were entirely or partially preserved in the transition to Cu(II), such effects would be predicted.

The main problem that this explanation faces is the apparent absence of a significant ESR signal. It requires the hypothesis of either spin coupling between Cu(II) atom pairs or very rapid resonance between the states indicated in Eq. (4). Well-documented cases of such behavior in Cu(II) complexes are known (*103*).

3. A third point of view rejects the question and holds that it is meaningless to assign an oxidation state to a metal in some circumstances (*103*). If Eq. (4) describes the situation, this may be the best view to take.

At present, the question must be described as still unsettled. The proposal for a distorted binding site has an attractiveness, for the rapid shifts between oxidation states which would be promoted by a copper binding site intermediate between those suitable for Cu(I) and Cu(II) should facilitate the facile oxygen exchange these proteins show. The fact that this is a Cu(I) protein that can accept oxygen without being irreversibly oxidized should not be overlooked. Whether or not the copper is oxidized during the presence of the oxygen, something compels it to revert to or remain in the Cu(I) state upon loss of the oxygen.

A final note of caution with respect to experimental results in this area is in order. It has been noted (see, for example, References *28* and *97*) that hemocyanins exhibit loss of oxygen-binding capacity upon aging. There are indications that the product of this process resembles the methemocyanin produced by H_2O_2 oxidation. It is possible that some of the confusion in the literature results from studies made on hemocyanin that had been modified by the aging process.

VII. SUMMARY

Phylogenetically, blood pigments represent a labile set of characters (*104*). Hemochromogens, being universally distributed in aerobic cells, serve as blood pigments in several unrelated groups of animals. Hemoglobins may have evolved numerous times; the heme is the same, but the globins differ greatly. Chlorocruorin has a porphyrin that differs from hemoglobin in only one side group of one pyrrole ring. Only one other iron pigment, hemerythrin, has been used in oxygen transport. Hemocyanin is a very different kind of molecule, being a large copper–protein of which several aggregation states are found within the same animal.

Most studies of blood pigments have concerned the hemoglobins. Structure–function relationships have been investigated by numerous groups of workers in connection with physicochemical properties. Compared to the hemoglobins, the hemocyanins have drawn only a little attention even though these proteins are widely distributed among many different invertebrate species. Our knowledge about hemocyanin is still very incomplete and based on an examination of relatively few species. For this reason we can raise many questions which still cannot be answered.

Hemocyanin definitely serves as a normal transporter of oxygen in the Arthropoda and Mollusca. The extent, however, to which it is essential for this activity probably varies greatly. In the Cephalopoda it is necessary for life, whereas the more sluggish forms, both arthropod and molluscan, could possibly do without its aid in oxygen transport, especially when not under the stress of greater-than-normal activity (*105*). The concentration of hemocyanin in the hemolymph is seasonally variable; Zuckerkandl (*106*) found a complete absence of hemocyanin for the spider crab *Maja squinado* during a certain period of its molt cycle.

Why are the hemocyanin molecules so large and why do they occur in different aggregation states? An osmotic regulation by this dissociation–association process is difficult to understand since the total contribution of the hemocyanin particles to the osmotic pressure is small.

Binding of oxygen changes the state of aggregation of some hemocyanins (Sections IV and VI). Physiologically, this probably means that the animal can influence the oxygen-binding of its blood by changing the aggregation state of the hemocyanin through environmental factors. The cooperative binding of oxygen in the *H. pomatia* α-hemocyanin 9×10^6 dalton structure appears to occur only through interactions between the 4.5×10^5 dalton subunits.

The reason for the occurrence of α- and β-hemocyanin in the hemolymph of some Gastropoda species is unclear. The definition is based on behavior in 1 M NaCl, the α-hemocyanin being dissociated into halves in this medium. For *H. pomatia* the pH-dependent shift of the oxygenation curves was shown to be opposite for the two components (*24*). However, recent studies on *P. leopoldvillensis* hemocyanin showed a β-character with respect to 1 M NaCl and behavior such as α-hemocyanin in binding oxygen (*107*). As all marine Gastropoda seem to possess β-hemocyanin, the only significance of this component may have to do with stability in an environment with a high salt concentration.

The biosynthesis of hemocyanins is still obscure. The hepato-pancreas has been suggested as the possible site of synthesis (*108*). The studies of Fahrenbach (*73*) reinforce this idea.

The occurrence of hemocyanin has been shown in only certain classes of Mollusca and Arthropoda. Its possible presence in other classes is uncertain in many cases simply because they have not been investigated. It is certainly worthwhile to repeat and expand the older studies with more sophisticated detection methods, especially since the properties of hemocyanin are now much better defined than in the past.

The amino acid composition generally corresponds to that of proteins built from a large number of subunits. The carbohydrate content is large, its possible role unknown. A thorough study of this problem, however, is under way.

Copper in biological macromolecules draws wide attention at the present time. Nevertheless, many fundamental questions about copper in hemocyanin are still unanswered. Even its valence (if it has one valence) is not determined with absolute certainty. Strong copper bands in the hemocyanin absorption spectrum at 350 and 400–800 nm are only visible in the oxygenated state. Deoxygenated hemocyanin is colorless and shows a typical protein spectrum. Only monovalent copper can be reintroduced in apohemocyanin. Perhaps ESR studies combined with ORD and CD measurements under carefully controlled circumstances can provide additional data. Recent studies of Andree and co-workers (*98*) indicate that the so-called aging of *H. pomatia* hemocyanin is not the result of a change in oxidation state of the copper. The same weak ESR signal is found in all situations, indicating that the phenomenon of aging is attributable to a change of the ligands in hemocyanin. Histidine, tyrosine, and cysteine have been suggested as possible candidates for the ligands. The removal of copper with KCN is certainly incomplete for Molluscan hemocyanins; here part of the copper seems to be bound in a different way. For *H. pomatia* hemocyanin the copper clearly plays a structural role (*24*), either by its influence on the protein conformation, or by a possible role as a "copper bridge".

How closely related are hemocyanins from Arthropoda and Mollusca? The copper content, the CD spectra, and the subunit organization are completely different. The amino acid composition shows a certain relationship. The meaning of the latter is open to argument, since a certain degree of similarity in amino acid composition is common to all proteins organized from subunits. Clearly, the answer must come from amino acid sequence studies, which at least for *H. pomatia* α-hemocyanin are turning out to be painstaking and time-consuming because of many unexpected and unusual problems.

Arthropod hemocyanins are probably built from identical 37,000-dalton protein chains, each chain having one copper, or perhaps from nonidentical 25,000-dalton protein chains, with one out of three chains lacking the copper. In any event, these chains form a 75,000-dalton (5 S) unit from which the larger aggregates are built. Association from and dissociation to these (functional?) units is a reversible process. The detailed organization of these aggregation states is discussed in Sections IV and V.

For the molluscan hemocyanin the situation is different. It is only possible to carry reversible dissociation to a subunit size of about 450,000 daltons. Dissociation of these 450,000-dalton units to protein chains of 50,000 daltons (and partially to 25,000 daltons) was shown to be possible but irreversible for *H. pomatia* α-hemocyanin. This dissociation was always accompanied by a loss of the copper and the functionality. The copper content of Molluscan hemocyanin corresponds to one copper per 25,000 dalton. Small differences between gastropod and cephalopod hemocyanins are observed both in ultracentrifugation and electron microscopy.

The microheterogenity of hemocyanin molecules shown during dissociation and association of the aggregation states presents an intriguing problem. It is unknown whether or not it is also present in a sample from one animal.

Why does the size of the hemocyanin molecules differ so depending on biological origin? This is another unsolved problem, although we know that within a given animal the aggregation state must be determined by the environment.

Our knowledge of hemocyanin is steadily increasing, but the progress is slow. A wide-range attack with all the available techniques of biochemists, physical chemists, the structural chemists will be needed to provide answers to the many unsolved fundamental problems.

VIII. NOTE ADDED IN PROOF

In the period since this manuscript was prepared, a number of advances have been made in the study of hemocyanins.

Our knowledge of hemocyanin structure has been considerably advanced by a low angle x-ray scattering study by I. Pilz and O. Kratky [*Z. Naturforsh.*, **B25**, 587 (1970)]. The results are particularly important, since they show that the dimensions of *H. pomatia* hemocyanin in solution are in very good agreement with those found for negatively-stained specimens by electron microscopy. Another Molluscan hemocyanin (from *Pila leopoldvillensis*) has been investigated by the light scattering technique [F. G. Elliott and J. Hoebke, *Comp. Bioch. & Physiol.*, **36**, 71 (1970)]. The new technique of determining diffusion coefficients by laser light scattering has been employed in a study of *M. trunculus* hemocyanin [R. Ford,

E. Jakeman, P. J. Oliver, E. R. Pike, R. J. Blagrove, E. Wood, and A. R. Peacocke, *Nature*, **227**, 242 (1970)].

Dr. G. Kegeles and coworkers have been investigating the association-dissociation equilibria of *H. americanus* hemocyanin by ultracentrifugation and temperature-jump light scattering. Two papers on this work have been accepted for publication in the Archives of Biochemistry and Biophysics, and should have appeared by the time this article reaches press.

Our understanding of the nature of the copper binding site is increased by other recent studies. N. Shaklai and E. Daniel [Biochemistry, **9**, 564 (1970)] have shown that fluorescent energy transfer from tryptophan to the Cu-O chromophore occurs in the hemocyanin of the mollusc *Levantina hierosolima*. An infrared study of the carbon monoxide complex of the hemocyanin of the key-hole limpet *Megathura crenulata* is reported by J. O. Alben, L. Yen, and N. J. Farrier [*J. Amer. Chem. Soc.*, **92**, 4475 (1970)]. Their data is interpreted as indicating that the carbon monoxide molecule is bound to only *one* copper atom per site.

ACKNOWLEDGMENT

We wish to take this opportunity to thank a number of persons who helped in our collection of the data for this manuscript. Our special thanks go to Drs. M. Gruber, J. Dijk, and W. N. Konings of the Laboratorium voor Structuurchemie, Groningen, to Dr. D. Teller, Department of Biochemistry, University of Washington, and to Dr. H. A. DePhillips, Department of Chemistry, Trinity College. These individuals have generously allowed us to examine manuscripts prior to publication and to quote their unpublished results.

REFERENCES

1. A. C. Redfield, *Biol. Rev.*, **9**, 175 (1934).
2. C. R. Dawson and M. F. Mallette, *Advan. Protein Chem.*, **2**, 179 (1945).
3. J. R. Redmond, *J. Cell. Comp. Physiol.*, **46**, 209 (1955).
4. C. Manwell, in *Oxygen in the Animal Organism* (F. Dickens and E. Neil, eds.), I.U.B. Symposium Series, Vol. 31, Pergamon Press, Oxford, 1964, p. 49.
5. F. Ghiretti, in *The Oxygenases* (O. Hayaishi, ed.), Academic Press, New York, 1962, p. 530.
6. F. Ghiretti, in *Physiology of Mollusca* (K. M. Wilbur and C. M. Yonge, eds.), Academic Press, New York, 1964, p. 233.
7. J. Swammerdam, *Historia Generalis Insectorum ofte Algemeene Verhandeling van de Bloedeloose Dierkens*, M. van Dreunen, Utrecht, 1669.
8. W. Blasius, *Z. Ration. Medicin*, **26** (1866).
9. L. Frédéricq, *Arch. Zool. Exptl. J.*, **7**, 535 (1878).
10. B. Lustig, T. Ernst, and E. Reuss, *Bioch. Z.*, **290**, 95 (1937).
11. M. E. Q. Pilson, *Biol. Bull.*, **128**, 459 (1965).
12. W. Decleir, *Arch. Intern. Physiol. Biochim.*, **76**, 2 (1968).
13. G. Duchateau and M. Florkin, *Bull. Soc. Chim. Biol.*, **36**, 295 (1954).
14. K. Heirwegh, H. Borginon, and R. Lontie, *Arch. Intern. Physiol.*, **67**, 514 (1959).
15. E. Phillipi, *Z. Physiol. Chem.*, **104**, 88 (1919).
16. J. B. Conant, F. Dersch, and W. E. Mydans, *J. Biol. Chem.*, **107**, 755 (1934).

17. G. Felsenfeld, *J. Cell. Comp. Physiol.*, **43**, 23 (1954).
18. W. Stricks and I. M. Kolthoff, *J. Am. Chem. Soc.*, **73**, 1723 (1951).
19. F. Kubowitz, *Bioch. Z.*, **299**, 32 (1938).
20. E. Bayer and H. Fiedler, *Annalen*, **653**, 149 (1962).
21. A. Ghiretti-Magaldi, C. Nuzzolo, and F. Ghiretti, *Biochemistry*, **5**, 1943 (1966).
22. A. Ghiretti-Magaldi and G. Nardi, in *Protides in Biological Fluids* (H. Peeters, ed.), Elsevier, Amsterdam, 1963, p. 507.
23. R. Lontie, V. Blaton, M. Albert, and B. Peeters, *Arch. Intern. Physiol. Biochim.*, **73**, 150 (1965).
24. W. N. Konings, R. van Driel, E. F. J. van Bruggen, and M. Gruber, *Biochim. Biophys. Acta*, **194**, 55 (1969).
25. I. M. Klotz and T. A. Klotz, *Science*, **121**, 477 (1955).
26. M. Lontie, *Clin. Chim. Acta*, **3**, 68 (1958).
27. L. C. G. Thomson, M. Hines, and H. S. Mason, *Arch. Biochem. Biophys.*, **83**, 88 (1959).
28. W. N. Konings, Ph.D. Thesis, Groningen, 1969.
29. E. J. Wood and W. H. Bannister, *Biochim. Biophys. Acta*, **154**, 10 (1968).
30. Y. Engelborghs, R. Witters, and R. Lontie, *Arch. Intern. Physiol. Biochem.*, **76**, 6 (1968).
31. K. E. van Holde, *Biochemistry*, **6**, 93 (1967).
32. M. Gruber, in *Physiology and Biochemistry of Haemocyanins* (F. Ghiretti, ed.), Academic Press, New York, 1968, p. 49.
33. R. Witters and R. Lontie, in *Physiology and Biochemistry of Haemocyanins* (F. Ghiretti, ed.), Academic Press, New York, 1968, p. 61.
34. J. Dijk and F. P. Schröder, unpublished results, 1969.
35. D. Teller, private communication, 1969.
36. J. Dijk, J. J. Dekker, M. Brouwer, and M. Gruber, unpublished results, 1969.
37. F. P. Schröder and J. Dijk, unpublished results, 1969.
38. I. B. Eriksson-Quensel and T. Svedberg, *Biol. Bull.*, **71**, 498 (1936).
39. T. Svedberg and K. O. Pedersen, *The Ultracentrifuge*, Oxford Univ. Press, Oxford, 1940.
40. W. N. Konings, R. J. Siezen, and M. Gruber, *Biochim. Biophys. Acta*, **194**, 376 (1969).
41. H. D. Ellerton, D. Carpenter, and K. E. van Holde, *Biochemistry*, **9**, 2225 (1970).
42. R. M. Condie and R. B. Langer, *Science*, **144**, 1138 (1964).
43. K. E. van Holde and L. B. Cohen, *Biochemistry*, **3**, 1803 (1964).
44. E. Stedman and E. Stedman, *Biochem. J.*, **22**, 889 (1928).
45. S. Brohult and K. Borgman, in *The Svedberg, 1884–1944*, Almquist and Wiksell. Uppsala, 1944, p. 429.
46. H. Heirwegh, H. Borginon, and R. Lontie, *Biochim. Biophys. Acta*, **48**, 517 (1961).
47. R. Lontie, G. Brauns, H. Cooreman, and A. van Clef, *Arch. Biochem. Biophys. Suppl.*, **1**, 295 (1962).
48. L. DiGiamberardino, *Arch. Biochem. Biophys.*, **118**, 273 (1967).
49. L. B. Cohen and K. E. van Holde, *Biochemistry*, **3**, 1809 (1964).
50. H. A. DePhillips, K. W. Nickerson, M. Johnson, and K. E. van Holde, *Biochemistry*, **8**, 3665 (1969).
51. K. E. van Holde, H. A. DePhillips, and K. Nickerson, *Federation Proc.*, **28**, 535 (1969).
52. H. A. DePhillips, K. W. Nickerson, and K. E. van Holde, *J. Mol. Biol.*, **50**, 471 (1970).

53. W. Stanley and T. Anderson, *J. Biol. Chem.*, **146**, 25 (1942).
54. G. L. Clark, M. L. Quaife, and R. B. Baylar, *Biodynamica*, **4**, 153 (1943).
55. A. G. Polson and R. W. G. Wyckoff, *Nature*, **160**, 153 (1947).
56. G. Schramm and G. Berger, *Z. Naturforsch.*, **7b**, 284 (1952).
57. M. A. Lauffer and L. G. Swaby, *Biol. Bull.*, **108**, 290 (1955).
58. C. E. Hall, *J. Biophys. Biochem. Cytol.*, **1**, 1 (1955).
59. H. E. Huxley, *Proc. Stockholm Conf. Electron. Microscopy Stockholm, 1956*, p. 260,
60. S. Brenner and R. W. Horne, *Biochim. Biophys. Acta*, **34**, 103 (1959).
61. H. Fernández-Morán, E. F. J. van Bruggen, and M. Ohtsuki, *J. Mol. Biol.*, **16**, 191 (1966).
62. B. von Borries and F. Lenz, *Proc. Conf. Electron Microscopy, 1956*, p. 60.
63. A. C. van Dorsten and H. F. Premsela, *Proc. 6th Intern. Congr. Electron Microscopy*, **1**, 21 (1966).
64. O. Levin, *Arkiv Kemi*, **21**, 15 (1963).
65. O. Levin, *Arkiv Kemi*, **21**, 29 (1963).
66. O. Levin, Ph.D. Thesis, Uppsala, 1963.
67. E. F. J. van Bruggen, V. Schuiten, E. H. Wiebenga, and M. Gruber, *J. Mol. Biol.*, **7**, 249 (1963).
68. L. di Giamberardino and C. de Haën, *Rapp. Lab. Fis. Inst. Super. Sanit.*, **1965**, ISS 65/5.
69. M. Wibo, Ph.D. Thesis, Louvain, 1966.
70. E. F. J. van Bruggen, in *Physiology and Biochemistry of Haemocyanins* (F. Ghiretti, ed.), Academic Press, New York, 1968, p. 37.
71. D. Eleveld and E. F. J. van Bruggen, unpublished results, 1969.
72. E. F. J. van Bruggen, *Proc. 3rd European Reg. Conf. Electron Microscopy, Prague, 1964*, Vol. B, p. 57.
73. W. H. Fahrenbach, *J. Cell Biol.*, **44**, 445 (1970).
74. T. Wichertjes, E. F. J. van Bruggen, and W. H. Fahrenbach, in preparation.
75. W. J. Bartels, A. P. Bois d'Enghien, F. G. Elliott, and E. F. J. van Bruggen, in preparation.
76. E. F. J. van Bruggen, E. H. Wiebenga, and M. Gruber, *J. Mol. Biol.*, **4**, 1 (1962).
77. W. N. Konings, J. E. Mellema, and T. Wichertjes, unpublished results, 1969.
78. V. Bloomfield, W. O. Dalton, and K. E. van Holde, *Biopolymers*, **5**, 135 (1967).
79. V. Bloomfield, K. E. van Holde, and W. O. Dalton, *Biopolymers*, **5**, 149 (1967).
80. Ch. Dhéré, *J. Physiol. Pathol. Gen.*, **18**, 503 (1919).
81. A. C. Redfield, *Biol. Bull.*, **58**, 238 (1930).
82. K. E. van Holde, *Biochemistry*, **6**, 93 (1967).
83. H. Takesada and K. Hamaguchi, *J. Biochem. (Tokyo)*, **63**, 725 (1968).
84. K. W. Nickerson, H. D. Ellerton, and K. E. van Holde, unpublished results, 1970.
85. A. C. Redfield, T. Coolidge, A. C. Hurd, *J. Biol. Chem.*, **69**, 475 (1926).
86. A. C. Redfield and E. D. Mason, *Am. J. Physiol.*, **85**, 401 (1928). results, 1969.
87. A. C. Redfield and R. Goodkind, *Brit. J. Exptl. Biol.*, **6**, 340 (1929).
88. A. C. Redfield and E. N. Ingalls, *J. Cell. Comp. Physiol.*, **1**, 253 (1932).
89. A. C. Redfield and E. N. Ingalls, *J. Cell. Comp. Physiol.*, **3**, 169 (1933).
90. H. P. Wolvekamp, *Z. Vergl. Physiol.*, **25**, 541 (1938).
91. J. Wyman, *J. Mol. Biol.*, **39**, 523 (1969).
92. W. N. Konings, J. Dijk, T. Wichertjes, E. C. Beuverg, and M. Gruber, *Biochim. Biophys. Acta*, **188**, 43 (1969).
93. N. Greenfield and G. Fasman, *Biochemistry*, **8**, 4108 (1969).

94. S. Beychock, *Science*, **154**, 1288 (1966).
95. R. W. Noble, *J. Mol. Biol.*, **39**, 479 (1969).
96. T. Nakamura and H. S. Mason, *Biochem. Biophys. Res. Commun.*, **3**, 297 (1960).
97. K. Heirwegh, V. Blaton, and R. Lontie, *Arch. Intern. Physiol. Biochem.*, **73**, 149 (1965).
98. P. J. Andree, A. J. M. Schoot Uiterkamp, R. van Driel, and J. Dijk, unpublished results, 1969.
99. W. E. Blumberg, in *The Biochemistry of Copper* (J. Peisach, P. Aisen, and W. E. Blumberg, eds.), Academic Press, New York, 1966, p. 473.
100. J. F. Boas, J. R. Pilbrow, G. J. Troup, L. Moore, and T. D. Smith, *J. Chem. Soc. A*, 965 (1969).
101. G. Felsenfeld, *Arch. Biochem. Biophys.*, **87**, 247 (1960).
102. W. E. Blumberg, in *The Biochemistry of Copper* (J. Peisach, P. Aisen and W. E. Blumberg, eds.), Academic Press, New York, 1966, p. 49.
103. G. Morpugo and R. J. P. Williams, in *Physiology and Biochemistry of Haemocyanins* (F. Ghiretti, ed.), Academic Press, New York, 1968, p. 113.
104. C. L. Prosser and F. A. Brown, *Comparative Animal Physiology*, Saunders, Philadelphia, 1961, p. 232.
105. J. R. Redmond, in *Physiology and Biochemistry of Haemocyanins* (F. Ghiretti, ed.), Academic Press, New York, 1968, p. 20.
106. E. Zuckerkandl, *Compt. Rend. Soc. Biol.*, **151**, 524 (1957).
107. W. J. Bartels, A. P. Bois d'Enghien, F. G. Elliott, E. F. J. van Bruggen, in preparation.
108. W. Johnston and A. A. Barber, *Comp. Biochem. Physiol.*, **28**, 1259 (1969).
109. M. Brunori, *J. Mol. Biol.*, **46**, 213 (1969).
110. J. Loehr, unpublished results, 1968.
111. C. H. Moore, R. W. Henderson, and L. W. Nichol, *Biochemistry*, **7**, 4075 (1968).
112. F. J. Joubert, *Biochem. Bliphys. Acta*, **14**, 127 (1954).
113. C. Bancroft and K. E. van Holde, unpublished results, 1967.
114. M. P. Printz, *Federation Proc.*, **22**, 291 (1963).
115. S. Brohult, *J. Phys. Colloid Chem.*, **51**, 206 (1947).
116. R. Lontie and R. Witters, in *Biochemistry of Copper* (J. Peisach, P. Aisen, and W. E. Blumberg, eds.), Academic Press, New York, 1966, p. 456.
117. F. E. Elliot and H. van Baelen, *Bull. Chem. Biol.*, **47**, 1979 (1965).
118. T. Omura, T. Fujita, F. Yamada, and S. Yamamoto, *J. Biochem. (Tokyo)*, **50**, 400 (1961).
119. J. P. Redmond, *Biol. Bull.*, **122**, 252 (1962).
120. S. M. Pickett, A. F. Riggs, and J. L. Larimer, *Science*, **151**, 1005 (1966).
121. E. F. J. van Bruggen, unpublished results, 1969.
122. W. Vanneste and H. S. Mason, in *The Biochemistry of Copper* (J. Peisach, P. Aisen, and W. E. Blumberg, eds.), Academic Press, New York, 1966, p. 465.

CHAPTER 2

HEMERYTHRIN

Irving M. Klotz

BIOCHEMISTRY DIVISION, DEPARTMENT OF CHEMISTRY
NORTHWESTERN UNIVERSITY, EVANSTON, ILLINOIS

I. INTRODUCTION

Nowadays it is widely recognized that structurally dissimilar protein molecules may actually perform the same biological function. As examples one might cite the different ribonucleases (pancreatic and T_1) or, in a perhaps more limited sense, any one of the many isoenzyme systems. It is of interest historically that this possibility was foreseen almost a century ago in the oxygen-carrying pigments. This recognition was possible long before anything was known about the molecular structure of these substances, and at a time when even the most elementary chemical facts were obscure, because the oxygen-carrying proteins are visibly different: in the oxygenated form, hemoglobin is red, hemerythrin is violet-pink, and hemocyanin is blue.

In the deoxygenated state, hemoglobin retains a deep color (somewhat more purple than that of the oxygenated protein), but hemerythrin and hemocyanin become completely colorless. In fact, the last two pigments show no absorption (*1*) in the visible region of the spectrum or even in the near ultraviolet down to almost 3000 Å (see, e.g., hemerythrin (*2*), Fig. 1).

55

The colors and the changes in color are a reflection of the chemical constitution (Table I) at an elementary level. Hemoglobin and hemerythrin contain iron atoms, whereas hemocyanin contains copper. In all of these macromolecules the metal in the deoxygenated protein exists in its lowest common oxidation state, II for iron, I for copper. In hemoglobin, however, the metal is set in a porphyrin ring to form the heme group, whereas (despite their names) hemerythrin and hemocyanin do not contain a heme group. In the last two proteins, therefore, the metal must be coordinated directly to protein side chains.

TABLE I

Comparison of Some Properties of Oxygen-Carrying Proteins

Property	Hemoglobin	Hemerythrin	Hemocyanin
Metal	Fe	Fe	Cu
Oxidation state of metal in deoxy protein	II	II	I
Metal:O_2	Fe:O_2	2Fe:O_2	2Cu:O_2
Color when oxygenated	Red	Violet-pink	Blue
Color when deoxygenated	Red-purple	Colorless	Colorless
Coordination of iron	Porphyrin ring	Protein side chains	Protein side chains
Molecular weight	65,000	108,000	400,000–20,000,000

Although there are exceptions, the hemoglobins in general have molecular weights near 65,000. Similarly, all known hemerythrins seem to have molecular weights near 100,000. Inside the organism these two pigments are carried within erythrocytes, as one might expect for proteins of such relatively low molecular weight. Hemocyanin, in contrast, is carried directly in the plasma and has molecular weights ranging from 400,000 to 20,000,000 depending on the species.

Enormous strides have been taken in working out the structure of hemoglobin within the past decade so that now precise molecular details are available at a level of 3 Å (3,4). It would obviously be of interest to learn to what extent a protein structure can be varied and still perform the same function, for example, carrying oxygen. It is with this objective in mind that investigators have been examining the molecular structure of hemocyanin and hemerythrin. Hemerythrin has proved to be the much more tractable substance and much more detail has been revealed in regard to its molecular structure. Almost all of the studies described in this review have been carried out with hemerythrin from the sipunuclid worm *Golfingia gouldii*. A few comments comparing pigments of other species also are reserved for the end of this chapter.

II. SIZE AND SUBUNIT CONSTITUTION

In many respects the molecular and behavioral characteristics of hemerythrin parallel those of hemoglobin. In addition to having a deoxy and an oxygenated form, both proteins may also be converted to a *met* state in which the iron has been oxidized from the oxidation state II in the deoxy form to an oxidation state of III in the met form (Fig. 2). The most

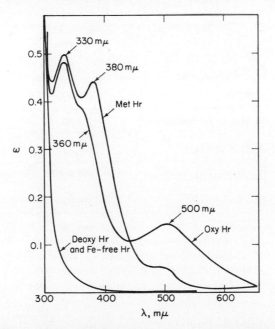

Fig. 1. Spectra of hemerythrin in oxy, deoxy, and met states.

extensive studies of the molecular weight of hemerythrin have been carried out with this met form since this is the only state that can be maintained without contamination from the other forms.

The iron content of hemerythrin, 0.82% (*2,5,6*), leads immediately to a value of 6800 for the minimum weight of a structural unit in this protein. Knowing that it takes two iron atoms to hold one oxygen molecule (*1,2,7*), we can conclude further that a weight of 13,600 must correspond to the minimum functional unit. Any multiple of this number could be the molecular weight of the naturally occurring form. Extensive measurements have been carried out by a variety of thermodynamic and hydrodynamic methods (8). The average value from all these experimental observations for the molecular weight of methemerythrin (Table II) is 107,000.

Since the sedimentation coefficient s of methemerythrin (7 S) is close to that observed for oxyhemerythrin and for deoxyhemerythrin (Fig. 2), it seems apparent that the molecular weight of the last two forms is also 107,000.

TABLE II

Molecular Weight of Hemerythrin
from Hydrodynamic and
Thermodynamic Measurements

Method^a	Molecular weight
s and D	102,000
Archibald	115,000
Sedimentation equilibrium	100,000
Osmotic pressure	100,000
Light scattering	107,000
$[\eta]$, s and β	115,000
Average	107,000

^{a} s, Sedimentation coefficient; D, diffusion coefficient; $[\eta]$, intrinsic viscosity; β, Scheraga-Mandelkern parameter.

Hemerythrin of this molecular weight, containing 16 iron atoms and 8 oxygen molecules in the oxy form, must be constituted of eight functional units. It becomes of interest, therefore, to determine whether this protein is actually constituted of eight subunits, linked by noncovalent bonds, as is the case with the four subunits in hemoglobin, or whether the eight oxygen-binding sites are distributed on one large polypeptide chain or on covalently cross-linked chains.

In fact, hemerythrin can be dissociated into eight subunits by a number of different procedures. Reagents such as sodium dodecylsulfate (*8*) or urea (*9*), which do not break covalent bonds, readily dissociate hemerythrin into species of approximately 13,500 molecular weight (Fig. 3). Alternatively, if a very large charge is introduced into the protein, for example by succinylation at neutral pH, the hemerythrin will dissociate into eight subunits (*10*). The most interesting procedure, from a molecular viewpoint,

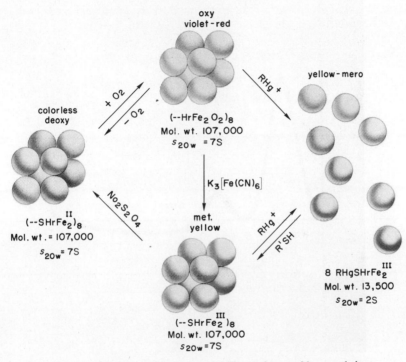

Fig. 2. Interrelationships between different forms of hemerythrin.

is that involving the use of organic materials (*11*) in which the reaction is precisely stoichiometric (see Fig. 4). There are eight SH groups in hemerythrin octamer (*11*); when 8 moles of mercurial have been added, dissociation is complete. Nevertheless, the process is reversible; if a small molecule mercaptan such as cysteine ethyl ester is added to the mercurial monomer, octamer is regenerated (Fig. 5).

Of particular interest is the all-or-none nature of the reaction with mercurial. For example, if the amount of mercurial added corresponds to 50% of the SH groups present in the original octameric hemerythrin, then a sedimentation velocity experiment shows 50% of the protein in the monomeric form and the remaining 50% completely in the octameric form.

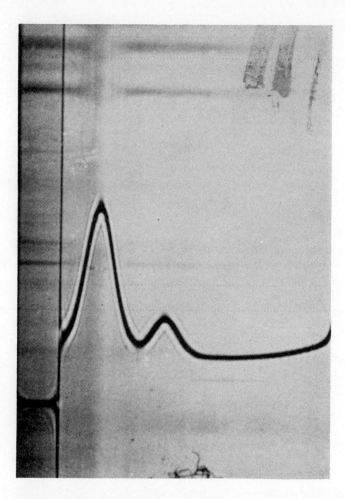

Fig. 3. Schlieren photograph of methemerythrin dissolved in sodium dodecyl sulfate and sedimented at 59,780 rpm. Change of shading near middle of slower-sedimenting peak corresponds to yellow color to right of peak which is associated with subunit as well as the large macromolecule. Similar patterns were obtained in succinylation, relative areas of peaks changing in parallel with increasing addition of anhydride.

No intermediates between these two extremes reveal themselves in the ultracentrifuge; sedimentation patterns (*11*) resemble that shown in Fig. 3.

One possible explanation of the all-or-none reaction is illustrated in Fig. 6. If the octamer is always in equilibrium with a very small amount of monomer and if some state of the monomer has an SH group in a condition more accessible to mercaptan-blocking reagents, then it is clear that the added organic mercurial would react essentially exclusively with the subunit form and continue to remove it from the octamer–monomer equilibrium. As a result, the octamer would continue to replace the used up monomer until a point was reached at which all of the mercurial added had combined with subunit SH groups. At this point the normal octamer subunit equilibrium would be reestablished, but the octamer would not dissociate further. The net result would be a distribution of protein between only two states: mercurial subunits and untouched octamer.

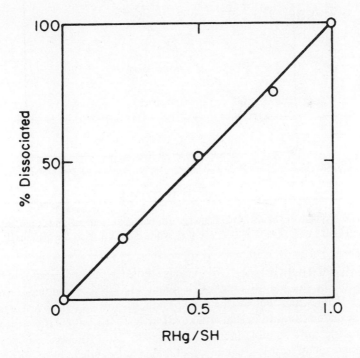

Fig. 4. Stoichiometric nature of dissociating effect of organic mercurial, salyrganic acid, in its blocking of mercaptan groups of hemerythrin.

This explanation is not the only one possible, however. A conformational rearrangement (without dissociation) of the octamer might be generated by the binding of the first mercurial, and this conformation might be much more reactive toward additional mercurial reagent.

It seemed essential, therefore, if the first explanation of the all-or-none reaction is correct, to demonstrate the presence of subunits even though their concentration might be extremely small, too small to be detected by conventional ultracentrifugal or osmotic measurements. The procedure

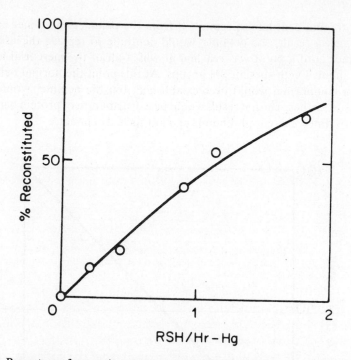

Fig. 5. Percentage of reconstituted hemerythrin obtained from subunits upon addition of cysteine ethyl ester in increasing stoichiometric ratios to the mercurial.

adopted first paralleled that used in early studies with hemoglobin: hybridization of the normal oligomer with a suitably labeled oligomer. Marine worms from which hemerythrin is obtained, however, are not easy to culture under conditions that might allow biosynthesis of the protein in the presence of a suitably labeled amino acid. Nevertheless, it is possible to prepare an appropriate chemically labeled hemerythrin by reaction with a small amount of succinic anhydride (12). Succinylated octameric hemerythrin was obtained by either of two methods (Fig. 7). In the first, the SH group of the protein was protected by a mercurial. The resulting monomer was succinylated and the modified octamer was regenerated by removal of the mercurial. The modified octamer was also obtained by succinylating native hemerythrin with a small amount of succinic anhydride in relation to the total amount of protein. Either procedure yields a marked hemerythrin with substantially larger negative charge than that of the unmodified protein. The small extra charge does not interfere with the aggregation process but does provide an aggregate with a much greater electrophoretic mobility than that of the native protein.

Using the succinylated octamer, one can perform an experiment (*12*) (Fig. 8) to test for the presence of a very small amount of subunit in equilibrium with the octamer. If native hemerythrin is in equilibrium with a small amount of its subunit, and succinylated hemerythrin is in equilibrium with a small amount of its monomer, a mixture of the two octameric proteins should generate hybrid octamers with various proportions of the two types of subunits. These hybrids will have different charges, hence their mobility should be different from either of the parent octamers. They should be easily detectable by electrophoresis experiments.

Such an experiment in a starch gel matrix has been carried out (*12*), and the results are shown in Fig. 9. The position of succinylated octameric hemerythrin is shown in column 1 of Fig. 9. The corresponding solution of native hemerythrin subjected to the same electric field in the same gel moved very little, as is demonstrated in column 3. A third solution containing a 1 : 1 mixture of native and succinylated hemerythrins which had been allowed to stand for 24 hours at 4°C shows the pattern of column 2 in Fig. 9. In column 2 very little density is found at the position of either original component hemerythrin. However, species of all intermediate electrophoretic mobilities are strongly visible. Since succinylation produces a statistical distribution of hemerythrins of modified charge and labeled and unlabeled monomers are randomly combined, this result is exactly what would be expected from the hypothesis of octamer–monomer equilibrium but would not be expected if the protein does not dissociate reversibly.

These experiments thus established the existence of subunits and the presence of a subunit equilibrium in the octameric protein.

III. STRUCTURE WITHIN SUBUNIT

A. Primary Structure

For a protein of molecular weight near 13,500, determination of the amino acid sequence is feasible with now classic techniques. The primary structure of hemerythrin has recently been completely determined (*5,6,13,14*) and is shown in Fig. 10.

Among the particularly interesting features is the presence of a single Cys residue (position 50) near the center of the polypeptide chain. The blocking of this SH group leads to dissociation of octamer, as described

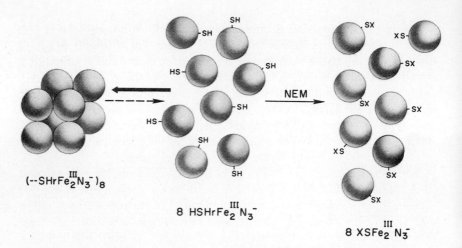

$(--SHrFe_2^{III}N_3^-)_8$

$8\ HSHrFe_2^{III}N_3^-$

$8\ XSFe_2^{III}N_3^-$

NEM

Fig. 6. An explanation of all-or-none reaction in terms of octamer–subunit equilibrium of hemerythrin. NEM represents N-ethylmaleimide, a mercaptan-blocking reagent.

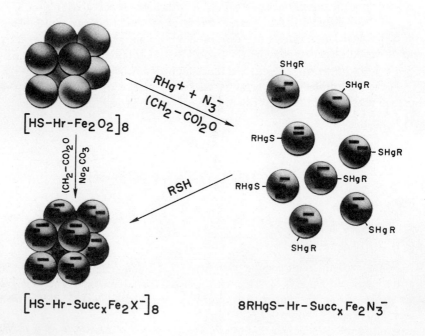

$\left[HS-Hr-Fe_2O_2\right]_8$

$RHg^+ + N_3^-$
$(CH_2-CO)_2O$

$(CH_2-CO)_2O$ Na_2CO_3

RSH

$\left[HS-Hr-Succ_xFe_2X^-\right]_8$

$8RHgS-Hr-Succ_xFe_2N_3^-$

Fig. 7. Schematic diagram of procedures used to prepare chemically labeled hemerythrin.

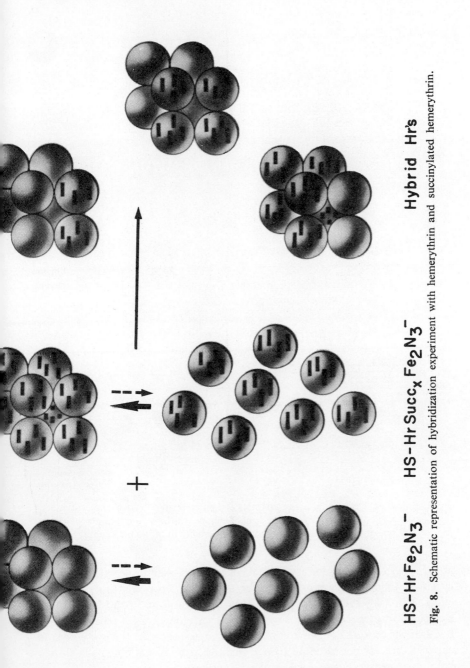

$HS-HrFe_2N_3^-$ $HS-HrSucc_xFe_2N_3^-$ Hybrid Hr's

Fig. 8. Schematic representation of hybridization experiment with hemerythrin and succinylated hemerythrin.

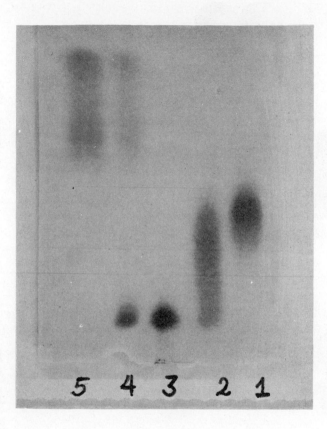

Fig. 9. Starch gel electrophoresis of mixtures of hemerythrins. 1, Succinylated hemerythrin octamer; 2, a mixture of 1 and 3; 3, native hemerythrin; 4, a mixture of 3 and 5; 5, heavily succinylated hemerythrin irreversibly in monomeric state.

above. Especially striking is the presence of Arg-Arg in positions 48 and 49 since it has been reported earlier that the mercurials that are most effective in dissociating hemerythrin (*11*) contain anionic groups. Evidently, electrostatic interactions play a role in attracting mercurials to the neighborhood of the SH group. It is tempting also to assume that the cysteine side chain is directly coordinated to one or both iron atoms but, as is shown below, spectroscopic observations rule out this assumption.

The only other residue that appears but once in the chain is Met. Whether it serves some special function has not yet been determined. Again it is tempting to speculate (by analogy to cytochrome ʋ) that the thioether group may provide one of the ligands to iron.

Some groupings of residues in the sequence may also be mentioned. The subunit chain contains four prolines, and these are all in the first tryptic peptide, at positions 3, 5, 7, and 12. Since peptide chains are synthesized from the N-terminal end, this grouping may condition the folding of the chain in a special way.

The lysines of the chain are grouped toward the carboxyl terminus. The first lysine occurs at position 26. The remaining ones are at positions 53, 74, 75, 88, 92, 95, 103, 108, 110, and 112. Some concentrations of dicarboxylic amino acid residues can also be noted. One such is at residues 22–24, another at residues 38–46, and a third at residues 71–78.

All eight polypeptide chains in octameric hemerythrin have essentially identical primary structures. There are, however, two sites of amino acid interchange, substitution of a Thr for a Gly at residue 79 and of an Ala for a Ser at residue 96 (*13,14*). These substitutions had already been suggested by the amino acid composition of the protein (*5*) in which the residue ratios for these four amino acids are closer to half than to integral residues. These substitutions cannot be responsible for the electrophoretic variants described by Manwell (*9*) since Gly → Thr and Ser → Ala mutations do not lead to changes in protein charge.

B. Optical Rotatory Properties and Helix Content

Hemerythrin shows optical absorption bands in the visible and in the UV region of the spectrum (*2,15*). It would be expected, therefore, that Cotton effects and circular dichroism (CD) bands would be found for this protein. In the visible region the rotatory bands should reflect the coordination environment of the iron atoms at the active site, and their description should be correlated with the optical spectra. This is discussed, therefore, in the next section describing the active site iron. Rotatory behavior in the UV range, however, arises primarily from the peptide bonds and from amino acid side chains of the protein itself.

Near-UV CD spectra of several methemerythrin complexes (*16*) are shown in Fig. 11a. All exhibit Cotton effects at 293, 289, 287, and 268–272 nm. These clearly arise from optically active absorption bands of electronic transitions in side-chain chromophores such as tyrosine and tryptophan; hemerythrin has no disulfide bonds (*13,14*).

In the vicinity of $n \to \pi^*$ and $\pi \to \pi^*$ peptide absorptions, fairly strong bands are found for methemerythrin chloride (*16*). Negative ellipticity bands (Fig. 11b) are observed at 222 and 209 nm, whereas a positive band appears at 197 nm. Helical polypeptides show bands at 222, 206–207, and 190–192 nm (*17*). The most distinguishing difference of the β-structure from α-helix is that the characteristic forked spectrum of the latter (extrema at 222 and 206 nm of approximately equal intensity) is replaced by a single band at 217–218 nm (*18*). Thus the spectrum of hemerythrin resembles that

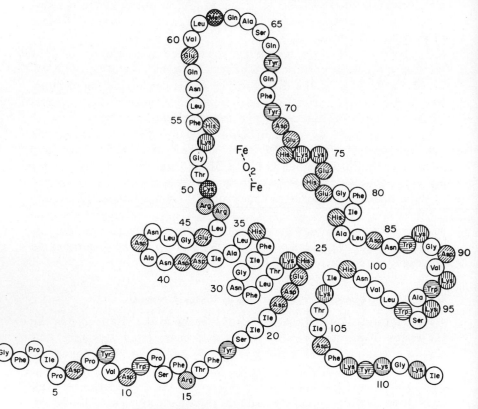

Fig. 10. Primary structure of hemerythrin.

of an α-helix rather than the β-form of secondary structure. If this spectrum is interpreted entirely in terms of α-helix (*17*), then hemerythrin must be about 75% helical. One cannot help noting that the oxygen carriers hemoglobin and myoglobin have helix contents in the same range (*3,4*).

C. State of Iron at Active Site

In regard to the active site, the iron locus, it seems reasonable to assume from the stoichiometry of oxygen uptake, $2Fe/O_2$, that the iron atoms are near each other and that the oxygen molecule forms a bridge between them

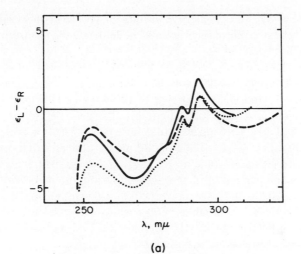

(a)

Fig. 11a. CD spectra of methemerythrin complexes at pH 7.0. ($\cdots$) Hemerythrin chloride; (- - -) hemerythrin thiocyanate; (—) hemerythrin azide.

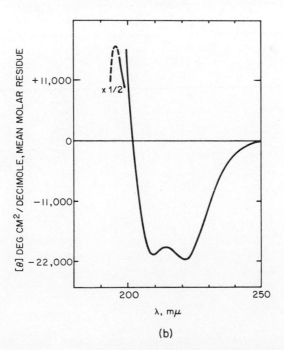

(b)

Fig. 11b. CD spectrum in amide bond region of methemerythrin chloride at pH 7.0. Buffer was 0.01 M tris–cacodylate containing 0.01 M Cl$^-$. Iron concentration was 3.5×10^{-5} M.

(Fig. 10): Fe—O_2—Fe. Extensive studies of spectroscopic and magnetic properties tend to confirm this view and to delineate the electronic structure of the iron more precisely.

Methemerythrin, with iron in the Fe(III) oxidation state forms complexes with a variety of small anions, including N_3^-, NCS^-, NCO^-, Br^-, Cl^-, CN^-, F^-, and OH^-. For two of these, azide (*15*) and thiocyanate (*19*), the stoichiometry of iron to ligand, L, has been determined and again found to be 2Fe/L. This suggests that here too L forms a bridge between the two irons: Fe—L—Fe. Each ligand also produces a rich optical spectrum in its methemerythrin complex (Fig. 12). A detailed analysis of these bands and a comparison of their wavelengths and intensities with those of model iron chelates has been made (*20*) and is summarized in Table III. It is remarkable how similar are the spectra of model oxo-bridged iron complexes, X_nFe(III)—O—Fe(III)X_n, to those of methemerythrin complexes, as well as to that of oxyhemerythrin.

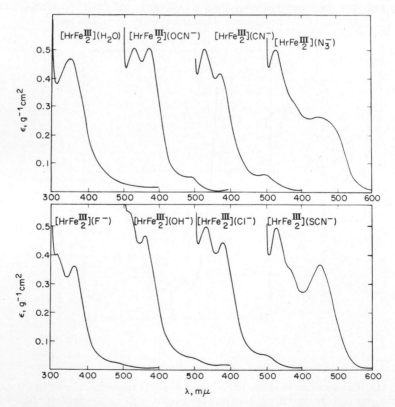

Fig. 12. Absorption spectra of coordination complexes of methemerythrin.

TABLE III

The Absorption Spectra of Hemerythrin Complexes and of Model Oxo-Bridged Complexes

Complex and ligand[a]	Stoichiometry[b]	Absorption spectra[c]															
		λ, mμ	εm	λ, mμ	εm	λ, mμ	εm	λ, mμ	εm	λ, mμ	εm	λ, mμ	εm	λ, mμ	εm	λ, mμ	εm
Oxyhemerythrin																	
O2	1	750sh	200	—	—	*500*	*2,200*	*360sh*	*5,450*	330	6,800	—	—	—	—	—	—
Methemerythrin	1 (?)																
SH−	1	680	190	—	—	*446*	*3,700*	*380sh*	*4,300*	326	6,750	—	—	—	—	—	—
NCS−	1	674	200	—	—	*452*	*5,100*	*370sh*	*4,900*	327	7,200	—	—	—	—	—	—
NCO−	1	650	166	—	—	*480sh*	*700*	*377*	*6,500*	334	6,550	—	—	—	—	—	—
Br−	1	677	165	—	—	*505sh*	*950*	*387*	*5,400*	331	6,500	—	—	—	—	—	—
Cl−	1	656	180	—	—	*490sh*	*750*	*380*	*6,000*	329	6,600	—	—	—	—	—	—
CN−	1	695	140	—	—	*493*	*770*	*374*	*5,300*	330	6,400	—	—	—	—	—	—
F−	1 or 2	595sh	200	—	—	*480sh*	*400*	*362*	*5,000*	317	5,600	—	—	—	—	—	—
OH−	2	597	160	—	—	*480sh*	*550*	*332*	*5,900*	320	6,800	—	—	—	—	—	—
H2O	2	580sh	220	—	—	*480sh*	*600*	—	—	355	6,400	—	—	—	—	—	—
H2O/I−	2	550sh	800	—	—	*480sh*	*1,200*	—	—	340	6,300	—	—	—	—	—	—
[Fe3O(Ac)6(H2O)3]2−[d]	—	973	15.7	515sh	124	464	199	*406sh*	630	335	5,010	—	—	—	—	—	—
[Fe(H2OB)2O]4+[d]	—	876	7.2	495	160	465	172	—	—	364	9,360	—	—	*274*	*15,756*	—	—
[FeEDTA)2O]4−	—	878	6.6	540sh	113	475	199	*405sh*	810	335	10,700	*303*	*14,100*	*267*	*14,100*	*240*	*17,200*
[Fe(−)PDTA)2O]4−	—	880	8.4	540sh	115	475	197	*400sh*	920	335	10,600	*305*	*13,200*	*267*	*14,700*	*240*	*16,500*
[FeHEDTA)2O]2−	—	875	6.5	540sh	104	474	196	*405sh*	830	335sh	11,400	*305*	*15,400*	*270*	*16,400*	*237*	*17,600*
[Fe(phen)2)2O]4+[e]	—	~970	—	580sh	—	—	—	—	—	352sh	—	*310sh*	—	*273*	—	*225*	—
[Fe(bipy)2)2O]4+[e]	—	960	—	575sh	—	525sh	—	*410*	—	345sh	—	*310sh*	—	*281*	—	*234*	—
[Fe terpy)2O]4+[f]	—	~950	—	590sh	—	525	—	—	—	—	—	—	—	—	—	—	—
[Fe salen)2O]e	—	~1,200	—	—	—	500sh	—	—	—	—	—	*290sh*	—	*256sh*	—	*226*	—
[Fe(salen)Cl)2]e	—	~1,250	—	—	—	525	—	*435*	—	377	—	*322*	—	*265*	—	*225*	—

a Ligands forming complexes with hemerythrin. Iodide does not coordinate to the iron but binds nearby.

b The stoichiometries indicated have been proved only in the case of oxygen, azide, and thiocyanate. Those of cyanate, chloride, and hydroxide have been inferred from the interaction constants. The remainder are those indicated by studies in Reference 20.

c The extinction coefficients are calculated per dimeric iron unit. Those bands indicated in italics are charge transfer bands arising from ligands other than the oxo bridge.

d Data recalculated from literature (20).

e Absorption spectra obtained from diffuse reflectance data (20).

f Estimated from the diffuse reflectance data in literature (20).

CD spectra in the visible range have also been examined (*16,20,21*). A typical spectrum of the iron-dependent bands in the visible and near UV is illustrated in Fig. 13. A tabulation of bands for a variety of hemerythrin complexes is assembled in Table IV.

EPR spectra of oxyhemerythrin have been reported (*22,23*). The spectrum consists of an absorption at $g = 2.00$, which is reduced in intensity when the protein is deoxygenated. However, the signal intensity cannot be correlated with the iron in the sample or with the optical spectrum. In fact, the EPR signal is eliminated upon addition of only a fraction of the oxidant required for titration of all the iron (*23*).

The signal at $g = 2$ could be a free radical produced by Fe(III) oxidation of the protein. This would account for the reduction of the intensity upon deoxygenation. The absence of an easily observable EPR signal attributable to iron is not surprising since as will be evident later the same processes that eliminate magnetic hyperfine interaction in the Mössbauer spectra described below also tend to broaden the EPR spectra.

Signals at $g = 2$ have also been observed with methemerythrins. These cannot account for more than 10% of the iron present.

Because the iron is imbedded in a sea of diamagnetic protein, it is very difficult to obtain absolute magnetic susceptibilities of iron in hemerythrin. However, *changes* in magnetic susceptibility are more accessible and have been measured for conversions of hemerythrin among the following forms (*23*),

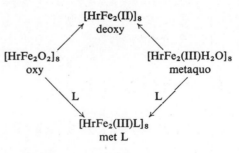

where L represents added ligand.

No appreciable changes in susceptibility occurred when oxyhemerythrin was converted into the metazide or metfluoride form, or when metaquohemerythrin was changed to the azide, cyanide, or thiocyanate complex. However, the magnetic susceptibility increased when deoxyhemerythrin was prepared from either oxyhemerythrin or metaquohemerythrin (Table V).

The magnetic susceptibility data show that oxyhemerythrin and the derivatives of methemerythrin exhibit a lower effective magnetic moment

TABLE IV

The Circular Dichroism of Hemerythrin Complexes[a]

Complex and ligand	Stoichiometry	λ, mμ	Δε	λ, mμ	Δε	λ, mμ	Δε	λ, mμ	Δε	λ, mμ	Δε	λ, mμ	Δε	λ, mμ	Δε	λ, mμ	Δε	λ, mμ	Δε
Oxyhemerythrin																			
O_2	1	—	—	—	—	520	−2.48	430	−0.70	—	—	—	—	336	−3.72	—	—	—	—
Methemerythrin																			
N_3^-	1	—	—	646	+0.06	500	−4.30	450sh	−2.44	370	−8.92	—	—	—	—	—	—	320sh	−1.64
NCS^-	1	—	—	623	+0.42	482	−4.88	—	—	425	−6.96	365sh	+2.74	347	+4.30	—	—	310	−2.10
NCO^-	1	—	—	622	+0.26	487	−0.80	439	+0.04	379	−7.16	—	—	338	+1.48	—	—	302	−2.96
CN^-	1	—	—	650	+0.50	496	−0.70	443	+0.38	378	−2.60	—	—	(332)[d]	—	—	—	305	−4.12
Br^-	1	—	—	638	+0.60	490sh	−1.42	—	—	409	−7.40	—	—	350	+2.52	—	—	297	−6.96
Cl^-	1	—	—	623	+0.54	492	−0.62	444	+0.08	384	−6.56	—	—	333	+1.72	—	—	303	−0.86
F^-	1[b]	—	—	~605	+0.10	500	−0.60	434	+0.04	367	−2.10	—	—	(330)[d]	—	—	—	302	−2.56
F^-	2[c]	670	−0.04	594	+0.04	530	−0.08	482	+0.28	386	+0.94	343	+1.50	(330)[d]	−1.28	—	—	304	−3.32
OH^-	2	—	—	640	+0.16	495	−0.50	470	−0.30	410	+0.18	364	+0.12	(335)[d]	−1.36	—	—	302	−3.12
H_2O	2	680	−0.04	—	—	491	−1.58	450sh	−1.46	415sh	−1.78	367	−1.42	335	+2.18	314	+1.38	298	−2.02
H_2O/I^-	2	—	—	680	+0.14	494	−1.88	430	—	417	−2.80	368	—	335	−1.00	(314)[d]	−3.40	296	−3.80

[a] The optically active bands are given in terms of the absorption maximum λ and the molar extinction coefficient Δε calculated per dimeric iron unit (i.e., per protein subunit).
[b] Inferred stoichiometry of fluoride complex extreme 1 (20).
[c] Inferred stoichiometry of fluoride complex extreme 2 (20).
[d] Position of CD minimum corresponding to a positive band in other complexes.

TABLE V

Changes in Magnetic Susceptibility on Conversion of
Hemerythrin from One Form to Another

Initial state	Final state	$\Delta\chi, \times 10^6$ per mole of iron
$[HrFe_2O_2]_8$	$[HrFe_2(II)]_8$	7500
$[HrFe_2O_2]_8$	$[HrFe_2(II)]_8$	6900
$[HrFe_2O_2]_8$	$[HrFe_2(II)]_8$	7600
$[HrFe_2(III)H_2O]_8$	$[HrFe_2(II)]_8$	7400
$[HrFe_2O_2]_8$	$[HrFe_2(III)N_3]_8$	700^a
$[HrFe_2O_2]_8$	$[HrFe_2(III)F]_8$	700^a
$[HrFe_2(III)H_2O]_8$	$[HrFe_2(III)N_3]_8$	-400^a
$[HrFe_2(III)H_2O]_8$	$[HrFe_2(III)CN]_8$	-900^a
$[HrFe_2(III)H_2O]_8$	$[HrFe_2(III)SCN]_8$	$\pm 500^a$

^{a} Experimentally these are not significantly different from zero.

than does deoxyhemerythrin, which contains Fe(II). The low effective
magnetic moment in the Fe(III) compounds could in principle be ascribed
either to spin pairing attributable to a strong ligand field or to an antiferromagnetic interaction between the pair of iron atoms in each protein
subunit. Mössbauer results, described below, support the latter interpretation.

The difference in susceptibility between deoxyhemerythrin and oxyhemerythrin is slightly lower than that found in high-spin Fe(II) compounds
and may be attributable either to a small magnetic moment in oxyhemerythrin or to some antiferromagnetic interaction between iron atoms
in deoxyhemerythrin (23).

Mössbauer spectra have been recorded for hemerythrin and for model
iron chelates (23), and pertinent parameters, δ (the isomer shift) and Δ (the
quadrupole splitting), are recorded in Table VI.

Deoxyhemerythrin shows a simple quadrupole split pair of lines at all
temperatures in the absence of a magnetic field (Fig. 14), a pattern indicating that there is but one iron environment. The relatively high value of
the isomer shift δ, 1.19 mm sec^{-1}, and the large quadrupole splitting Δ,

TABLE VI

Zero-Field Isomer Shift and Quadrupole Splitting in Mössbauer Spectra

Compound[a]	Temperature, °K	Δ, mm sec^{-1} (± 0.05 mm sec^{-1})	δ, mm sec^{-1} (± 0.05 mm sec^{-1})
[HrFe(II)$_2$]$_8$	195	2.75	1.11
	77	2.81	1.19
	4.2	2.89	1.20
[HrFe(III)$_2$NCS]$_8$	77	1.81	0.52
[HrFe(III)$_2$OH$_2$]$_8$	77	1.57	0.46
[HrFe$_2$O$_2$]$_8$			
Outer pair	77	1.93	0.51
	4.2	1.92	0.54
Inner pair	77	1.03	0.48
	4.2	1.09	0.51
[Fe(salen)Cl]$_2$	300	1.38	0.39
	77	1.39	0.48
	4.2	1.38	0.51
[Fe(salen)Cl]	300	1.32	0.37
Monomer + dimer	4.2	1.39	0.45
[(Fe{phen$_2$}Cl)$_2$O]Cl$_2$ · 6H$_2$O	77	1.70	0.46
	4.2	1.50	0.47
[Fe(salen)]$_2$O	77	0.84	0.41
	4.2	0.85	0.46
enH$_2$[(Fe{HEDTA})$_2$O]	298	1.63	0.40

[a] Salen, N, N'-bissalicylideneethylenediamine; en, ethylenediamine; HEDTA, N-hydroxyethylethylenediaminetriacetate; phen, 1, 10-phenanthroline.

2.81 mm sec^{-1}, show the compound to be high-spin Fe(II). In comparison, Fe(NH$_4$)$_2$ (SO$_4$)$_2$ · 6H$_2$O gives $\delta = 1.40$ mm sec^{-1} and $\Delta = 2.70$ mm sec^{-1} at 77°K (24). The lower value of δ in hemerythrin suggests that nitrogen as well as oxygen ligands could be involved in binding.

Methemerythrin, both as the aquo and thiocyanato complexes, also gives a simple quadrupole split spectrum (Fig. 15), indicating that here, as in the deoxy form, there is only one type of iron site in each of the compounds. The thiocyanate is formed from the aquo by the combination of 1 mole of SCN$^-$ with two iron atoms. It is most likely, therefore, that the thiocyanate acts as a bridging group, for addition of SCN$^-$ to only one

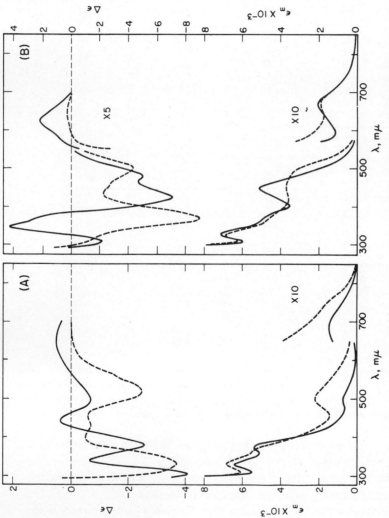

Fig. 13. CD spectra (upper) and absorption spectra (lower) of iron-dependent bands of: (A) Metcyano-hemerythrin, $[HrFe_2^{III}(CN)]_8$ (—); oxyhemerythrin, $[HrFe_2(O_2)]_8$ (- - -). (B) Metisothiocyanatohemerythrin, $HrFe_2^{III}(NCS)]_8$ (—); metazidohemerythrin, $[HrFe_2^{III}(N_3)]_8$ (- - -).

of the two iron atoms would be expected to split the Mössbauer signal. These data together with the values of the isomer shifts (0.46 mm sec^{-1} for aquo- and 0.52 mm sec^{-1} for thiocyanatohemerythrin at 77°K) and the quadrupole splitting (1.57 and 1.81 mm sec^{-1}, respectively), which are

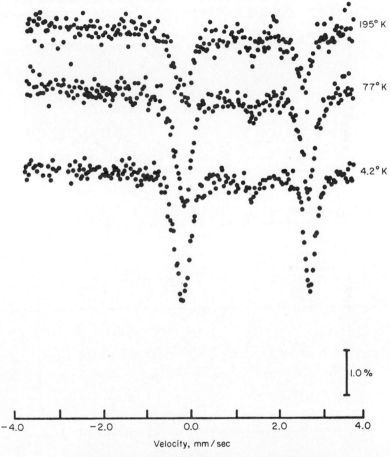

Fig. 14. Mössbauer spectra of deoxyhemerythrin.

very similar to the values for the high-spin binuclear Fe(III) model compounds, strongly indicate that the simple quadrupole split spectrum of the Fe(III) hemerythrin compounds is that of a pair of coupled high-spin Fe(III) ions. An alternate possibility is that the iron atoms are isolated (and that the observation of one simple quadrupole spectrum for the two atoms in both compounds is coincidental). This can be eliminated by an examination of the Mössbauer spectrum in a magnetic field. No magnetic hyperfine structure is seen in the signal (23). This result also eliminates the possibility that the iron atoms are *two* isolated low-spin Fe(III) cations. A final possibility, that the Fe(III) atoms are low-spin interacting cations is also most unlikely in view of the high isomer shift observed typical of that for high-spin cations and higher than that of any known low-spin iron complex. In toto chemical susceptibility and Mössbauer measurements

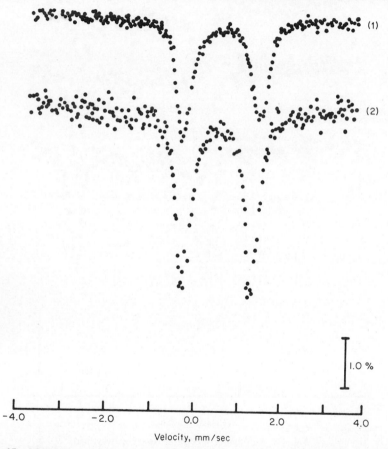

Fig. 15. Mössbauer spectra of methemerythrin at 77°K: (1) thiocyanato complex; (2) aquo complex.

lead to the conclusion that there is strong antiferromagnetic interaction between the iron atoms that are in a high-spin state.

Oxyhemerythrin, which is of low susceptibility (compared with deoxyhemerythrin), gives rise to a spectrum consisting of two quadrupole split pairs of lines superimposed on each other in the absence of a magnetic field (Fig. 16).

Combining the chemical, magnetic, Mössbauer, and spectroscopic information we arrive at the following structural conclusions. Deoxyhemerythrin with its high magnetic moment, Mössbauer parameters typical of high spin ($s = 2$) Fe(II), absence of absorption or CD spectra (from 300 to 1000 mμ), and absence of ESR bands must contain its two irons (per subunit) in essentially equivalent environments but as noninteracting high-spin Fe(II) atoms, presumably, therefore, with the structure

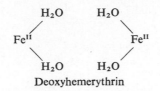

Deoxyhemerythrin

Water is presumed to occupy the open coordination positions since there is no evidence of an oxo bridge in deoxyhemerythrin. It should be noted that no oxo-bridged Fe(II) model complexes are known either (20).

In contrast, for methemerythrin all the electromagnetic properties examined are consistent with the presence of an oxo ($-O^{2-}-$) bridged pair of antiferromagnetically coupled high-spin ($s = 5/2$) Fe(III) cations, with the second ligand L forming a second bridge, at least for L = Cl$^-$, Br$^-$, NCS$^-$, N$_3{}^-$, CN$^-$, and NCO$^-$. These methemerythrin complexes may therefore be represented by the structure

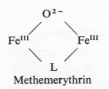

Methemerythrin

There is some evidence (20) that when L = OH$^-$, H$_2$O, or F$^-$ two ligands are coordinated to each site, one to each iron, and thus the second bridge through L is cleaved.

Turning finally to oxyhemerythrin, we note that its properties strongly resemble those of methemerythrin. It has a low magnetic susceptibility, no EPR, charge transfer bands in the absorption spectrum characteristic of an Fe^{III}—O—Fe^{III} system and CD similar to Fe(III) model compounds (20). These properties all point to a μ-oxo—μ-peroxo double-bridged iron dimer (20). However, the Mössbauer spectrum shows two pairs of quadrupole split peaks (Fig. 16), and consequently the two iron atoms must be nonequivalent in some respect. It seems unlikely that oxyhemerythrin consists of an equilibrium mixture of a ferrous oxygen complex and a ferric peroxo complex since the Mössbauer spectrum in high magnetic fields and its dependence on temperature indicates that both pairs of doublets should be attributed to high-spin Fe(III) dimers. It seems most likely that nonequivalence arises from a nonsymmetrically bridged peroxo group, for example one in which the peroxo group is hydrogen-bonded to a donor from the surrounding protein. Such a structure might be represented as

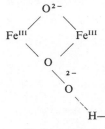

Oxyhemerythrin

This structure accounts for the nonequivalence reflected in Mössbauer spectroscopy and yet is consistent with all the other characteristics of an oxo-bridged high-spin Fe(III) dimer (*20*).

In comparing the structure of deoxy with oxyhemerythrin, it can be seen that the uptake of oxygen must be accompanied by a substantial rearrangement in the steric position of the iron atoms. It seems likely therefore that in hemerythrin, as in hemoglobin, oxygenation is accompanied by conformational changes in the protein.

D. Coordination of Iron to Protein

Since there is no heme or other prosthetic group in hemerythrin, the iron atoms presumably are held at the active site by direct coordination to side chains from amino acid residues. Assuming that each iron in methemerythrin or oxyhemerythrin has two coordination sites participating in the double bridge, each metal must be involved in four (or conceivably five) linkages to the protein.

Hemerythrin contains only a limited number of amino acid residues capable of participating in metal coordination. Coordination affinities of ligands are usually related to acid-base properties, that is, to their tendency to bind a proton. In general, groups with very high or very low pK values do not form bonds with metal cations.

The hydroxyl groups of threonine and serine, the guanido group of arginine, the indole nitrogen of tryptophan, and the amide groups of asparagine and glutamine all have pK values in the extreme range, hence are unlikely to be involved in iron binding. The free amino groups of lysine and the N-terminal glycine are potential ligands. However, the affinity of

Fe(III) for nitrogen is low compared to its affinity for other ligands containing sulfur or oxygen. Methionine, cysteine, tyrosine, and histidine are all known to coordinate with iron, in cytochrome c, ferredoxin, transferrin, and hemoglobin, respectively. Although the carboxylate groups of aspartic and glutamic acid are only weakly basic, they are also capable of providing ligands to iron. From these considerations, it can be concluded that the following 43 of the 113 amino acids per subunit in hemerythrin are potential ligands to the irons: 11 lysines, 1 glycine, 1 cysteine, 1 methionine, 7 histidines, 5 tyrosines, 11 aspartic acids, and 6 glutamic acids.

Group-specific reagents have been used extensively to probe the tertiary structure of a number of proteins. Amino acids that react with these reagents under nondenaturing conditions are usually considered "free", that is, accessible to solvent; those that do not react are assumed to be submerged within the protein matrix. In the case of hemerythrin, nucleophilic groups bound to the irons should be blocked from reaction with typical electrophiles. However, groups that can undergo ring substitution at positions not directly involved in iron coordination, such as histidine and tyrosine, may participate in chemical reactions.

Modification of hemerythrin by a number of group-specific reagents has yielded some information about its active site. The single cysteine of hemerythrin can be blocked with typical SH reagents without any accompanying effect on the visible spectrum of the protein (*11*). Such behavior indicates that the SH of cysteine does not provide an iron ligand. In addition, Klotz and Keresztes-Nagy (*8*) were able to succinylate hemerythrin exhaustively without affecting the iron locus. Although no attempt was made to quantitate this reaction, the absence of any changes in the spectrum of hemerythrin on succinylation suggests that lysines are not coordinated to the irons. More recently, Fan and York (*25*) have shown that all 11 lysines and the N-terminal glycine can be blocked with trinitrobenzenesulfonic acid, a behavior which again indicates strongly that amino groups are not coordinated to the irons. Fan and York also found that three of the seven histidines react with diazonium-1*H*-tetrazole (DHT) to form the di-DHT derivative, whereas the remaining four histidines form only the mono-DHT derivative. This differentiation in reactivity suggests that the latter four histidines may be bound to the irons.

Finally, we note that Rill and Klotz (*26*) have found that treatment of hemerythrin with tetranitromethane, a reagent that selectively modifies tyrosine (and cysteine) residues (*27*), leads to a release of approximately 50% of the iron in the protein, as well as to a complete loss of the CD spectra arising from the metal at the active site. Their observations indicate that tyrosine residues do provide ligands to the iron. Moreover, since the percentage drop in CD ellipticity increases approximately twice as fast as the loss of iron, it seems evident that the loss of a single iron in a pair at each site is sufficient to abolish the spectrum. The spectra and chemical results also indicate that the two iron atoms at each active site are not in identical molecular environments. One iron of each pair is evidently released more readily since even with nitration of all tyrosines only half of the total iron is lost. It does not follow, however, that tyrosine provides ligands for only one iron in each pair, for we cannot tell at present what the other protein ligands to each iron are and how much they contribute to retaining the metal within the protein.

Summarizing, we see that there is evidence of participation of histidine and tyrosine residues in direct binding of the two irons at the active site. It is totally unclear, however, how many of each ligand are bonded to each iron, and whether other groups such as carboxylates also coordinate with the metals. All chemical evidence from group-specific reagents is also subject to a residual uncertainty that modification of a particular residue in the primary structure leads to a conformational adaptation which in turn affects the active site. Conclusive evidence for the structure of the active site will become available only from a complete X-ray diffraction analysis of the tertiary structure of hemerythrin.

E. Oxygen Equilibria

The first quantitative studies of the oxygen uptake of hemerythrin were carried out by Marrian (28), and this work was further extended by Florkin (29). More recent studies (30–33) have been directed toward evaluating the strength of site–site interactions and the thermodynamic parameters of the oxygenation reaction. In general, the n of the Hill equation has been found to be only slightly above 1 (32,33), or in thermodynamic terms the free energies of site–site interaction range from zero to only a few hundred calories (30,31). Thus site–site interaction is much weaker than in hemoglobins. Also, in contrast to hemoglobins, generally no Bohr effect appears in the oxygenation of hemerythrin (29,32,33). This has been confirmed by direct pH measurements of proton release on oxygenation (34). It should be mentioned, however, that hemerythrin from the brachiopod *Lingula* behaves differently, showing both a Bohr effect and a strong site–site interaction (35).

There is a strong temperature dependence for the oxygenation reaction. Observed heats of oxygenation (30,31,33,35) vary from about -12 to -20 kcal/mole, substantially greater than that found for hemoglobin.

IV. GEOMETRY OF SUBUNIT ARRANGEMENT

The actual spatial arrangement of subunits within the octamer also requires an X-ray diffraction analysis for definitive answers. Nevertheless, there are certain restrictions in geometries that can be stated before-hand (36).

The number of possible geometries for an oligomer increases rapidly as the number of subunits increases. However, there are reasonable restrictions that can be imposed at the outset. Since monomers of hemerythrin are essentially identical, it can be assumed that each subunit in the octamer is in an equivalent (or pseudoidentical) environment. In essence, any linear array or permutations of a linear structure are eliminated in this way. Second, since there is little if any tendency of the octamer to aggregate, it can be assumed that the interface binding regions of the subunits are all saturated. Thus closed sets of geometries must prevail, and the subunits must be regularly packed around a central point, that is, must have point-group symmetry (36).

For an octamer these considerations limit the number of arrangements to three: (1) monomers arranged at corners of a regular octagon, with cyclic symmetry (C_8); (2) monomers arranged in dihedral symmetry (D_4) at vertices of a cube; and (3) monomers arranged dihedrally as a square antiprism (Fig. 17). The octagon would have eight bonding regions, all identical. A cubic arrangement would possess 12 bonding regions, of two different types. In the square antiprism arrangement, there are 16 bonding regions, of three different types (36).

Other things being equal, the array with the largest number of subunit interface contact regions would be expected to be the most stable. For hemerythrin this predicts the square antiprism (Fig. 17). In actuality, of course, it is not only the number of bonds formed but the strengths of these bonds that determine the stability and geometric arrangement of the subunits of an oligomer, and these bond energies are not well established. It is of interest in this regard to note that although most tetrameric proteins whose structures are known are found to be tetrahedra, with six intersub-unit bonding regions, some are found in a square array, with only four bonding interfaces (36). Nevertheless, we conclude that for hemerythrin, compact dihedral symmetry (Fig. 17) is likely and a square antiprism the most probable geometric arrangement of subunits.

V. ENERGETICS OF SUBUNIT INTERACTIONS

Two methods have been developed to evaluate association–dissociation equilibrium constants for hemerythrin. The first of these has not been widely used although the principle is an old one going back to Gibbs (37) and having been elaborated upon for many protein situations by Wyman (38,39). In essence, it says that if an equilibrium is difficult to follow experimentally it can be coupled with a reaction that is easier to follow. The linked equilibria must also have intertwined equilibrium constants. If one of these constants can be measured under a series of different conditions, then generally the other can be readily extracted from the data.

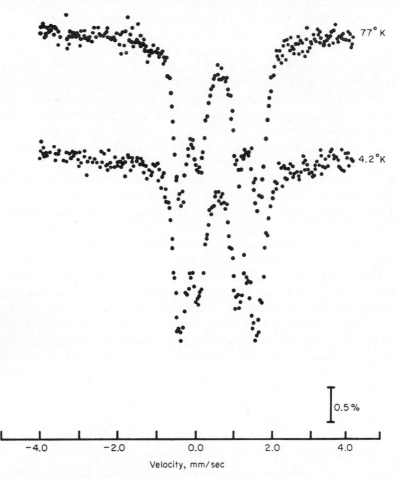

Fig. 16. Mössbauer spectra of oxyhemerythrin.

For hemerythrin dissociation the "indicator" reaction that has been used is the binding of various small ligands by the iron site. As described in Section III, the Fe(III) protein, methemerythrin, combines with a variety of iron-coordinating ligands. Coordination of the protein iron results in characteristic spectroscopic changes (Fig. 12). Thus the binding affinity between methemerythrin and the ligands can be determined readily by spectrophotometric means. The binding reaction may now be coupled to the octamer–monomer dissociation equilibrium

$$
\begin{array}{ccc}
& K_p & \\
\mathrm{Hr_8} & \rightleftharpoons & \mathrm{8Hr} \\
+ & & + \\
\mathrm{8L} & & \mathrm{8L} \\
k_L \Updownarrow & & \Updownarrow K_L \\
\mathrm{Hr_8L_8} & \rightleftharpoons & \mathrm{8HrL} \\
& K_{pL} &
\end{array}
\tag{1}
$$

$$
\text{Fraction liganded} = \frac{8(\mathrm{Hr_8L_8}) + (\mathrm{HrL})}{8(\mathrm{Hr_8L_8}) + 8(\mathrm{Hr_8}) + (\mathrm{HrL}) + (\mathrm{Hr})}
\tag{2}
$$

If octamer H_8 binds L with a different affinity than does monomer Hr, that is, if k_L and K_L are not identical, then it is possible to evaluate K_p, the parameter of central interest in this linked system of equilibria (19).

Brought more closely to experimental terms, this means that if the protein disaggregates on dilution, the apparent binding constant for ligand will change with protein concentration. Let us assume, as is true in actual observation, that the monomer, Hr, has a larger affinity for the ligand. In that case as the protein solution is diluted, the proportion of the strong binding form, the monomer, is increased and consequently the fraction of the protein that is in the liganded state [Eq. (2)] should increase. In practice then, it is necessary to measure the binding of ligand by hemerythrin over a wide range of protein concentration.

Typical results with SCN⁻ as the ligand (19) are shown in Fig. 18. The ordinate is essentially a measure of the fraction of protein bonded ot the ligand, and the abscissa is the concentration of free ligand in equilibrium with this system. Each line refers to a set of experiments over a range of SCN⁻ concentrations, but for a single concentration of protein. As is apparent from the legend to Fig. 18, the protein concentration decreases for the lines that are higher in the figure. Quite clearly, then, the fraction of hemerythrin in the liganded form increases with increasing dilution of the protein. A quantitative treatment of these data (19) leads to an equilibrium constant for the association reaction

$$
\mathrm{8Hr = Hr_8}; \quad K_8 = \frac{(\mathrm{Hr_8})}{(\mathrm{Hr})^8} = K_p
\tag{3}
$$

of $K_8 = 10^{38}\ M^{-7}$.

An alternate approach to an evaluation of the association–dissociation constant depends on a very elementary chemical principle, Le Chatelier's rule. For an equilibrium of the type shown in Eq. (3), it is obvious that more of the monomeric species Hr must be produced as the total concentration of the protein is diluted. In practice this was difficult to follow when only schlieren optics were available in the ultracentrifuge. With the development of Rayleigh optical methods and photoelectric scanner systems (40–43), it has proved possible to examine protein solutions at very high dilution.

Some results obtained with Rayleigh interference methods for hemerythrin in the presence of azide ions (44) are summarized in Fig. 19. As is apparent from this graph, the weight-average molecular weight remains

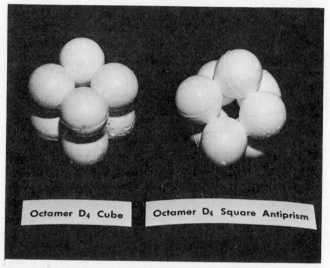

Fig. 17. Arrangement of subunits in a cube or a square antiprism.

close to that of the octamer as the protein concentration is decreased from high concentrations down to approximately 0.04%. Further dilution, however, results in a steep drop in the molecular weight, and at a concentration slightly below 0.01% the value for monomer is reached. The experimental points can be fitted very well by a simple thermodynamic equation, which assumes that only monomer and octamer are present in significant quantities at equilibrium throughout the concentration range examined [Eq. (3)]. The line shown in Fig. 19 is one calculated on this basis (44), with a single equilibrium constant of $3.4 \times 10^{36} \, M^{-7}$.

It is perhaps instructive to present the results of the energetics of the dissociation process in terms of a free-energy diagram (Fig. 20). For comparative purposes among proteins with different numbers of subunits, it is most convenient to present the diagram for the formation of 1 mole of subunit. As Fig. 20 indicates, for hemerythrin the monomer is approximately 6 kcal per mole higher in free energy in the dissociated state than in the octameric aggregate.

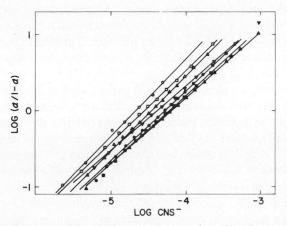

Fig. 18. Dependence of SCN⁻ binding on concentration of methemerythrin. Protein concentration in moles (of subunit) per liter: (○), 0.221×10^{-5}; (□), 0.876×10^{-5}; (△), 1.86×10^{-5}; ▽, 3.59×10^{-5}; (●), 7.29×10^{-5}; (■), 13.4×10^{-5}; (▲), 23.3×10^{-5}.

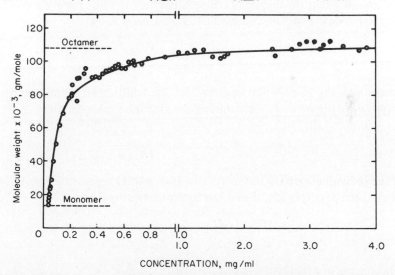

Fig. 19. Weight-average molecular weights of methemerythrin–N_3^- as obtained from equilibrium ultracentrifugation (at 5°C).

Furthermore, qualitative studies (*44*) indicate that hemerythrin binding other ligands dissociates to still different degrees. It seems apparent, therefore, that one should show not a set of single energy levels for the dissociation process, but rather a band of energy levels (Fig. 20) to represent properly the multiplicity of states that must be available in order to account for the difference in $\Delta G°$ values observed for hemerythrin carrying different ligands.

One of the most noteworthy features of the subunit interactions of hemerythrin is that a *single* equilibrium constant, K_8, fits the dissociation behavior very well. Since a single equilibrium constant does describe the dissociating hemerythrin system, it is apparent that this equilibrium involves overwhelmingly only the monomer and octamer species [Eq. (3)]. In other words, equilibrium centrifugation studies do not reveal the presence of any significant quantities of other oligomeric intermediates such as dimer or tetramer. Hemerythrin thus behaves differently from, for example, hemoglobin (*42,45–47*), in which dimers play an important role, or from lactic dehydrogenase (*48*) in which dimers and tetramers are intermediates between monomer and octamer. Since only the final octamers are detectable, association of monomers in hemerythrin must be a highly cooperative process. It therefore seems relevant to ask how this strong cooperativity need be to reduce intermediate oligomeric species to negligible concentration. This question has been approached quantitatively in the following way (*44*).

Assuming that the degree of cooperativity is the same for each associating step, one may define a cooperativity parameter α, by

$$\alpha = \frac{k_{i+1}}{k_i} \tag{4}$$

where k_i is the equilibrium constant for the addition of a monomeric A molecule to oligomer A_{i-1} containing $(i-1)$ monomeric subunits

$$A + A_{i-1} = A_i; \qquad k_i = \frac{A_i}{(A)(A_{i-1})} \tag{5}$$

If the individual steps in the successive associations represented by Eq. (5) are written out explicitly, then it can be shown readily that

$$\frac{A_i}{A^i} = \prod_2^i k_j \tag{6}$$

Making use of α from Eq. (4), one may write

$$k_3 = \alpha k_2$$
$$k_4 = \alpha k_3 = \alpha^2 k_2 \tag{7}$$
$$\vdots \qquad \vdots \qquad \vdots$$
$$k_i = \alpha k_{i-1} = \alpha^{i-2} k_2$$

Combination of the relations of Eq. (7) with Eq. (6) leads to

$$A_i = (\Pi k_j)(A)^i = k_2^{i-1}(\alpha \cdot \alpha^2 \cdots \alpha^{i-2})A^i = k_2^{i-1}\alpha^\sigma A^i \tag{8}$$

where

$$\sigma = [1+2+\cdots(i-2)] = \frac{(i-1)\,(i-2)}{2} \tag{9}$$

An experimentally fitted single association constant for a monomer i-mer equilibrium would be written as

$$K_i = \frac{A_i}{A^i} \tag{10}$$

From Eqs. (8) and (10) it follows that

$$k_2^{i-1}\alpha^\sigma = K_i \tag{11}$$

Starting with an experimental determination of K_i, one can treat α as a parameter, calculate k_2, and then each k_i. With this information a curve

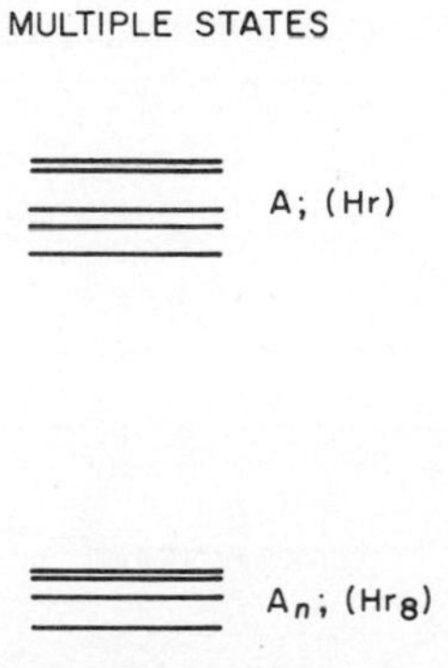

MULTIPLE STATES

Fig. 20. Free-energy diagram illustrating multiplicity of free energies of hemerythrin with different bound ligands.

for the concentration dependence of M_w as well as graphs of the mole fraction of each oligomeric species can be computed.

This method of analysis has been applied to the hemerythrin system (44) using $K_8 = 3.4 \times 10^{36}$. Figure 21 shows that the calculated molecular weight curves quickly converge for $\alpha \geq 4$, giving a limiting curve that fits well with the experimental one (Fig. 19). For $\alpha = 4$, computed curves for the mole fraction (in monomeric units) of monomer and octamer (Fig. 22) agree well with corresponding curves calculated from the experimental observations on the assumption that only a monomer–octamer equilibrium exists (no intermediate species), with $K_8 = 3.4 \times 10^{36}$ (Fig. 23). The insert in Fig. 22 illustrates how very trivial the equilibrium concentrations of dimer, trimer, and intermediate oligomers are for a system with a cooperativity factor $\alpha = 4$.

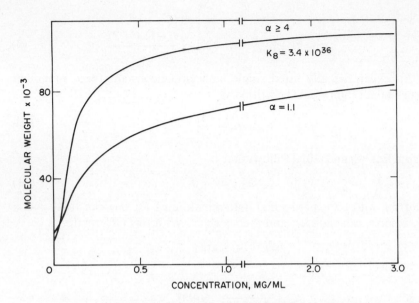

Fig. 21. Weight-average molecular weight vs. concentration of hemerythrin; computed for different values of cooperativity parameter, α.

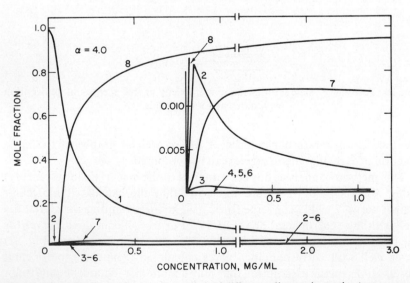

Fig. 22. Mole fractions (in monomer units) of different oligomeric species i vs. concentration of hemerythrin, computed for $\alpha = 4$. The integer labeling each curve gives i, the number of subunits in the respective oligomer (1 = monomer, 2 = dimer, and so on). The insert shows the same data for $2 \leq i \leq 7$ on an expanded scale to illustrate how low the concentration of these species is.

Such a value for α corresponds to a very small free-energy *increment* for the $(i+1)$th association as contrasted to the ith

$$\frac{k_{i+1}}{k_i} = \alpha$$

$$\Delta(\Delta G^\circ) = RT \ln \frac{k_{i+1}}{k_i} = RT \ln 4 \tag{12}$$

$$= 0.8 \text{ kcal}$$

Thus a very minor increase in interface contact area between incoming monomer and each successive oligomeric species leads to only a minor increment in cooperativity free energy; and yet such a minor increase in contact interface would be reflected as a steeply rising association curve.

The very crucial role that a minor change in structure may play in the energetics of associating proteins can also be seen in another way. For comparison of proteins with different numbers of subunits, it should be convenient to distribute the free energy of association of the highest oligomer equally among the monomers and to define a free energy of formation of monomer, ΔG_m°, as follows (*44*)

$$n\text{A} = \text{A}_n; \quad \frac{\Delta G^\circ}{n} = -\Delta G_m^\circ = -\frac{1}{n} RT \ln K_n \tag{13}$$

ΔG_m° may be viewed as the difference in (free) energy levels between the monomeric and oligomeric states of a subunit. For hemerythrin, for example,

$$\Delta G_m^\circ = \frac{1}{8} RT \ln K_8 = 5.8 \text{ kcal/mole monomer} \tag{14}$$

We may now examine the effect of small changes in ΔG_m° on the association–dissociation equilibrium. Typical calculations for hemerythrin are summarized in Fig. 24. A change of only 2 kcal in ΔG_m° (from 6.8 to 4.8 in Fig. 24) would convert this quaternary structure into one that would appear completely monomeric in the range below 0.5 mg/ml, and overwhelmingly dissociated (whereas actually it is almost completely in the octameric state) up to very high concentrations of protein. Even a change of 0.8 kcal would markedly affect the molecular weight vs. concentration curve (Fig. 24).

Although obvious, it should be mentioned explicitly that ΔG_m°, being a difference, can be changed by variation of G° for either the initial or final state. Changes in ΔG_m° may arise either from a small perturbation in the interactions of side chains of neighboring monomers in the interface area within the oligomer or from a modification of interactions of such side chains in the dissociated monomer with the surrounding solvent.

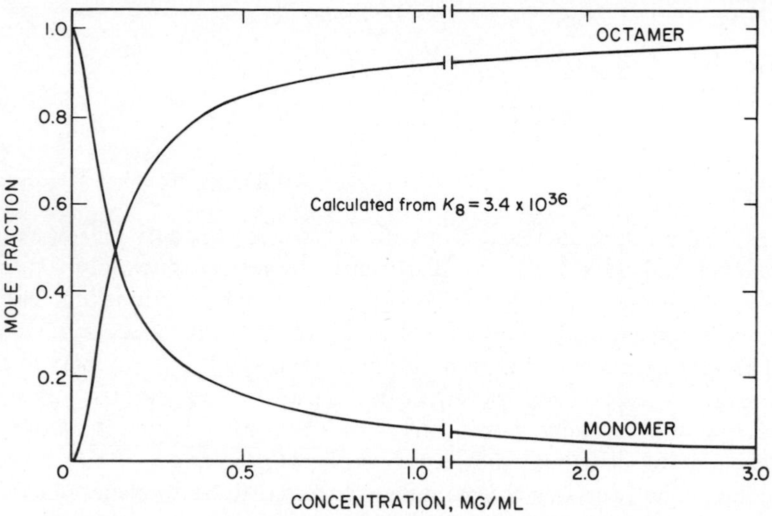

Fig. 23. Mole fraction of monomer and of octamer calculated from experimental data for weight-average molecular weight on the assumption that no intermediate species exist in significant concentration and that $K_8 = 3.4 \times 10^{36} M^{-7}$.

VI. CHEMICAL AND OPTICAL EVIDENCE OF MULTIPLE CONFORMATIONAL STATES

It has also been of interest to see whether or not more direct evidence can be obtained, beyond thermodynamic indications, of the existence of different conformational forms of monomeric hemerythrin and perhaps even of octameric hemerythrin. This question can be defined somewhat more precisely, for the specific case of the monomer, in the following terms. Monomeric hemerythrin can be obtained by dilution of Hr_8 or by reaction of octamer with a mercurial, RHgX. Are the two monomers, Hr and Hr', essentially the same in so far as the protein conformation is concerned?

Chemical and optical methods indicate quite clearly that Hr differs from Hr'. For example, UV difference spectra have been obtained (49) for the monomer in which the mercaptan group has been blocked (for convenience in the spectroscopic observations, with cysteine rather than with a mercurial) against a reference solution of the octamer, and also for monomer obtained by dilution against a reference solution containing the octamer. As is apparent from Fig. 25, these two difference spectra are in themselves markedly different from each other in the region of the aromatic group absorption. Evidently, these aromatic group internal indicators are in significantly different environments in the two different forms of monomeric hemerythrin.

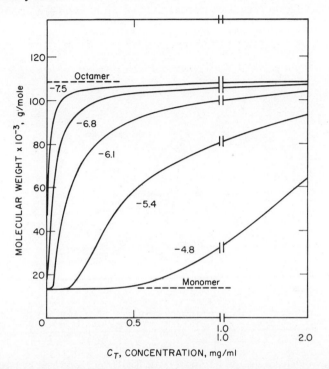

Fig. 24. Concentration dependence of weight-average molecular weight of hemerythrin computed for different values of $-\Delta G°_m$, the difference in free-energy levels between monomeric and octameric state of the subunit.

Similar differences between monomer hemerythrins are also apparent in CD spectra (*49*).

Furthermore, there are also differences in chemical reactivity. These are perhaps best illustrated in studies of the kinetics of the reaction of the monomers in the Fe(III) state with the colorimetric reagent Tiron:

$$HrFe_2^{III}(H_2O)_2 + 2Tiron \rightleftharpoons HrFe_2^{III}(Tiron)_2 + 2H_2O \qquad (15)$$

Tiron:

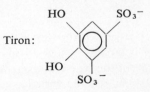

As is illustrated in Fig. 26, the mercaptan-blocked monomer reacts at a very appreciable rate with Tiron. In contrast, the dilution monomer shows no detectable rate of reaction over the same period of time, approximately 3 hr. Again it is obvious that the two monomers are of a different nature.

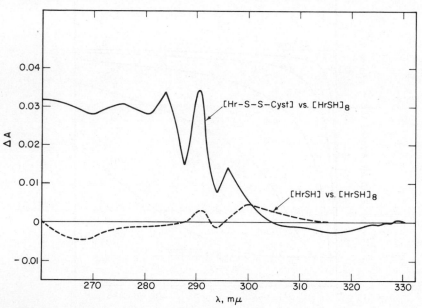

Fig. 25. UV difference spectra. For monomeric hemerythrin obtained by blocking the mercaptan group (with Cys) (—) and for monomeric hemerythrin obtained by dilution of the octamer (- - -).

Similar effects have been observed with monomeric deoxyhemerythrin (49), in which the iron had been reduced to the Fe(II) oxidation state, in its reaction with o-phenanthroline, a colorimetric reagent for Fe(II):

$$HrFe_2^{II}(H_2O)_4 + 2\ o\text{-Phen} \leftrightharpoons HrFe_2^{II}(o\text{-Phen})_2 + 4H_2O \qquad (16)$$

o-Phen: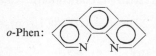

Chemical observations of complex formation with Tiron also provide evidence of different types of octameric hemerythrin. The results (49) are perhaps best described in terms of the schematic diagram shown in Fig. 27. If the mercurial monomer Hr' has reacted with Tiron to form Hr'(Tiron) this complex may now be converted to octamer by removal of the mercurial with cysteine ethyl ester. The octamer so formed is found to retain the Tiron ligands. However, this state $[Hr(Tiron)_2]_8$, is not directly accessible from natural hemerythrin, Hr_8, as the observations represented in Fig. 26

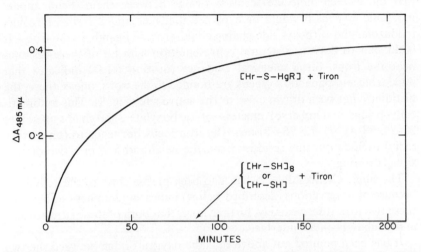

Fig. 26. Kinetics of reaction of Tiron with different forms of hemerythrin. Hr—S—HgR represents monomer obtained by blocking of SH group with mercurial reagent. Hr—SH represents monomer obtained by dilution of octamer, (Hr—SH)$_8$.

clearly show. Thus the chemical evidence indicates that not only are there a number of different states of monomeric Hr but also a number of different states of octameric Hr$_8$ (*49*). The representation of the transition from Hr$_8$ to Hr (Fig. 20) by bands of energy levels is therefore confirmed by chemical evidence also.

VII. STRUCTURAL-MOLECULAR ASPECTS OF SUBUNIT INTERACTIONS

An ultimate explanation of subunit interactions at the molecular level depends on the determination of a full three-dimensional X-ray structure of this protein molecule. Only then will it be possible to see which side chains appear at the interfaces between the subunits. At the moment only a one-dimensional structure is available, that is, the primary sequence. Nevertheless, combining this information with that obtained from a variety of chemical reactions, some reasonable guesses can be made as to the character of the interaction at the molecular levels.

The primary structure of hemerythrin from *G. gouldii* is shown in Fig. 10. This distorted cloverleaf representation has been constructed to reflect a variety of suggestive physical and chemical facts about this oxygen-carrying pigment. It is known that each subunit of hemerythrin must provide side chains for the attachment of two iron atoms. Mössbauer and magnetic studies (20,23) show that the iron atoms are near each other and that the oxygen molecule forms a bridge between them. From model inorganic systems it would be expected that mercaptan, lysine, tyrosine, imidazole, and carboxyl side chains could provide ligands to the irons. It is difficult to draw a structural representation with all of these residues near the irons. Since chemical evidence (Section III.D) indicates that imadazole and phenolic oxygens are bonded to the irons, one of these, the histidines, has been drawn close to the active site (Fig. 9). This automatically brings a number of clusters of carboxylate side chains (residues 22–24, 38–45, 71–72, 76–78) near the iron locus but tends to (unwarrantedly) exclude tyrosine residues from the neighborhood in a two-dimensional drawing.

The single Cys, residue 50, has also been placed close to the active site because of observations (described below) indicating interactions between these two loci. It seems likely, furthermore, that both of these sites are also at the subunit contact interface.

It has been pointed out above that the oxygen bridge between the two iron atoms can be replaced by other ligands such as N_3^-, SCN^-, Cl^-, and so on. These different ligands to the irons have very different effects on the reactivity of the SH of the Cys residue (15). For example, if a mercaptan-blocking reagent N-ethylmaleimide is added to a solution of aquohemerythrin, no dissociation will occur in the period of time necessary to make a sedimentation experiment, that is, only 6 S particles are found. However, if to a portion of this solution, azide ion is added in addition, then the sedimentation experiment shows 2 S particles. Clearly, the mercaptan-blocking reagent was not effective in dissociating hemerythrin in the absence of liganding ion. Its ineffectiveness might be attributable to either of two causes, however: (1) it did not combine with protein SH groups; (2) it combined with SH groups but dissociation did not proceed until ligand (N_3^-) was bound. To settle this point the following experiment was carried out (15).

Part of the solution containing aquomethemerythrin and N-ethylmaleimide was dialyzed to remove unreacted maleimide. If any reagent had reacted with protein mercaptan, it would not dialyze out since a covalent S—C bond is formed. After dialysis, azide was added to the solution. If maleimide had reacted with protein but needed the azide to dissociate hemerythrin, 2 S particles should have appeared in the ultracentrifuge. If maleimide had not reacted in the absence of azide, it should have been removed during the dialysis, and the subsequent addition of azide should have had no effect on particle size, since addition of azide to undissociated aquomethemerythrin merely leads to the octameric azide complex. The actual sedimentation experiment, after dialysis and subsequent addition of azide, showed only 6 S particles (*15*). Clearly *N*-ethylmaleimide had not reacted with aquo protein in the original solution in the absence of azide. Similar effects are observed with other mercaptanblocking reagents.

Even more striking effects are seen in direct experiments on the rates of reaction of a mercurial with methemerythrin (*50*) in different liganded states (Fig. 28). Whereas the aquo form $(Hr—H_2O)_8$, reacts slowly over

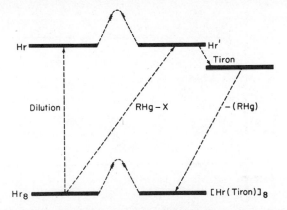

Fig. 27. Comparison of chemical behavior with Tiron of different forms of hemerythrin. Barriers are indicated between (approximately) equipotential energy states which are not mutually accessible.

a period of hours, the azide complex reacts extremely rapidly. Hemerythrins with more weakly bound ions such as chloride or fluoride react at intermediate rates.

Thus we see that the nature of the environment at the locus of the iron atom affects reactivity at the SH locus of Cys 50.

Interestingly enough there are reciprocal effects. These have been revealed particularly in studies on the influence of a number of specific ions, not bound at the iron locus, on the reactivity of the SH group.

Specific ions such as ClO_4^- and NO_3^- have a marked effect on the reactivity of the SH group in hemerythrin with mercurials (*51*). For example, if a solution of aquohemerythrin is divided into two parts, to one of which approximately 10^{-3} M ClO_4^- ion is added and to both of which a quantity of salyrganic acid stoichiometric with SH groups is added, then, after standing overnight the hemerythrin samples give the results shown in Fig. 29 in a sedimentation velocity experiment. Whereas the sample without perchlorate is about 80% dissociated (into 2 S particles), that with perchlorate shows only nondissociated octameric protein. In other experiments essentially complete protection of the sulfhydryl was found when the concentration of perchlorate was the same as that of cysteine in the protein. This indicates that a single perchloric-binding site (per subunit) provides the locus for binding ClO_4^- strongly and producing the effects observed.

Titrations of the sulfhydryl group with chloromercuribenzoate have also shown markedly reduced rates of reaction in the presence of perchlorate (51). In contrast to the effects of ligands bonded to the iron, such

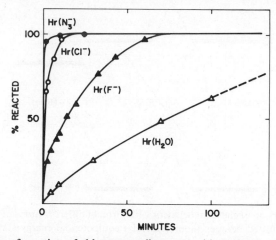

Fig. 28. Rates of reaction of chloromercuribenzoate with methemerythrin containing different ligands.

as azide or chloride, perchlorate ion produces a decrease in reactivity of the SH group rather than the marked enhancement produced by iron-coordinating ligands. From studies of simple inorganic model systems, ClO_4^- or NO_3^- would not be expected to be coordinated to the iron. To make certain that this is also true in hemerythrin, Darnall et al. (51) examined the CD spectra of solutions of hemerythrin in the presence and absence of ClO_4^- ion. In the presence of 0.1 M sodium perchlorate the CD spectrum is essentially identical to that of metaquohemerythrin (Fig. 30). Since CD spectra are particularly sensitive to changes at the iron locus, it is clear that the species formed in the presence of perchlorate is simply metaquohemerythrin and not an iron perchlorate coordination complex.

Further studies of the effect of perchlorate on the CD spectra of metaquohemerythrin over a range of pH values uncovered some very unexpected and interesting phenomena (51). At high pH values where the aquo form

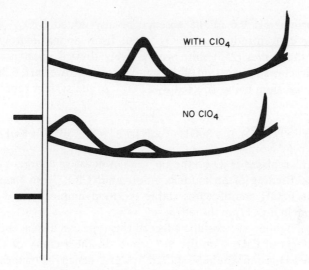

Fig. 29. Sedimentation of metaquohemerythrin. Upper contains 9×10^{-4} M NaClO$_4$.

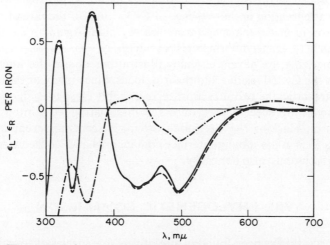

Fig. 30. CD spectra of metaquohemerythrin and methydroxohemerythrin. — . —, pH 8.5, 94% methydroxo; - - - - -, pH 8.5, 0.09 M NaClO$_4$, 93% metaquo; ——— pH 7, 87% metaquo.

goes to methydroxohemerythrin the addition of perchloric ion to the solution shifts the CD spectrum back to one identical with the metaquohemerythrin state. The same effect on the equilibrium, a shift from the hydroxo to the aquo form is also seen very strikingly in studies of the absorption spectra of these two species at pH values which both can exist together (Fig. 31). Curve 1 shows the absorption spectrum of hydroxohemerythrin at pH 8.6 in the absence of any added ClO_4^- ion. With increasing concentrations of ClO_4^- ion, from approximately 10^{-4} to 10^{-2} M, the curves (2–5) shift progressively toward spectrum 6 which is the characteristic spectrum of the aquo form at a lower pH, 6.36.

The equilibrium being affected by ClO_4^- in the system represented by Fig. 31 is:

$$[HrFe^{III}(H_2O)_m]_8 + 8nOH^- = [HrFe_2^{III}(OH^-)_n]^{8n-} + 8mH_2O \quad (17)$$

When OH^- replaces H_2O on the Fe^{III}, a net negative charge is created at this locus. Binding of an anionic, nonligand, ClO_4^- ion near the iron makes the FeOH complex less stable thermodynamically and shifts the equilibrium in Eq. (17) to the left.

A near-neighbor electrostatic effect of this type can also account for the kinetic effects of ClO_4^- on the SH group. In the vicinity of ClO_4^- the pK_a of the SH group would be shifted upward, hence the concentration of the reactive species in mercaptan-blocking reactions, S^-, would be markedly reduced. A site–site interaction of this nature would also explain the marked attentuation in the presence of ClO_4^- ion, of the rate of reaction of the iron in metaquohemerythrin with N_3^- ion. Again an electrostatic repulsion between neighboring sites is evident.

Summarizing, we have seen many illustrations of site–site interactions between the Cys 50 residue and the iron locus. Since many of these effects are electrostatic, the two sites must be reasonably close to each other. The only chemically specific procedure for dissociating Hr_8 into subunits depends on blocking the mercaptan group. It seems apparent therefore that Cys 50 is in the subunit interface contact area and, consequently, that the iron locus must also be nearby.

VIII. PHYLOGENETIC COMPARISON

Hemerythrin has been found in a relatively few species of four different phyla (52), the sipunculids, polychaete annelids, priapulids, and brachiopods. Pure hemerythrins have been isolated from species of three different genera of sipunculids, *Phascolosoma*, *Sipunculus*, and *Dendrostomum*, and one brachiopod, *Lingula*. Early molecular weight determinations (of hemerythrin from *Sipunculus nudus*) were incorrect. The most detailed modern determinations of the molecular weight have been made with samples of protein from *G. gouldii* (see Section II of this review). Sedimentation experiments have also been carried out recently with hemerythrins from *S. nudus* (33), *Dendrostomum pyroides* (53), and *Lingula reevi* (54),

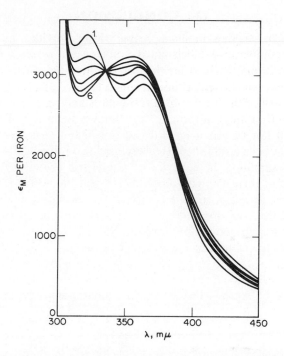

Fig. 31. Shift in absorption spectrum of hemerythrin from hydroxo to aquo form with added increments of ClO_4^- ion. 1, Absorption spectrum of hydroxohemerythrin, pH 8.6; 2–5, spectra at pH 8.6 with added increments of ClO_4^- to give 3.75×10^{-4} M, 1.00×10^{-3} M, 3.75×10^{-3} M and 2.50×10^{-2} M, respectively; 6, aquohemerythrin, pH 6.36.

and all show sedimentation coefficients near 6 S. It is likely therefore that all these hemerythrins have molecular weights near the 108,000 found for *G. gouldii*. At least in this property this respiratory pigment isolated from animals as geographically separated as Hawaii, the California coast, the New England coast and the Mediterranean belongs in one macromolecular class. This is particularly striking in a comparison of *Lingula* (from Hawaii) with the others, for the brachiopods are systematically isolated from the sipunculids.

The phylogenetic and paleontological significance of relationships between hemerythrins will be obscure until further work is carried out on the molecular properties of these proteins. Comparisons of primary structure would be particularly relevant, in view of the wealth of insights obtained in corresponding comparisons for cytochrome c, hemoglobins, and the immunoglobulins (*55*).

IX. CONCLUSION

The oxygen-carrying pigment hemerythrin provides an interesting system for comparative biochemical studies. It illustrates an alternate chemical evolution, aside from that of the heme pigments, of an iron-containing oxygen carrier. Although much has been learned of the structural chemistry of the active site, a full understanding of the unique properties of the iron locus in terms of the known behavior of model small chelates will not be attained until the full ligand structure and stereochemistry has been established. Hemerythrin also provides a very tractable system for examining the behavior of oligomeric proteins. Although the architecture of this supermacromolecule is still unknown, it is clear that the geometric arrangement is a compact one, probably of dihedral symmetry, and that the dissociation–association equilibrium involves only the end species, monomer and octamer. Shifts in equilibrium and in conformation of the monomer and octamer can be produced by a variety of small molecules. In turn, the properties of the iron locus seem to be linked to these conformational changes so that a wide range of subtle changes in behavior may be expressed under different environmental conditions.

REFERENCES

1. I. M. Klotz and T. A. Klotz, *Science*, **121**, 477 (1955).
2. I. M. Klotz, T. A. Klotz, and H. A. Fiess, *Arch. Biochem. Biophys.*, **68**, 284 (1957).
3. M. F. Perutz, H. Muirhead, J. M. Cox, L. C. G. Goaman, F. S. Mathews, E. L. McGandy, and L. E. Webb, *Nature*, **219**, 29 (1968).
4. M. F. Perutz, H. Muirhead, J. M. Cox, and L. C. G. Goaman, *Nature*, **219**, 131 (1968).
5. W. R. Groskopf, J. W. Holleman, E. Margoliash, and I. M. Klotz, *Biochemistry*, **5**, 3779 (1966).
6. W. R. Groskopf, J. W. Holleman, E. Margoliash, and I. M. Klotz, *Biochemistry*, **5**, 3783 (1966).
7. E. Boeri and A. Ghiretti-Magaldi, *Biochim. Biophys. Acta*, **23**, 465 (1957).
8. I. M. Klotz and S. Keresztes-Nagy, *Biochemistry*, **2**, 445 (1963).
9. C. Manwell, *Science*, **139**, 175 (1963).
10. I. M. Klotz and S. Keresztes-Nagy, *Nature*, **195**, 900 (1962).
11. S. Keresztes-Nagy and I. M. Klotz, *Biochemistry*, **2**, 923 (1963).
12. S. Keresztes-Nagy, L. Lazer, M. H. Klapper, and I. M. Klotz, *Science*, **150**, 357 (1965).
13. A. R. Subramanian, J. W. Holleman, and I. M. Klotz, *Biochemistry*, **7**, 3859 (1968).
14. G. L. Klippenstein, J. W. Holleman, and I. M. Klotz, *Biochemistry*, **7**, 3868 (1968).
15. S. Keresztes-Nagy and I. M. Klotz, *Biochemistry*, **4**, 919 (1965).
16. D. W. Darnall, K. Garbett, I. M. Klotz, S. Aktipis, and S. Keresztes-Nagy, *Arch. Biochem. Biophys.*, **133**, 103 (1969).
17. G. Holzwarth and P. Doty, *J. Am. Chem. Soc.*, **87**, 218 (1965).
18. S. N. Timasheff and M. J. Gorbunoff, *Ann. Rev. Biochem.*, **36**, 13 (1967).
19. M. H. Klapper and I. M. Klotz, *Biochemistry*, **7**, 223 (1968).

20. K. Garbett, D. W. Darnall, I. M. Klotz, and R. J. P. Williams, *Arch. Biochem. Biophys.*, **135**, 419 (1969).
21. D. D. Ulmer and B. Vallee, *Biochemistry*, **2**, 1335 (1962).
22. H. Beinert, W. Heinen, and G. Palmer, *Brookhaven Symp. Quant. Biol.*, **15**, 229 (1962).
23. M. Y. Okamura, I. M. Klotz, C. E. Johnson, M. R. C. Winter, and R. J. P. Williams, *Biochemistry*, **8**, 1951 (1969).
24. P. R. Brady, J. F. Duncan, and K. F. Mok, *Proc. Roy. Soc. (London)*, **A287**, 343 (1965).
25. C. C. Fan and J. L. York, *Biochem. Biophys. Res. Commun.*, **36**, 365 (1969).
26. R. L. Rill and I. M. Klotz, *Arch. Biochem. Biophys.*, **136**, 507 (1970).
27. M. Sokolovsky, J. F. Riordan, and B. L. Vallee, *Biochemistry*, **5**, 3582 (1966).
28. G. F. Marrian, *Brit. J. Exptl. Biol.*, **4**, 357 (1927).
29. M. Florkin, *Arch. Intern. Physiol.*, **36**, 247 (1933).
30. I. M. Klotz, *Science*, **128**, 815 (1958).
31. W. L. Peticolas, Ph.D. Dissertation, Northwestern University, Evanston, Illinois, 1954.
32. C. Manwell, *Ann. Rev. Physiol.*, **22**, 191 (1960).
33. G. Bates, M. Brunori, G. Amiconi, E. Antonini, and J. Wyman, *Biochemistry*, **7**, 3016 (1968).
34. M. Y. Okamura and I. M. Klotz, unpublished observations, 1969.
35. C. Manwell, *Science*, **132**, 550 (1960).
36. I. M. Klotz, N. R. Langerman, and D. W. Darnall, *Ann. Rev. Biochem.*, **39**, 25 (1970).
37. J. W. Gibbs, *Trans. Conn. Acad.*, **3**, 108 (1875–1876); **3**, 343 (1877–1878).
38. J. Wyman, Jr., *Advan. Protein Chem.*, **4**, 407 (1948).
39. J. Wyman, Jr., *Advan. Protein Chem.*, **19**, 223 (1964).
40. H. K. Schachman, L. Gropper, S. Hanlon, and F. Putney. *Arch. Biochem. Biophys.*, **99**, 175 (1962).
41. K. Lamers, F. Putney, I. Z. Steinberg, and H. K. Schachman, *Arch. Biochem. Biophys.*, **103**, 379 (1963).
42. H. K. Schachman and S. Edelstein, *Biochemistry*, **5**, 2681 (1966).
43. D. A. Yphantis, *Biochemistry*, **3**, 297 (1964).
44. N. R. Langerman and I. M. Klotz, *Biochemistry*, **8**, 4746 (1969).
45. S. Vinograd and W. O. Hutchinson, *Nature*, **187**, 216 (1960).
46. G. Guidotti and L. C. Craig, *Proc. Natl. Acad. Sci. U.S.*, **50**, 54 (1963).
47. M. A. Rosemeyer and E. R. Huehns, *J. Mol. Biol.*, **25**, 253 (1967).
48. D. B. Millar, V. Frattali, and G. E. Willick, *Biochemistry*, **8**, 2416 (1969).
49. D. W. Darnall and I. M. Klotz, unpublished experiments, 1967.
50. S. Keresztes-Nagy, private communication, 1968.
51. D. W. Darnall, K. Garbett, and I. M. Klotz, *Biochem. Biophys. Res. Commun.*, **32**, 264 (1968).
52. F. Ghiretti, in *Oxygenases* (O. Hayaishi, ed.), Academic Press, New York, 1962, Chapter 12.
53. G. L. Klippenstein and I. M. Klotz, unpublished experiments, 1966.
54. J. Swaney and I. M. Klotz, unpublished experiments, 1968.
55. C. Nolan and E. Margoliash, *Ann. Rev. Biochem.*, **37**, 727 (1968).

CHAPTER 3

PHYCOCYANINS AND DEUTERATED PROTEINS

Donald S. Berns

DIVISION OF LABORATORIES AND RESEARCH
NEW YORK STATE DEPARTMENT OF HEALTH

AND

DEPARTMENT OF BIOCHEMISTRY
ALBANY MEDICAL COLLEGE, ALBANY, NEW YORK

I. INTRODUCTION

Phycocyanin† is a protein of extensive intrinsic interest, a fact realized by a minority of scientists. Svedberg and co-workers (*2–4*) found phycocyanin

† The several varieties of phycocyanin and their sources are discussed by Ó hEocha (*1*). In general, we are concerned only with phycocyanin arbitrarily designated by Svedberg et al. (*2–4*) as C-phycocyanin and which is found in the large majority of blue-green and red algae. "Phycocyanin" as it appears in the text indicates C-phycocyanin unless specifically stated otherwise.

to be a very interesting protein primarily because of the advantageous absorption spectrum of the protein which permitted convenient use of the absorption optical system on the early analytical ultracentrifuge. The distinction of being one of the first proteins thus investigated, however, did not prevent it from remaining only a laboratory curiosity. Indeed, early investigators of Sephadex and column chromatography were drawn to phycocyanin and phycoerythrin because of their advantageous spectral properties (5,6), but little detailed work on these proteins was undertaken.

Phycocyanin is a chromoprotein that contains a linear tetrapyrrole as the prosthetic group, the identity of which has been elucidated in recent work by Siegelman et al. (7), Ó hEocha (8), and Rüdiger and Ó Carra (9). Phycocyanin is found in blue-green and red algae and represents as much as 25% of the total weight of the algal cell (10). Its function is reported to be that of an energy transfer system, part of the so-called photosystem II. The attractiveness of the protein stems from several factors. First, it is found in blue-green algae which grow under all extremes of environmental conditions as halophiles, psychrophiles, thermophiles, and mesophiles. Perhaps no other microorganisms are as ubiquitous as the Cyanophyta. Immunochemical studies (11) have demonstrated that phycocyanins extracted from a very large variety of blue-green algae are all antigenically related. Isolation and purification of the protein can be accomplished by simple procedures unlikely to produce artifacts. The characterization of an aggregating protein from such diverse environmental sources is particularly suited for contributing information about forces and specific reciprocal influences involved in protein–protein interaction.

Photosynthetic organisms, blue-green algae in particular, also offer a unique opportunity for obtaining large amounts of fully deuterated proteins (12). Cultivation of autotrophic microorganisms in D_2O and inorganic nutrients is a relatively simple matter, permitting the investigation of proteins fully deuterated at normally nonexchangeable positions. The complete substitution of deuterium for hydrogen represents the largest isotopic perturbation that can be achieved. This provides a unique method for probing the forces involved in stabilizing secondary, tertiary, and quaternary structure of proteins.

Red algae which also contain phycocyanin are eucaryotic, whereas blue-green algae are procaryotic organisms. This difference represents the most significant dichotomy in all organelle development. This situation is ideal for the application of the immunochemical method for tracing phylogenetic relationships. One tentative conclusion supported by such studies is that red algae are derived from blue-green algae or from some common ancestral cell type (11). Another aspect of organelle development of particular interest is associated with the location of phycocyanin in

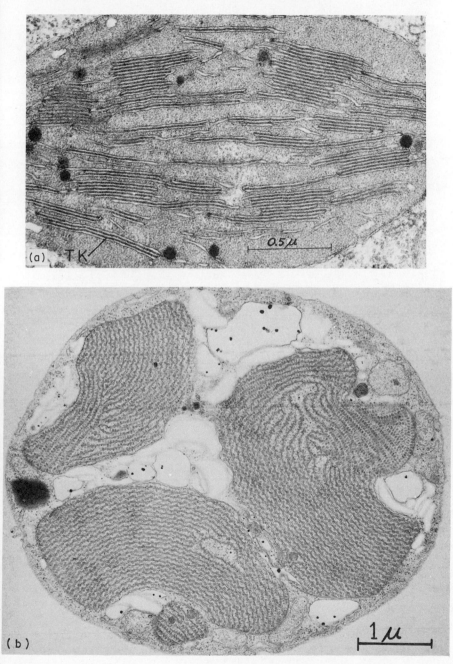

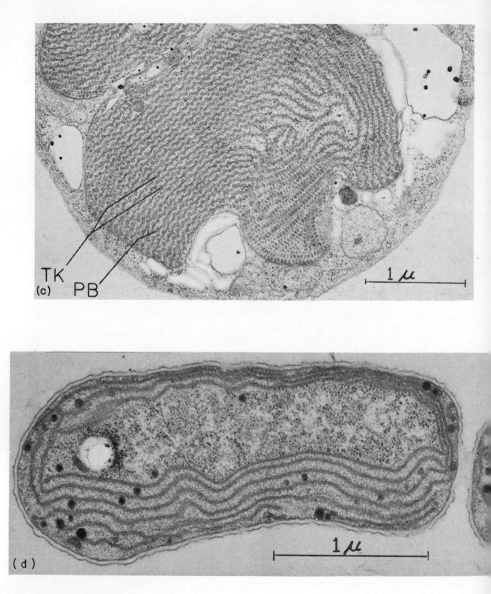

TK

PB

(c)

$1\,\mu$

(d)

$1\,\mu$

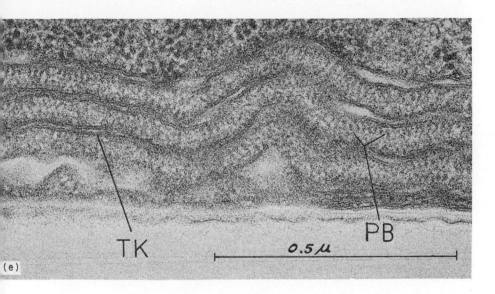

(e)

Fig. 1. Electron micrographs of stroma and grana structure in widely varying photosynthetic cell types. (a) Chloroplast from higher plant type, *Paeonia albiflora*, a eucaryotic cell with stacked grana. Note little if any stroma space between thalykoids (TK) in the grana stack. No phycocyanin is found in this organism. 60,000 ×. (b) Whole cell of red alga, *Porphyridium cruentum*, eucaryotic cell type with *no* stacked grana. Thalykoid membranes are well defined with definite stroma spaces. The cobblestone appearance in stroma space is the site of phycocyanin (phycobilisome). 20,000 ×. (c) A higher magnification of Fig. 1b to better demonstrate the fine structure between thalykoids (TK) and the phycobilisomes (PB). 40,000 ×. (d) Whole cell of blue-green alga, *Synechococcus lividus*, a procaryotic cell with *no* stacked grana. The thalykoids are well defined and a stroma space is the site of phycocyanin. 55,000 ×. (e) A higher magnification of a portion of an axial view of *Synechococcus lividus* cell. This shows with good definition the thalykoid membranes (TK) and the stroma space with the phycobilisomes (PB), the site of phycocyanin in the cell. 160,000 ×.

blue-green and red algae. Phycocyanin is always located in the stroma area between the thalykoid membranes, as indicated in part by the work of Gantt and associates (*13–15*) on phycobilisomes. In no phycocyanin-containing organisms examined have stacked grana been observed (Fig. 1). Organisms containing phycocyanin perform photosynthetic functions with an efficiency equal to those algae with highly developed organelles and stacked grana. If large amounts of membranes are needed for embedding and orienting the constituent parts of the photosynthetic system, then an analogous function is probably played by phycocyanin in the organisms lacking grana. We have suggested that phycocyanin represents an interesting prototype of a membrane system (*16*).

The progress made in investigating the phycocyanin system can best be described by first delineating the characterization of the subunit or monomer and then the more complex aggregates.

It is necessary to delineate the physical and chemical characteristics of the protein subunits of phycocyanin by employing a variety of experimental techniques. With the knowledge of the identity and properties of the phycocyanin subunit, it is possible to investigate the presence of phycocyanin aggregates, including demonstration of the kinds of aggregation present and how various perturbations affect aggregation. Evaluation of the observed properties of the aggregating system permits postulation of mechanisms to explain the aggregation processes. It is then extremely important to present evidence implicating these aggregates in an in vivo physiological role.

Since phycocyanin can be obtained in a fully deuterated form, a protein with deuterium substituted for hydrogen at all sites is available. It is then possible to investigate the changes in physical properties brought about by this largest possible stable isotopic perturbation of a protein. These experiments may give valuable insight into the relative importance of specific forces in protein aggregation, such as dispersion interactions. The effect of deuterium oxide on protein aggregation can also uncover interesting correlations concerning the nature of the stabilizing influences in protein aggregation. Finally, it is necessary to consider the prospects for continued research in areas related to phycocyanin and its aggregation properties.

II. SUBUNITS

A. Chemical Composition

A major decision in beginning work with a protein is the adoption of a method of purification. In most cases a long history of methodology generally prejudices all decisions concerning purification. Such is the case with phycocyanin (5,6,17–19). We have carefully investigated the various methods in the literature and adopted those that we feel are least deleterious to the preservation of the native structure and ability to aggregate (20). It is possible to prepare homogeneous and monodisperse phycocyanin by a number of procedures (5,6,17–19) that inhibit the retention of large aggregates. We have found that the chemical composition and properties of C-phycocyanin isolated in the highly aggregated state are essentially identical to the homogeneous monodisperse preparations.

Phycocyanin is a chromoprotein and the chemical composition of the chromophore has been characterized as a linear tetrapyrrole (1). The exact composition of the chromophore has been described by Siegelman et al.

(7) to be identical for all phycocyanins studied. The most probable chemical structure for the native chromophore (7–9) is indicated in Fig. 2. A metal is *not* reported to be associated with the tetrapyrrole. The chromophore peptide linkage is under active study; however, little conclusive evidence has been presented as to the nature of the linkage. Present estimates indicate that there are most probably two chromophores per 30,000-molecular-weight subunit (21,22).

1. Amino Acid Content

The amino acid composition of phycocyanin is an important consideration in the initial characterization of this protein. The amino acid composition of phycocyanin isolated from a number of blue-green and red algae is listed in Table I. Calculation of the number of residues in a particular phycocyanin is based simply on the assumption that the amino acid

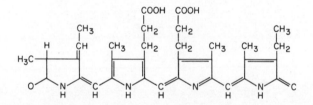

Fig. 2. Structure of phycocyanobilin, the chromophore present in all C-phycocyanins and allophycocyanins.

present in the least abundance must have at least one whole residue which in most cases turns out to be one whole cystine residue or one histidine. The total residues present, as calculated by this method, indicate an arbitrary subunit weight of about 30,000 for the phycocyanins analyzed (23). The comparative amino acid analysis indicates that most phycocyanins from blue-green algae are quite similar in amino acid composition (*Plectonema calothricoides*, *Phormidium luridum*, *Nostoc muscorum*, *Anacystis nidulans*, and *Synechococcus lividus*). Immunochemical studies (11) also indicate great similarity among phycocyanins from blue-green algae. Phycocyanin from the red algae (*Cyanidium caldarium* and *Porphyra tenera*) is also similar in amino acid composition to that of the Cyanophyta. The important observation is the similarity of composition and subunit size inferred from the amino acid analysis even though the phycocyanin is isolated from diverse algal types, such as filamentous and unicellular forms, as well as from those that differ widely in organelle development (procaryotic Cyanophyta and eucaryotic Rhodophyta). The data for *P. calothri-*

TABLE I

Amino Acid Composition of Some Phycocyanins

	Plectonema calothricoides[a]	Phormidium luridum[a]	Nostoc muscorum[b]	Anacystis nidulans[c]	Synechococcus lividus[a]	Cyanidium caldarium[a]	Porphyra tenera[d]
Polar residues							
Acidic							
Asp	28	29	31	38	28	33	36
Glu	18	19	27	18	30	28	38
Basic							
Lys	10	13	10	11	11	10	11
His	2	2	2	1	2	1	1
Arg	14	17	16	16	19	17	19
Hydroxyl							
Thr	14	17	17	15	14	14	18
Ser	16	20	19	20	10	13	32
Nonpolar residues							
Aliphatic							
Gly	20	24	23	22	10	22	29
Ala	40	43	36	48	26	47	52
Val	16	20	14	17	17	18	25
Ile	14	18	13	15	16	17	18
Leu	22	25	22	26	23	25	36
Pro	8	9	9	9	10	11	10
Met	8	9	2	3	8	8	11
Cys	1	1	1	1	1	1	3
Aromatic							
Tyr	10	14	11	12	16	8	17
Phe	8	11	7	10	10	5	7
Total molecular weight of residues	28,080	31,186	27,666	30,000	28,902	30,209	38,807

[a] Reference 23. [b] Reference 70. [c] Reference 21. [d] Reference 71.

coides and *P. luridum* cited in Table I also represent amino acid analyses performed on fully deuterated phycocyanins extracted and purified from these algae grown in 99.8 % D_2O. The amino acid analysis is identical with that for the protio protein (9).

2. Cyanogen Bromide Cleavage

Very few studies of peptide analysis of phycocyanin have been made, but several workers report cyanogen bromide cleavage experiments (21,24). Cyanogen bromide cleavage at methionine bonds makes it a convenient method for corroborating the monomer molecular weight and the number of peptide chains. The maximum number of peptides that should result from treatment with cyanogen bromide is equal to one plus the number of methionines in the monomer unit. Obvious exceptions would be the presence of uncleaved disulfide linkages and the possiblity of methionine as a chain-terminating group, or redundant methionines. We have exposed phycocyanin from *P. calothricoides* to cyanogen bromide and analyzed the products by several methods, including polyacrylamide disc electrophoresis (24), that can separate seven or eight peptides. Paper chromatography and electrophoresis give similar results. Amino acid analysis of phycocyanin from *P. calothricoides* indicates the presence of eight methionines per 28,000 molecular weight, demonstrating that a maximum of nine different peptides should result from cyanogen bromide treatment if the monomer molecular weight approximates 28,000. The results, which indicate seven or eight peptides, preclude the assignment of any molecular weight based on amino acid analysis and cyanogen bromide cleavage other than 28,000. More extensive cyanogen bromide cleavage experiments with phycocyanin from *A. nidulans* has been reported by Neufeld and Riggs (25) in which the amino acid analysis shows three methionines per 30,000 molecular weight and four peptides.

B. Immunochemistry

It has been repeatedly demonstrated that rabbit antisera prepared against native phycocyanin contain antibodies directed toward at least three different antigenic species (monomer, trimer, and hexamer) (11,20). Experiments with fully deuterated phycocyanin (26) have established clearly the absence of any detectable antigenic difference between the deuterio and protio proteins. Ouchterlony double diffusion gives lines of complete identity (Fig. 3) which fluoresce red under long-wavelength UV light. This phenomenon is a natural indicator of the presence of phycocyanin. The presence of multiple precipitin lines is interpreted as resulting from the difference in antigenicity of the several aggregates and is useful in discern-

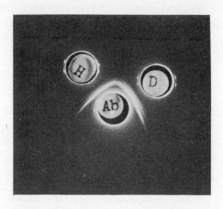

Fig. 3. Ouchterlony immunodiffusion performed in 1.5% bacto-agar and phosphate buffer, pH 6.0, 0.1 ionic strength. Ab, Antisera to protiophycocyanin from *P. calothricoides;* H, the antigen protiophycocyanin from *P. calothricoides;* D, the antigen deuterio (fully deuterated) phycocyanin from *P. calothricoides.* Note multiple lines of identity. All lines fluoresce red in the presence of long-wavelength UV light.

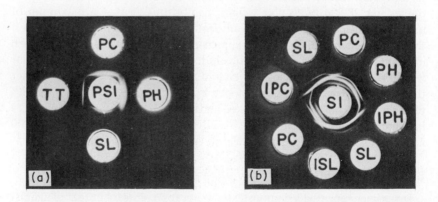

Fig. 4. Ouchterlony immunodiffusion performed in 1.5% bacto-agar and phosphate buffer, pH 6.0, 0.1 ionic strength. All lines fluoresce red in the presence of long-wavelength UV light. (a) PSI, antisera to phycocyanin from *P. luridum* injected in the presence of 1% sodium dodecyl sulfate; PH, antigen, phycocyanin from *P. luridum;* PC, antigen, phycocyanin from *P. calothricoides;* SL, antigen, phycocyanin from *S. lividus;* TT, antigen, phycocyanin from *Tolypothrix tenuis.* (b) SI, Antisera to phycocyanin from *S. lividus;* PC, PH, SL, as in Fig. 4a; IPC, IPH, and ISL are the corresponding antigens in the presence of 1% dodecyl sulfate. (Reprinted from Reference *11*, p. 1569, by courtesy of The American Society of Plant Physiologists.)

ing the difference in relative amounts of aggregates found in deuterio and protio proteins, an observation substantiated by physical studies (27). Analytical precipitin studies support the observation of antigenic identity of deuterio and protio proteins (26). The immunochemical evidence for identity of protio and deuterio proteins received more recent confirmation by the report of optical rotatory dispersion and circular dichroism analysis of these proteins (28).

The presence of multiple precipitin lines can confuse the comparison of phycocyanin monomers isolated from different algal sources. There are, however, several simple technical improvements that simplify the matter. Phycocyanin in the presence of 1% sodium docecyl sulfate is over 90% monomer (23). Antisera against this material should have a preponderance of antibody directed toward the monomer. This technique is moderately successful. Several phycocyanins examined with these antisera (Fig. 4) appear to have closely related monomer units. Another possibility for monomer production is the use of detergent-treated phycocyanin in the Ouchterlony double-diffusion experiments. These experiments were also somewhat successful and results indicated the partial antigenic identity of monomer and lower phycocyanin aggregates from the several algal sources.

C. Physical Characterization

1. Sedimentation

The earliest report of sedimentation properties of C-phycocyanin from Svedberg's laboratory (2–4) indicated that the protein was polydisperse under a variety of conditions, the major species present being 7 and 11 S. A slower sedimenting species was detected at a high pH (9.0 and above) and Svedberg pointed out that a subunit molecular weight of about 34,000 would be consistent with all the data. The suggestion of a subunit of this particular size was in agreement with the *then* startling Svedberg theory that protein molecular weights were equal to the molecular weight of ovalbumin, 34,500, multiplied by some integer. Unfortunately, no additional work concerning attempts to isolate the subunit was reported. Several other laboratories reported (17,29) sedimentation studies of phycocyanin, but without additional characterization of the subunit.

Our earliest attempts to characterize the monomer unit of phycocyanin began when it was found that near the isoelectric point, pH 4.7, sedimentation velocity patterns of phycocyanin contained a minor peak sedimenting at approximately 3.5 S (20) (Fig. 5). Samples of phycocyanin were then centrifuged in the presence of increasing concentrations of detergent, and at 1% sodium dodecyl sulfate approximately 90% of the protein present sedimented at 3.7 S. Similar experiments with increasing concentrations of

urea were severely limited since in solutions of a greater concentration than
2 M urea most of the phycocyanin precipitated. Even in 2 M urea there was
a substantial amount of 3.5 S sedimenting material. The sedimentation
experiments with the most convincing evidence for the existence of a
monomer unit of approximately 30,000 are those performed in the presence
of 5 M guanidine hydrochloride (24) by the method of Lapanje and Tanford
(30). Phycocyanin from *P. calothricoides* purified by salt fractionation,
hydroxyapatite column, or batch procedures, or treated with *p*-chloro-
mercuribenzoate, was all sedimented in the presence of guanidine hydro-

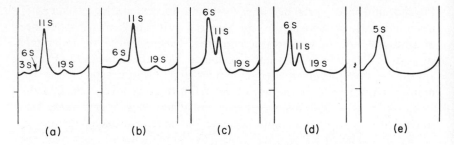

Fig. 5. Sedimentation study as a function of pH of phycocyanin from *P. calothricoides*.
All frames are at 32 min. Sedimentation is from left to right and speed is 59,780 rpm
at a rotor temperature of ~23°C. All buffers are 0.1 ionic strength. (a) pH 5.0, acetate;
(b) pH 6.0, phosphate; (c) pH 7.0, phosphate; (d) pH 8.0, phosphate; (e) pH 9.0, carbonate.

chloride and all data were consistent with a common *s*-vs.-concentration
plot (Fig. 6). The extrapolated *s*-value is 0.65 which may then be used in
the equation $s^0/(1-\phi'\rho)-0.286N^{0.473}$, where ϕ' is the apparent partial
specific volume, ρ is the solvent density, and N is the number of residues
in the polypeptide chain. The value of $N = 238$ is calculated from these
data. The amino acid analysis for this phycocyanin (Table I) indicates the
number of residues per minimal subunit of 28,000 to be 249. The sedimentation data with guanidine hydrochloride are therefore in excellent agreement with the monomer molecular weight suggested by the chemical
results.

High-speed sedimentation equilibrium experiments have been performed
by Dr. R. MacColl of this laboratory. A monochromator and photoelectric scanner on the Spinco model E ultracentrifuge permitted absorption at 620 and 278 mμ to be used in evaluating the molecular weight of
phycocyanin monomer from several algal sources (*P. calothricoides*, *Ana-
baena variabilis*, and *Lyngbya* sp.); all monomer molecular weights were
between 29,500 and 32,000. These experiments were performed in acetate
buffer, pH 3.9, 0.1 ionic strength, where only monomer is found, and at

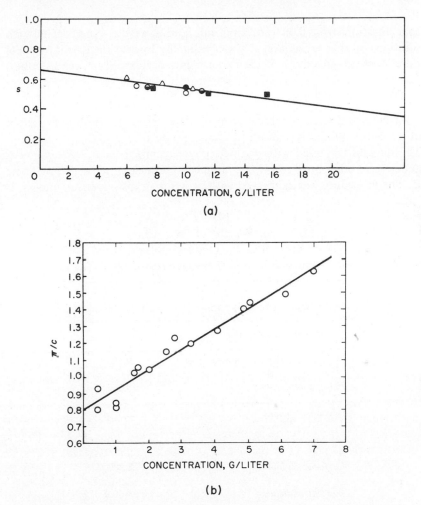

Fig. 6. (a) Plot of data from study of phycocyanin from *P. calothricoides* in the presence of 5 *M* guanidine hydrochloride. All studies were performed in double-sector synthetic boundary cells at rotor speeds of 59,780 rpm and a temperature of ~25°C. Phycocyanin samples used were from a variety of purifications with different chemical treatments and all fall on the same plot. △, Phycocyanin in phosphate buffer plus 5 *M* guanidine hydrochloride; ■, phycocyanin in water plus 5 *M* guanidine hydrochloride; ○, phycocyanin purified by hydroxyapatite batch procedure; ◑, monomer from *p*-chloromercuribenzoate treatment; ●, allophycocyanin fraction from *p*-chloromercuribenzoate treatment. (b) Plot of data from osmotic pressure study of phycocyanin from *P. calothricoides* in 6 *M* guanidine hydrochloride, and 0.1 *M* mercaptoethanol. The extrapolated value of π/c is used in the van't Hoff equation and a monomer molecular weight of ~30,600 is determined.

concentrations from 0.05 to 0.12 mg/ml. Studies were also performed with solutions of phycocyanin in 6 M guanidine hydrochloride and the results were identical with those of the experiments performed in acetate buffer.

2. Disc Electrophoresis

An empirical method of great utility in determining the molecular weight of protein subunits is sodium dodecyl sulfate disc electrophoresis (24). The method was used with purified phycocyanin from *P. calothricoides* and several other purified phycocyanins and crude extracts listed in Table II. The procedure was calibrated with proteins (trypsin, pepsin, and oval-

TABLE II

Molecular Weight of Phycocyanins
from Sodium Dodecyl Sulfate
Polyacrylamide Gel Electrophoresis

Source of phycocyanins	Molecular weight[a]
Plectonema calothricoides	28,500
Plectonema calothricoides-monomer[b]	29,400
Phormidium luridum	28,000
Tolypothrix tenuis	28,300
Synechococcus lividus	28,900
Anabaena variabilis	28,800
Anacystis nidulans	28,600
Porphyridium cruentum	28,700
Calothrix membranacea	28,800
Porphyridium aerugineum	29,000
Cyanidium caldarium	29,500

[a] These results are a compilation of the original published data (24) and many subsequent determinations using a microcomparator and a least-squares analysis for evaluation.
[b] This is a species separated by PCMB treatment and subsequent gel filtration (38).

bumin). All 10 different phycocyanins examined gave subunit molecular weights between 28,000 and 30,000.

3. Gel Filtration

Using Sephadex gel filtration experiments, Neufeld and Riggs (25) and Neufeld (21) determined the monomer molecular weight of phycocyanin from *A. nidulans*. At very low protein concentrations (<0.02 mg/ml) at pH 4.5, they were able to separate monomer from hexamer, and estimated from carefully calibrated G-200 Sephadex columns that the monomer molecular weight was 28,700. An estimate of the molecular weight of the monomer of this phycocyanin with crude extracts and sodium dodecyl sulfate gel electrophoresis was 28,600 (24), a value in respectable agreement with the other semiempirical method.

4. Osmometry

Osmotic pressure measurements were performed with samples of phycocyanin from *P. calothricoides* in 6 *M* guanidine hydrochloride in the presence of 0.1 *M* mercaptoethanol. The resulting osmotic pressure data were analyzed with π/c vs. c and also $(\pi/c)^{1/2}$ vs. c plots. The extrapolated values yielded a molecular weight of $30,500 \pm 1,000$, which is in excellent agreement with all other estimates of the monomer molecular weight.

III. AGGREGATION

A. Evidence for Several Protein Aggregates

The fact that phycocyanin is an aggregating protein was first indicated in the original sedimentation studies performed in Svedberg's laboratory (2–4). These studies demonstrated that phycocyanin exists as both 7 and 11 S species and, as mentioned, some suspicion of the existence of a slower sedimenting monomer of approximately 34,500 molecular weight. Svedberg and Katsurai (3) estimated the molecular weight of the 7 S species to be 104,000, a trimer, and of the 11 S to be 208,000, a hexamer. Experimental data used in estimating the molecular weight of the aggregates indicate that the trimer is approximately 90,000–100,000 daltons and the hexamer 180,000–200,000 daltons. The polydisperse nature of the system makes precise determination of molecular weights very difficult. Support for the existence of at least three distinct species is found in right-angle immunodiffusion studies (11,20) (Fig. 7). Approximate diffusion coefficients are reported for monomer, trimer, and hexamer (Table III). The most convincing direct evidence for the state of aggregation of phycocyanin is found

Fig. 7. Ouchterlony immunodiffusion, right-angle plates, in 1.5% bacto-agar. All lines fluoresce red in the presence of long-wavelength UV light. (a) 0.85% Saline; D-Ab., antiserum to phycocyanin from *P. calothricoides;* H-Ag, antigen, phycocyanin from *P. calothricoides.* (b) 0.85% Saline; S3 Ab., antiserum to phycocyanin; PH. Lur. (.1DS), phycocyanin from *P. luridum* in the presence of 0.1% sodium dodecyl sulfate. Note disappearance of precipitin lines in going from (a) to (b). (c) pH 7.0, phosphate; D-2Ab., antiserum to phycocyanin from *P. calothricoides;* HA *P. Cal.,* antigen, phycocyanin from *P. calothricoides* purified by hydroxyapatite chromatography.

TABLE III

Diffusion Coefficients of Phycocyanin from Immunodiffusion[a]

Species	$D \times 10^7$, cm^2/sec^2
Hexamer	4.2 ± 0.4
Trimer	7.5 ± 0.6
Monomer	13.2 ± 1.6

[a] Reference 20.

in the electron microscope studies of Berns and Edwards (31). The hexamer was demonstrated to be about 125 × 30 Å in size and consists of apparently spherical monomer units about 30 Å in diameter (Fig. 8). These micrographs provide ample indication that hexamers split into trimers and that isolated trimers and monomers exist. A higher state of aggregation (17–19 S), probably a dodecamer of approximately 360,000 molecular weight, is found in small amounts when the 11 S species is predominant (20). It is

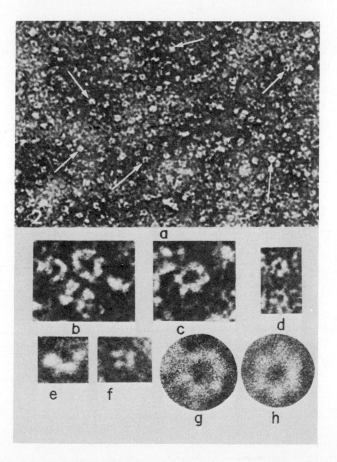

Fig. 8. Electron micrographs of phycocyanin from *P. calothricoides*. In all micrographs negative-stain technique with phosphotungstic acid was used at phosphate buffer pH 6.3, 0.1 ionic strength. (a) 200,000×, arrows point to molecular aggregates of phycocyanin; (b) and (c) 800,000 ×, hexamers and fragments of other aggregates; (d) 600,000×, obvious hexamers; (e) and (f) 800,000×, examples of probable trimer configurations; (g) and (h) 1,200,000×, highly enlarged micrographs demonstrating the appearance of monomers aggregated in rings of hexamers as shown in (b), (c), and (d). (Reprinted from Reference *31*, p. 511, by courtesy of Academic Press.)

possible to purify phycocyanin so that sedimentation analysis at pH 7.0 indicates the presence of 7 S material and little if any 11 S hexamer. At pH 5.0 the 11 S hexamer is present with 7 S only as some possible asymmetry of sedimentation peaks. The purification method used for these preparations is hydroxyapatite column or batch procedures (19), or ion-exchange cellulose chromatography. Purifications with results of this type appear to inhibit the reversible aggregation phenomena. Therefore, in work reported

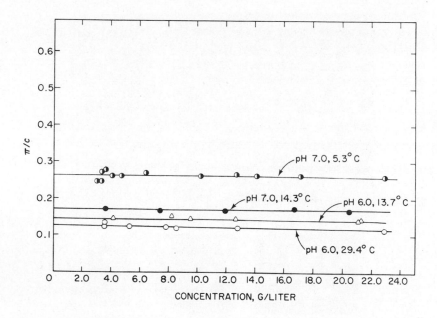

Fig. 9. Osmotic pressure studies of aggregation. The osmotic pressure of phycocyanin solutions in phosphate buffer pH 6.0 and pH 7.0, 0.1 ionic strength, was measured at different temperatures. Under all conditions the π/c vs. c plots have essentially zero slope indicative of no effect of concentration on state of aggregation.

from our laboratory, we have avoided these procedures. The retention of aggregation properties makes the characterization of the molecular weight of aggregates more difficult since the system is always polydisperse because of the simultaneous presence of monomer, trimer, hexamer, and dodecamer. It is possible to utilize conditions under which more than 80% of the total protein appears to be present as a particular species. Under these conditions osmotic pressure measurements have been used to estimate the trimer and hexamer molecular weights of phycocyanin from *P. calothricoides* (32) (Fig. 9). The number-average molecular weight from osmometry with more than 90% trimer (7 S) present is 89,400. In the presence of about 80%

hexamer, 17% trimer, and 3% dodecamer, as estimated from sedimentation velocity schlieren patterns, osmometry gives a molecular weight of 198,000 (*32*).

Neufeld and Riggs (*25*) made another direct measurement of the molecular weight of the hexamer on phycocyanin from *A. nidulans*, using calibrated G-200 Sephadex columns. At pH 4.5 the estimated molecular weight of the hexamer was 216,000 and at pH 6.5, 195,000. They believe that there is a monomer–dimer, hexamer–dodecamer equilibrium and have determined a dimer molecular weight of 67,100 at pH 6.5.

The existence of aggregates larger than dodecamer has been reported in sedimentation studies of phycocyanin purified by gentle methods (*33*). The exact size of these aggregates remains to be determined and only a preliminary estimate of size is possible from the available data (Table IV).

TABLE IV

Molecular Weights and Equilibria of Proposed Phycocyanin Aggregates

Aggregate	Estimated molecular weight	Method
Monomer	28,000–30,000	Sedimentation velocity in 5 M guanidine hydrochloride
		Sodium dodecyl sulfate polyacrylamide electrophoresis
		Sephadex gel filtration
		Amino acid analysis and cyanogen bromide cleavage
		Osmometry in 6 M guanidine hydrochloride
Dimer	67,000	Sephadex gel filtration
Trimer	90,000	Osmometry
Hexamer	190,000	Osmometry
		Sephadex gel filtration
Dodecamer and higher aggregates	360,000 540,000 720,000	None

Possible equilibria for phycocyanin from *Plectonema calothricoides* and *Phormidium luridum*:

$$M_6 \ (11\ S) \rightleftharpoons 6M \ (3\ S)$$
$$M_6 \ (11\ S) \rightleftharpoons 2M_3 \ (6\ S)$$
$$M_3 \ (6\ S) \rightleftharpoons 3M \ (3\ S)$$
$$M_{12} \ (19\ S) \rightleftharpoons 2M_6 \ (11\ S)$$

B. The Perturbation of Aggregation Equilibrium

The detailed investigation of the apparent equilibrium between trimer and hexamer elucidates several problems of central importance in protein chemistry in particular, and in molecular biology in general, namely, protein subunit assembly and the stability of protein aggregates. Aggregation phenomena have been investigated as a function of pH, ionic strength, temperature, and concentration (20). Sedimentation velocity studies by Scott and Berns (20) indicate that the relative amount of 11 S hexamer is greatest at constant ionic strength (0.1) at pH values close to the isoelectric point (pH 4.7). This finding has been confirmed by other sedimentation studies and the elegant gel filtration experiments of Neufeld and Riggs

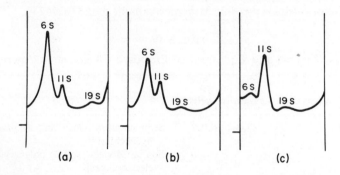

Fig. 10. The effect of ionic strength in sedimentation properties of phycocyanin from *P. calothricoides*. All sedimentation is from left to right at 59,780 rpm, at 40 min at ~25°C, and phosphate buffer pH 7.0. (a) 0.025 Ionic strength; (b) 0.1 ionic strength; (c) 2 *M* in sodium chloride.

(21,25). As the pH is increased, at constant ionic strength, the relative amount of hexamer decreases with a corresponding increase in 7 S trimer (Fig. 5).

Neufeld reports the same pH dependence of the stability of the hexamer (in *A. nidulans* phycocyanin), except he believes (21,25) that a dimeric species is favored at higher pH. Increasing the ionic strength at constant pH by addition of sodium chloride (Fig. 10) favored the 11 S hexamer at the expense of the trimer, while decreasing the ionic strength from 0.1 to 0.01 resulted in greater amounts of 7 S trimer (20). The relative amounts of each species were estimated from the sedimentation velocity measurements by integrating the area under the schlieren peaks. The effect of radial dilution was very small, and the area estimation was over a range of protein concentrations of 4–40 mg/ml. In this range there was no detectable concentration dependence of disaggregation (20), an observation that has been confirmed subsequently by osmotic pressure measurements (32) (Fig. 9). Studies using the photographic absorption system in the ultracentrifuge were performed at phycocyanin concentrations as low as 0.2 mg/ml. While it was difficult to quantitate the relative amount of each species present from such data, it was demonstrated that the larger aggregates were indeed

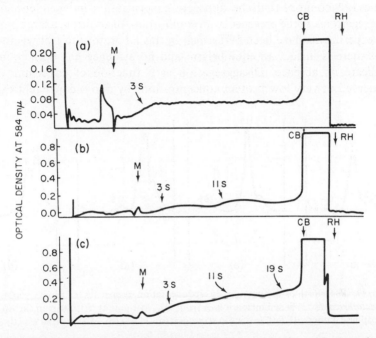

Fig. 11. Sedimentation patterns for phycocyanin from *P. luridum* with photoelectric scanning system and monochromator on Model E ultracentrifuge. 584-mμ light used for all scans. M, Location of meniscus; CB, location of cell bottom; and RH, location of outer reference hole on each scan. Phosphate buffer pH 6.0, 0.1 ionic strength, and 23°C for all samples. (a) 0.008 mg/ml, 32 min, indication of single peak; (b) 0.028 mg/ml, 36 min; (c) 0.056 mg/ml, 38 min. Note appearance of 11 and 19 S before any 6 S in both (b) and (c).

present (34). The monochromator-scanner system on the analytical ultracentrifuge has permitted sedimentation velocity studies at a concentration of 0.028 mg/ml by Drs. R. MacColl and J. J. Lee in this laboratory (35) (Fig. 11). Higher aggregates are still present at this concentration although there is a much larger amount of 3 S material. The observation that very low concentrations (< 0.02 mg/ml) of phycocyanin must be examined to obtain only monomers was confirmed by the gel filtration experiments of Neufeld and Riggs (25), who observed complete resolution of monomer from hexamer only at a concentration of approximately 0.02 mg/ml.

The observation that there is no detectable concentration dependence of the relative amounts of 11 S hexamer and 7 S trimer over a very extensive concentration region appears difficult to understand if true equilibria are present. At first it seems that either the observation is in error or that the system is not in equilibrium. Examination of the osmotic pressure properties (32) confirms both the apparent concentration independence of the aggregation and the presence of an equilibrium. In addition, a large number of experiments have been performed in this laboratory to test whether or not there is in fact an equilibrium, and no evidence has been found to indicate its absence. Disaggregation as a function of concentration is observed at very low protein concentrations by two independent experi-

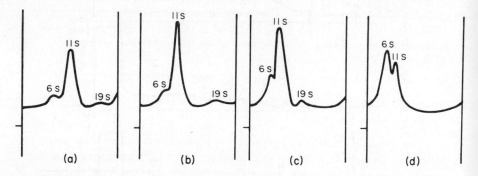

Fig. 12. The effect of temperature in sedimentation properties of phycocyanin from *P. calothricoides*. All sedimentation is from left to right at 59,780 rpm, at 40 min, phosphate buffer pH 6.0, 0.1 ionic strength. (a) 34°C; (b) 25°C; (c) 10°C; (d) 3°C.

mental techniques [gel filtration (25) and ultracentrifugation (35)]. From these results it is possible to estimate the pertinent association constants and to determine why the lack of concentration dependence of aggregation occurs. This analysis is presented in a later section.

The effect of temperature on the aggregation is such that at higher temperatures the relative amount of 11 S hexamer is increased (Fig. 12) at the expense of 7 S trimer. Lower temperatures favor the trimeric species. In all the pH, ionic strength, and temperature studies of aggregation, the effect on the concentration of 19 S material is difficult to delineate since it is a relatively minor component. In studies with phycocyanin from the thermophilic alga *S. lividus*, it was found that significant 19 S material appeared at a temperature of 49°C, which is close to the culturing temperature (36). Purification at a pH above 6.0 appears to be deleterious to the preservation of aggregates higher than 11 S (34). High salt concentrations do not seem to enhance aggregates other than 11 S.

The effect of specific chemical perturbants on the aggregation phenomena is also revealing. As already mentioned in Section I.C.1, the addition of increasing concentrations of detergent results in the enrichment of mono-mer units. At constant pH, urea can be added only up to a concentration of 2 *M*. Beyond that point the protein precipitates; however, at concentra-tions of 2 *M* in urea, there is almost complete degradation of 11 S hexamer to 7 S trimer (*20*). This is only 75 % reversible by dialysis away of the urea. Guanidine hydrochloride is a more efficient denaturant, and at concentra-tions of 5 *M* the monomer unit is present. Concentrations as low as 0.5 *M* in guanidine at pH 7.0 are efficient in degrading hexamer to trimer (*37*). It is curious that at very low concentrations of several guanidine salts (0.1 *M*) there is an enhancement of hexamer which is reversed by dialysis away of the guanidine (*36*). High concentrations of other salts, such as sodium thiocyanate, which are thought of as inhibiting hydrophobic inter-actions through destruction of water structure, also disaggregate phyco-cyanin hexamers.

C. Sulfhydryl Reagents

The reaction of sulfhydryl reagents with phycocyanin has been investi-gated in considerable detail. The earliest report (*29*) proposed that the reaction of *p*-chloromercuriphenyl sulfonic acid (PMPSA) with phyco-cyanin resulted in a reversible disaggregation to monomer units. The addition of reducing agent to the reaction product was sufficient to effect an apparent, almost complete, reaggregation as judged by spectral and sedimentation measurements. The conclusions from these data were that sulfhydryl groups are involved in the aggregation, they react with the PMPSA, and the product can be reduced to regenerate the native system. Careful studies with a variety of sulfhydryl reagents have cast considerable doubt on this explanation (*38*). The reaction of phycocyanin with *p*-chloro-mercuribenzoate (PCMB) appeared to give results identical to those with the PMPSA reagent, but several inconsistencies in the data were readily apparent. Although amino acid analysis indicated that phycocyanin from *P. calothricoides* has two sulfhydryl groups per 30,000 molecular weight, it was impossible to titrate clearly the groups with PCMB. The titration was never sharp and breaks in the titration course did not occur at repro-ducible ratios of reagent to protein concentration. When the reaction product was treated with a reducing reagent, such as dithiothreitol, regeneration of the original aggregation occurred. When the PCMB reac-tion product was exposed to G-100 gel filtration, two species were separated. Treatment of these species with reducing reagents had no effect, and recombination of the fractions prior to treatment with reducing agents did not serve to regenerate the native properties. It was evident that changes other than a reaction of sulfhydryl groups had occurred. Methyl mercuribromide reacted with phycocyanin in a manner identical to PCMB. N-Ethylmaleimide (NEM) was paradoxical in its reaction. The spectral properties of NEM were monitored and the attenuation of absorption associated with the reagent indicated a reaction; however, aggregation and spectral properties were unaffected. Addition of PCMB after NEM elicited a response identical to that from PCMB alone.

A relatively new sulfhydryl reagent, 4,4'-dithiodipyridine (39), was investigated and found to be capable of titrating sulfhydryl groups in a reproducible fashion. The titration was followed spectrally, and 2.2 sulfhydryl groups per 30,000 molecular weight were obtained with a wide range of protein concentrations. Spectral shifts were not of the same nature as those with PCMB. Aggregation properties were *not* affected in a dramatic fashion. The addition of PCMB after dithiodipyridine resulted in spectral changes. The dithiodipyridine reaction with sulfhydryl groups is analytical, does not cause dramatic changes in aggregation properties, and seems to confirm the suspicion that the reactions observed with PCMB and PMPSA and methyl mercuric bromide are not only reactions with sulfhydryl groups. Considerable care should be taken in assigning functions to sulfhydryl groups on the basis of studies with a single sulfhydryl reagent. Dramatic changes in aggregation properties may result from spurious side reactions of a particular reagent.

D. A Mechanism for the Observed Aggregation

The binding responsible for stabilizing the several equilibria in the aggregation of phycocyanin is obviously of a noncovalent character. The noncovalent bonds can be considered in two categories—those that involve intensely polar interactions, such as hydrogen bonds and salt linkages, and other types, such as the van der Waals or London dispersion forces and hydrophobic bonds described by Kauzmann (40). The hydrophobic bond does not represent a specific affinity of nonpolar groups for each other but, rather, as its name implies, results from the aversion of these groups to water. In a simple example, the formation of a soap micelle, the molecules coalesce because the negative free-energy change in transferring the aliphatic chains from an aqueous to a hydrocarbon environment is greater than the positive change involved in bringing the ionized carboxyl groups close together in the micelle surface. Favorable dispersion interactions quite likely keep the micelle together even with substantial charge repulsion between carboxyl groups.

The thermodynamics of hydrophobic interactions are such that the entropy change is positive and the enthalpy increment may also be positive (40,41). The positive entropy change is associated with the more random arrangement of water molecules that occur in the vicinity of the juxtaposed nonpolar groups. Hydrophobic bonds would be weakened by a decrease in the dielectric constant of the medium since nonpolar groups are more soluble under these conditions. Increasing the ionic strength of the solution tends to strengthen hydrophobic bonds by decreasing the solubility of nonpolar groups. The entropy change associated with the formation of salt linkages is also positive (40); however, their behavior in decreasing dielectric constant and increasing ionic strength would be the opposite of that of hydrophobic bonds. An obvious exception to the decrease of solubility of nonpolar groups by increasing ionic strength is the use of salts, which are thought to destroy water structure and consequently decrease the necessary entropic effect and therefore increase the solubility of nonpolar groups. Examples are sodium thiocyanate and guanidine hydrochloride.

Thermodynamic parameters were estimated for the phycocyanin trimer–hexamer aggregation from the temperature dependence of sedimentation studies by Scott and Berns (20) and the entropy, as well as the enthalpy, change for the aggregation is positive. The association to hexamer at pH 7.0 on raising the temperature from 5 to 30°C takes place as a result of the increase in entropy and in opposition to the increase in enthalpy. The magnitude of the entropy change is $+100$ eu and the $\Delta H = +24$ kcal/mole. Increasing salt concentration at constant pH enhances the hexamer aggregation. This observation is consistent, as are the thermodynamic data, with a hydrophobic interaction mechanism for formation of hexamer. More definitive evidence is provided by the fact that increasing pH at constant ionic strength, which charges up the subunits, favors depolymerization of the hexamer, as do high concentrations of urea and guanidine hydrochloride. The effect of urea and guanidine hydrochloride on proteins is now interpreted by Whitney and Tanford (42) in terms of ability to increase the solubility of nonpolar groups. Thus the disaggregation of the hexamer in high concentrations of urea, guanidine salts, and sodium thiocyanate is further support for a hydrophobic force mechanism for hexamer formation.

It is possible to be more specific in delineating a mechanism for hexamer formation. The gel filtration experiments of Neufeld and Riggs (25) indicate that very close to the isoelectric point the hexamer at very high dilution is disaggregated to monomer. The sedimentation experiments also indicate the presence of monomer and hexamer and no intermediates (20). In the sedimentation velocity experiments of Scott and Berns, the 3 S monomer is present at pH 5.0, but it is not evident at pH 6.0, 7.0, and 8.0. In gel filtration studies, Neufeld and Riggs do not observe disaggregation to monomer at pH 6.5; the disaggregation is to a dimer. All this evidence would be consistent with a mechanism suggesting that hexamer exists in equilibrium with monomer and that the monomer–hexamer equilibrium greatly favors the hexamer in a situation analogous to micelle formation. The critical micelle concentration for the hexamer is somewhat less than a protein concentration of 0.01 mg/ml or a monomer molarity close to 10^{-7}.

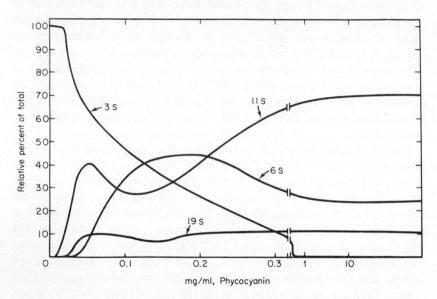

Fig. 13. Plot of relative amount of monomer (3 S), trimer (6 S), hexamer (11 S), and dodecamer (19 S) as a function of total phycocyanin concentration. These data are from overlapping studies of sedimentation velocity with absorption and schlieren optics. Multiple wavelengths (278, 420, and 584 mμ) were used in absorption studies. Note that 11 and 19 S aggregates appear before 6 S species.

The concentration profile for the relative amount of each of the aggregates is depicted approximately in Fig. 13. This information results from overlapping studies performed with the analytical ultracentrifuge using both the scanner and schlieren optical systems *(35)*. Below a concentration of 0.01 mg/ml, we observe essentially 100% monomer. We can assume this is the critical micelle concentration and then analyze the proposed equilibria.

$$6M \rightleftharpoons M_6; \quad K_1$$
$$M_6 \rightleftharpoons 2M_3; \quad K_2, \text{ and so on}$$
$$K_1 = \frac{[M_6]}{[M]^6}$$

Above a total phycocyanin concentration of 0.01 mg/ml, or with the monomer (M) molecular weight of 30,000, $\sim 3 \times 10^{-7}$ M, the concentration of monomer does not exceed 3×10^{-7} M. The association constant K_1 as determined by sedimentation measurements at concentrations close to 0.01 mg/ml is $\sim 10^{30}$ (35). With an association constant of 10^{30}, it is obvious that at concentrations above 0.01 mg/ml the hexamer predominates. One can then begin to understand the lack of monomer at concentrations of 4–40 mg/ml. The 6 S trimer appears in significant amounts only at concentrations of 0.05 mg/ml. When examined at a pH close to the isoelectric point (4.7), samples of 0.06 mg/ml exhibit only monomer and hexamer boundaries.

The hexamer forms as a result of hydrophobic bonding or coalescing of the subunits; once the monomers are in close proximity, dispersion forces serve to stabilize the hexamer. By raising the pH the subunits become charged, and a competition of forces results between the electrostatic repulsion of identically charged subunits and the dispersion interactions stabilizing the hexamer. Consequently, the disaggregation to trimer or dimer occurs above the critical micelle concentration for monomer-to-hexamer aggregation. This type of micelle formation, stabilized by dispersion forces, is probably quite common and is an ideal method for assembling polymeric building blocks for formation of membranes and other structural components. Methods for testing the suggested importance of hydrophobic forces and dispersion interactions in formation of the hexamer are presented in the studies of fully deuterated proteins and protein aggregation in D_2O.

E. Evidence for Phycocyanin Hexamers in Vivo

It is always important to ascertain whether or not a physically oriented study of biological material has any biological significance, that is, is the hexamer a species of possible biological importance in the function of phycocyanin? Are phycocyanin hexamers present in phycocyanins other than in those isolated from *P. calothricoides*, *P. luridum*, and *A. nidulans*? An ideal phycocyanin to study is *S. lividus* which is isolated from a thermophile which grows at 55°C in laboratory culture and as high as 80°C in the natural state. Phycocyanin isolated from mesophiles denatures at 55°C. Would the same temperature dependence of aggregation be observed for the hexamer–trimer aggregation in the thermophile? The investigation of this protein demonstrated that under comparable conditions there is far more 7 S trimer than in the mesophilic phycocyanin (36). Heating the thermophilic phycocyanin to 49°C is necessary to stimulate as much hexamer formation as found at 25°C with *P. calothricoides* phycocyanin

(Fig. 14). At lower temperatures 19 S material is absent, but at 49°C a considerable amount of 19 S is detectable. The *S. lividus* phycocyanin is antigenetically related to *P. calothricoides* phycocyanin in both trimeric and hexameric states (Fig. 15). From amino acid analysis there seems to be a greater proportion of charged and polar amino acids in the thermophile (Table I). The isoelectric point appears to be identical with that of other

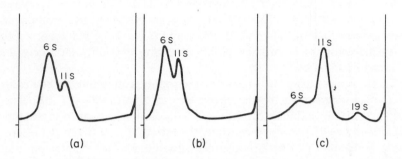

Fig. 14. Temperature study of sedimentation of phycocyanin from *S. lividus* in phosphate buffer pH 6.0, 0.1 ionic strength. Sedimentation is from left to right at 59,780 rpm. (a) At 8°C; (b) ~23°C; (c) 49°C.

Fig. 15. Ouchterlony double-diffusion studies of thermophilic phycocyanin. ASL, Antisera to *S. lividus* phycocyanin; SL, *S. lividus* phycocyanin; PC, *P. calothricoides* phycocyanin. All lines fluoresce red in the presence of long-wavelength UV light. Phosphate buffer, 0.1 ionic strength. (a) pH 6.0; (b) pH 7.0. (Reprinted from Reference *36*, p. 1528, by courtesy of the American Chemical Society.)

phycocyanins. However, greater migration of the thermophilic protein occurs in an electric field as the pH is increased (*36*). All these properties seem to be consistent with the proposed mechanism for hexamer aggregation.

In a study of phage-lysed blue-green algae, Luftig and Haselkorn (*43*) published an electron micrograph of what they believe is a hexameric array of phycocyanin. Related studies on phage-lysed blue-green algae by Padan et al. (44) also indicate semicrystalline arrays of hexamer similar to those reported from electron microscopy of Berns and Edwards (31). Fluorescence depolarization studies of crude phycocyanin extracts by Goedheer (45) offer interesting data as to the advantage of the hexamer configuration for the energy transfer function of phycocyanin. The depolarization of phycocyanin fluorescence is most efficient at pH values (4.0–6.0) where the hexamer is the predominant structure. The almost complete depolarization of fluorescence must be a result of internal inductive energy transfer since rotatory Brownian motion of such a large protein molecule would not permit complete depolarization. The hexamer structure certainly would favor internal energy transfer between subunits and subsequent fluorescence from the almost randomly oriented chromophores. At pH 7.0 and above, the depolarization decreases since the trimer and monomer eventually become more dominant. An array of hexamers appears to be an ideal structure for efficient energy transfer to chlorophyll. Studies of the efficiency of energy transfer to chlorophyll in cultures of A. nidulans reveal the importance of the proper ratio of phycocyanin to chlorophyll (46). Growth of cultures in strong orange light depresses the phycocyanin-to-chlorophyll a ratio, while growth in red light causes an increase in the ratio. A decrease in the efficiency of energy transfer occurs with a deviation of the phycocyanin-to-chlorophyll ratio from the normal value. In addition, cells grown in high-intensity orange light, which therefore have a lower phycocyanin-to-chlorophyll ratio, have a 7% polarized chlorophyll fluorescence. Cells grown in low-intensity orange light with a normal phycocyanin-to-chlorophyll ratio have only a 1% polarization of fluorescence. All these facts seem to provide good circumstantial evidence for the role of phycocyanin structure in efficient energy transfer. Studies of the relative fluorescence efficiency of purified phycocyanin fractions separated by sucrose density gradient sedimentation indicate that the fluorescence efficiency increases substantially with increased aggregation (33).

F. Evidence for Large Phycocyanin Aggregates

The presence of what appeared to be a 19 S species in purified phycocyanin from P. calothricoides and P. luridum was the first clue that higher aggregates of phycocyanin exist (20). From the accumulating evidence concerning the stability of the hexamer (20,47), it became obvious that purification of phycocyanin at a pH below 7.0 and avoiding exposure to chromatographic methods would permit the isolation of substantial amounts of higher aggregates (34). Crude extracts examined immediately after cell lysis demonstrated that more than 40% of the phycocyanin sedimented faster than 11 S (*33*) (Fig. 16). The work of Gantt and Conti (*13*) suggests that there are large aggregates of the biliproteins or "phycobilisomes". Sucrose density gradients confirm that the aggregation found above 11 S hexamer is phycocyanin (*33*). In addition, fractions from density gradient studies demonstrate the association of a red shift with increasing aggregation, as well as a marked increase in fluorescence efficiency with the presence of larger aggregates. The fluorescence excitation spectra of

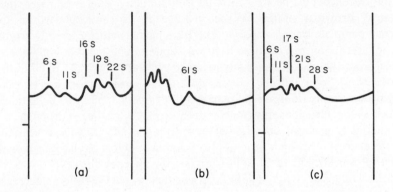

Fig. 16. Sedimentation velocity experiments with phycocyanin from *P. luridum*, purified carefully to preserve large aggregates (*33*). Sedimentation was at 59,780 rpm, at ∼23°C, in sodium phosphate buffer, 0.1 ionic strength. Wherever a pD is indicated instead of a pH, the samples were lyophilized and reconstituted in D_2O. (a) pH 6.0, 32 min; (b) pD 6.0, 16 min; (c) pD 6.0, 32 min.

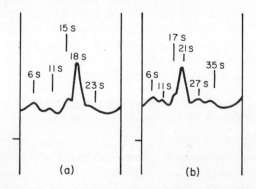

Fig. 17. Sedimentation velocity experiments with phycocyanin from *P. luridum*, purified carefully to preserve large aggregates (*33*). Sedimentation was at 59,780 rpm at ∼23°C, in sodium phosphate buffer, 0.1 ionic strength. Wherever a pD is indicated instead of a pH, the samples were lyophilized and reconstituted in D_2O. (a) pH 6.5, 32 min; (b) pD 7.0, 32 min.

fractions from sucrose density gradient studies are of interest in that the narrowest bandwidth of fluorescence excitation in the 620-mμ region also occurs with the largest aggregates. Gel filtration experiments indicate that the presence of the very large aggregates is not a simple concentration-dependence phenomenon. When fractions from a Sephadex G-200 column were examined by sedimentation velocity, it was demonstrated that the

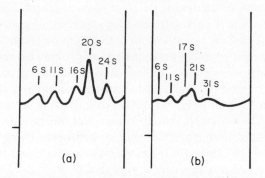

Fig. 18. Sedimentation velocity experiments with phycocyanin from *P. luridum*, purified carefully to preserve large aggregates *(33)*. Sedimentation was at 59,780 rpm at ~23°C, in sodium phosphate buffer, 0.1 ionic strength. Wherever a pD is indicated instead of a pH, the samples were lyophilized and reconstituted in D_2O. (a) pH 5.5, 32 min; (b) pD 5.5, 32 min.

species found in the leading Sephadex column fractions were not the largest, and considerable dissipation of the largest phycocyanin aggregates occurred. Reconstitution of the phycocyanin fractions did not result in reaggregation. The size of phycocyanin aggregates can be greatly enhanced by exposure to D_2O *(16,48)* (Figs. 17 and 18). The effect of deuterium oxide on the size of the largest aggregates is striking and is discussed in Section IV.

IV. DEUTERATED PROTEINS

A. Explanation

The term "deuterated protein" as used here must be clearly differentiated from that normally encountered, that is, the exposure of proteins to D_2O and the deuteration of exchangeable sites on the protein. Deuteration occurs in phycocyanin as a result of growth of alga in 99.8% D_2O. The extracted protein consequently has deuterium at all exchangeable and nonexchangeable sites. The purification of the protein in H_2O results in replacement of deuterium at exchangeable sites with protons. The non-

exchangeable sites (approximately 80% of the hydrogen) are still deuterated. For convenience, we refer to the deuterated protein, whether in D_2O
or H_2O, as deuterio protein and to the normally encountered protein as
protio protein.

B. The Possibilities

The substitution of deuterium for hydrogen at all available sites in a
protein is the largest possible stable isotopic perturbation of a protein. If
indeed the same protein is available in both protio and deuterio form, the
possibilities are opened for a unique method of probing the relative contributions to stabilizing secondary, tertiary, and quaternary protein structure.
The role of hydrogen bonding in protein structure is open to investigation,
as is the contribution of internal rotation to stabilizing structures and the
effect of dispersion interactions. All these contributions are significantly
affected by the substitution of deuterium for hydrogen. The first goal is the
establishment of the identity of the primary, secondary, and tertiary structure of deuterio and protio proteins. Amino acid analysis (12), qualitative
and quantitative immunochemistry (26), and optical rotatory dispersion
(ORD) and circular dichroism (CD) (28) have provided convincing evidence
for the identity of the deuterio and protio proteins. These data permit
comparison of the behavior of the two proteins and make relevant interpretations about the contribution of the several types of interaction to
the stabilization of protein structure.

C. Thermal Denaturation

The earliest comparative studies of deuterio and protio phycocyanin
were made by the analysis of thermal denaturation of the proteins (49).
The fluorescence of phycocyanin furnished a significant and convenient
method (50). It was reported that very fast denaturation studies with phycocyanin were reversible, and the fluorescence and absorption changes were
indicative of denaturation or unfolding effects (50). Careful studies in H_2O
(12) as a function of pH and ionic strength demonstrated that the deuterio
protein under comparable conditions always denatured at a lower temperature than the protio protein. No noticeable effect of ionic strength on
denaturation was observed (49). Related experiments in D_2O gave comparable results in that the deuterio protein always denatured at a lower
temperature (49). These results were later confirmed in considerable detail
(19). The most meaningful comparison of deuterio and protio proteins is
made in the same solvent under identical conditions (Fig. 19). The deuterio
proteins always denature at a temperature from 3–5°C lower than the
protio (Table V). Comparisons of the same protein in D_2O and H_2O are

difficult to interpret because of probable changes in pK values of the several amino acids in the proteins as characterized by the work of Appel and Yang (51) with synthetic polypeptides. Katz and co-workers (19) feel that their denaturation studies indicate a greater stability of protein structure in D_2O than in H_2O because of the difference between hydrogen and

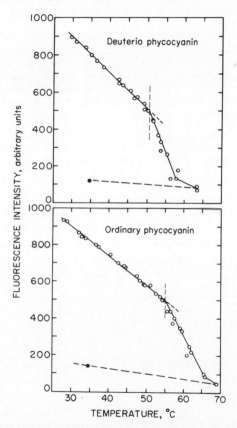

Fig. 19. Thermal denaturation of *P. calothricoides* phycocyanin in D_2O–acetate buffer. Break in curve is taken as transition temperature. ●, Descending temperature. (Reprinted from Reference 49, p. 1377, by courtesy of the American Chemical Society.)

deuterium bond strength. There is disagreement about the blanket applicability of this concept in the work of Berns (49). This point is minor, however, when uniform agreement is observed under comparable conditions in which the deuterio protein always denatures at a lower temperature than does the protio analog. The suggestion of a mechanism for the observed difference brings to light some disagreement. Berns has suggested (49) that the temperature increment involved in the denaturation of deuterio vs.

protio protein may be a reflection of the importance of internal rotation in the determination of protein structure. Immunochemical experiments (26), ORD and CD (28), have not uncovered any significant differences in secondary or tertiary structures in the deuterio and protio protein. Little is known about the actual polypeptide structure in phycocyanin. Differences that may exist in the ionization state of the protein induced by the presence

TABLE V

Thermal Denaturation of Deuterio and Protio Phycocyanins

Phycocyanin	Buffer	Critical denaturation temperature, °C	
Plectonema calothricoides			
Deuterio	Acetate–H_2O	44[a]	—
Protio		48	—
Deuterio	Acetate–D_2O	50	—
Protio		55	—
Deuterio	Phosphate–H_2O	45	42[b]
Protio		51	49
Deuterio	Phosphate–D_2O	45	53
Protio		48	55
Phormidium luridum			
Deuterio	Phosphate–H_2O	—	43
Protio		—	49
Deuterio	Phosphate–D_2O	—	52
Protio		—	54
Synechococcus lividus			
Deuterio	Phosphate–H_2O	—	66
Protio		—	68

[a] Temperatures in this column from Reference *49*.
[b] Temperatures in this column from Reference *19*.

of deuterium at nonexchangeable positions must be very small (*12*) and in the wrong direction to cause a destabilization of the protein because of charge repulsion. The suggestion of a decrease in nonpolar side-chain interactions resulting from substitution of deuterium for hydrogen is attractive (*19*) but difficult to substantiate by model compound studies. The available data have some basic inconsistencies difficult to explain by this mechanism. The ORD data in the UV range for protio and deuterio phycocyanin in D_2O and H_2O at pD = pH were studied and were found to be indistinguishable. Whatever shortcomings ORD may have for determining α-helix content, it has always been accepted as a very sensitive indicator of structural differences. Some slight structural opening might have been expected if hydrophobic forces were the dominant factor affected. It also might be expected that in D_2O there would be an enhancement of hydrophobic forces (see Section V) as indicated by the greater water structure (52) and by model compound studies including detergent micelle formation (53). The lack of an ionic strength or pH-induced charge effect on the relative difference between deuterio and protio denaturation temperatures is difficult to understand (49). One possibility is that the thermodynamic concept of an apolar interaction vis à vis water structure is not the factor to consider, but instead a perturbation of the dispersion forces between side chains (12,49). Another possibility is a change in the heat capacity term used to describe thermal denaturation in which it is assumed that the stable configurations of polypeptide chains and proteins are explicable in terms of hindered rotation about single bonds and the effect of secondary forces from within and without, for example, intramolecular and intermolecular hydrogen bonds and hydrophobic interactions that tend to compensate for the excess energy arising from the internal rotation potential (54). Tanford (55), in proposing a model that attempts to elucidate the stability of globular conformation of proteins on the basis of hydrophobic interaction, notes that the existence of potential energy barriers to rotation may be a contributing factor to the inability of his model to explain thermal denaturation. It can then be assumed that there is considerable hindered rotation when the protein is denatured or unfolded and that the stabilizing influence of the other structural forces (barrier to hindered rotation, for example, hydrogen and hydrophobic bonds) is the same in both deuterio and protio proteins. The observed difference in temperature of thermal denaturation in the case of deuterio and protio proteins in the same solvent and buffer may be explained in terms of a simple difference in moments of inertia of the C—D vs. the C—H. It should be possible to calculate at least semiquantitatively the type of temperature difference necessary to achieve the same conformational state in the two isotopically different proteins.

The entire argument can be cast in terms of the rotational contribution to the partition function. One can postulate that in the deuterio protein species the rotational energy levels will be more closely spaced than in the hydrogen protein and the heat capacity for the deuterio protein correspondingly larger, permitting the barrier to hindered rotation to be overcome at a lower temperature than in the protio protein. Unfortunately, no experimental test of any explanation for the isotope effect is readily forthcoming.

D. Protein Aggregation

The effect of deuteration on phycocyanin aggregation was studied as a function of pH, temperature, and ionic strength (27) in H_2O in a manner completely analogous to work with the protio protein (20). Dodecamer was completely absent in deuterio phycocyanin purified at pH 7.0 but was present in pH 6.0 purified protein. Sedimentation velocity studies at pH 5.0 on deuterio and protio proteins were similar (Fig. 20). Increasing the pH resulted in consistently smaller amounts of 11 S hexamer in the deuterio protein and larger amounts of the trimer (27). Increasing the ionic strength

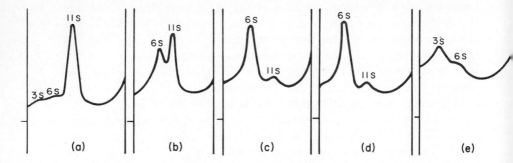

Fig. 20. Study of pH effect on sedimentation properties of phycocyanin from deuterio *P. calothricoides*. This figure should be compared with Fig. 5, which is a similar study for the protiophycocyanin from *P. calothricoides*. All sedimentation is from left to right at 59,780 rpm, at 40 min, ~25°C, and H_2O buffers at 0.1 ionic strength at each pH. (a) pH 5.0, acetate; (b) pH 6.0, phosphate; (c) pH 7.0, phosphate; (d) pH 8.0, phosphate; and (e) pH 9.0, carbonate.

at constant pH resulted in enhancement of hexamer as in the protio protein. Temperature studies at pH 5.0, 5.4, and 6.0 indicated that enhancement of hexamer is associated with increasing temperature as in the protio protein, except that the degree of enhancement of hexamer is smaller than in the protio phycocyanin. The effect of lower ionic strength at pH 6.0 is interesting in that the expected decrease in hexamer is observed as well as some evidence for an intermediate between hexamer and trimer (Fig. 21). Related studies on deuterio and protio phycocyanins have also been reported by Hattori et al. (56). These results also show a decrease in stability of hexamer in the deuterio protein without a specific increase in trimer, but with more emphasis on the appearance of monomer. Their report was not extensive in pH or ionic strength studies and did not include temperature data. The explanation for the decrease in stability of hexamer suggested by Hattori et al. (56) is again the decrease in hydrophobic forces. Unfortunately, there is no evidence to suggest that trimer is stabilized with respect to monomer by hydrophobic forces, although this is a possibility. In addition, solubility studies with model compounds such as amino acids, as suggested by the work of Krescheck et al. (*52*), are not available. Investigations of the solubility of some aromatic compounds by Olsson (*57*) and Erlenmeyer (*58*) give no clear indication of a general trend.

Deuterium substitution at nonexchangeable positions results in second-order effects dealt with in great detail in studies of simple organic molecules (*59*). The deuterium bonded to carbon is effectively more electropositive but less polarizable than hydrogen. Using several acids, Streitwieser and Klein (*60*) and Paabo et al. (*61*) characterized the increased electropositive

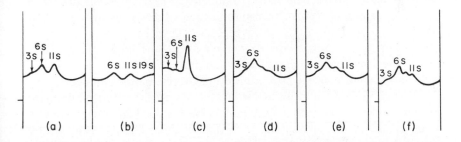

Fig. 21. Sedimentation study of phycocyanin from deuterio *P. luridum* purified at pH 7.0. All sedimentation is from left to right at 59,780 rpm, at 40 min; H_2O–phosphate buffers, pH 6.0, at ~ 25°C, and 0.1 ionic strength unless otherwise noted. (a) pH 6.0; (b) sample purified at pH 6.0; (c) pH 6.0, phosphate plus 2 *M* sodium chloride; (d) 0.01 ionic strength; (e) 0.025 ionic strength; (f) 0.05 ionic strength.

contribution of deuterium in this situation. The expected small increase in pK values of the several deuterium-substituted acids was found. Small increases in pK values of the constituent amino acids of the protein tend to decrease the charge of the protein at a particular pH. This decrease in charge should result in less trimer in the deuterio than in the protio protein since this acts effectively as a pH decrease. The prediction of a greater stability of the hexamer is contrary to the observed behavior and seems to eliminate consideration of the effect of deuterium substitution on pK as a prime consideration in understanding the destabilization of the hexamer.

A decrease in polarizability (*62*) or force constant (*63*) of a C—D vs. a C—H bond would result in a sizable decrease in dispersion forces. Deuterium substitution in simple hydrocarbons results in changes in molar volume and surface tension which may best be explained by a decrease in dispersion forces in the C—D vs. C—H bond. It is of course a considerable extrapolation from simple compounds to a summation over the hundreds of C—H bonds found in a protein. Van der Waals forces have been implicated by several investigators (64,65) as being of importance in biological systems. The dispersion force is the dominant contribution to the van der Waals forces in simple compounds (66) and the interaction energy does not decrease with increasing temperature. It should always be kept in mind that hydrophobic bonds and dispersion interactions are not mutually exclusive or identical. Hydrophobic forces, as delineated by Kauzmann (40) or Scheraga (41), arise from water structure as an entropic effect. Dispersion forces result from electron cloud interactions which become most effective with increasing molecular size and decreasing intermolecular distance.

In discussing the mechanism for hexamer stabilization, it was suggested by Scott and Berns (27) that hydrophobic forces pushed subunits together, but at a small distance there was competition between electrostatic repulsion of charged subunits and dispersion forces. At high pH repulsion caused the hexamer to be destabilized and more trimer was found. At low pH and high salt concentration, there was more efficient competition by the dispersion forces and hexamer was more stable. Studies on deuterio protein seem to support this hypothesis. If dispersion forces decrease as a result of deuterium substitution, then there is more effective competition by the electrostatic repulsion of the subunits and more trimer will appear, a phenomenon that has been observed. If the charge repulsion is neutralized by moving close to the isoelectric point, then behavior analogous to that in the protio protein would be expected since the competition with the dispersion forces has been minimized. The experiments at pH 5.0 in protio and deuterio proteins are almost identical. If there were a simple decrease in hydrophobic forces, one might expect destabilization of the hexamer even at pH 5.0. It might seem tempting to examine a correlation of the thermal denaturation data (49) with the change in stability of aggregates; however, at present no simple correlation is evident. For example, near the isoelectric point where aggregation is a maximum, no increase in thermal stability of phycocyanin is noted.

The conclusion drawn from the comparative studies of aggregation in deuterio and protio phycocyanins is that van der Waals forces are considered to make a significant contribution to protein aggregation. The total protein molecule may not necessarily be the contributing factor in this effect; it may be several side-chain interactions acting cooperatively. The contribution of other effects such as hydrophobic interactions should not be minimized. They may serve to bring the interacting groups close enough so that van der Waals forces can be most effective in stabilizing the aggregate.

V. EFFECT OF D₂O ON PROTEIN AGGREGATION

The critical micelle concentration of ionic detergents in D_2O has been reported to be smaller than that found in water (52,53). In D_2O there is a greater amount of localized structure than in H_2O and there should be a greater entropic effect with a consequent enhancement of hydrophobic

TABLE VI

Typical Data from H_2O and D_2O Sedimentation
Velocity Experiments with Phycocyanin at a Single
Protein Concentration of ∼ 15 mg/ml[a]

	H₂O			D₂O	
pH	$s_{20,w}$	Relative % of total area	pD	$s_{20,w}$	Relative % of total area
6.0	6.2	22	6.1	6.4	13
	10.3	55		10.8	69
	16.7	23		17.4	18
7.4	5.8	80	7.5	6.8	43
	10.4	16		11.6	41
	15.1	4		18.1	17
8.0	6.1	85	8.0	6.6	66
	11.1	11		11.4	29
	17.8	4		18.0	5

[a] Reference 16.

interactions (52). Several reports have been made of the effect of D_2O on protein aggregation (16,48,67,68) and at least one on an in vivo system (69). The studies with protio and deuterio phycocyanins are especially revealing. In previous work it was clearly demonstrated that a particular aggregate, the phycocyanin hexamer, was apparently enhanced by conditions favoring hydrophobic bonds. In studies of the effect of D_2O on hexamer stability, it was first established that the same species existed in sedimentation in D_2O and H_2O (Table VI). Sedimentation studies as a function of pD were performed with both deuterio and protio proteins. The effect of D_2O was the definite enhancement of the hexamer aggregation at the expense of the trimer (Table VI). No noticeable increase in dodecamer was reported in these studies, which included only samples with monomer, trimer, hexamer, and dodecamer. A comparison of the effect of D_2O on protio and deuterio phycocyanins seems to indicate that the percent increase in hexamer over that originally present is approximately the same. Thus the evidence supports the postulate that the hexamer results in part from hydrophobic forces and there is little if any difference between the enhancement of the entropic effect in the protio or deuterio proteins going from H_2O to D_2O.

TABLE VII

Absorption and Fluorescence Studies

Approximate $s_{20,\,w}$	Maximum of visible absorption peak, mμ	OD_{620}/OD_{280}	Relative fluorescence efficiency
Sucrose density gradient experiment, pH 6.0[a]			
> 22	629	3.0	29
19	623	5.4	45
16	622	5.4	49
10	619	3.6	20
8	619	3.5	18
< 6	617	2.0	14
Sucrose density gradient experiment, pD 6.0[b]			
Sample in D_2O before sedimentation			55
> 39	626	4.0	54
28	623	5.6	82
23	622	5.1	43
16	625	4.5	42
< 10	620	2.9	34

[a] Reference *33*.
[b] Reference *16*.

The effect of D_2O on hydrophobic bonds was confirmed by studies on other proteins whose ability to aggregate is unaffected by D_2O and in which hydrophobic forces have never been implicated as a mechanism (trypsin, soybean trypsin inhibitor, and catalase) (48). There is an effect of D_2O on α_s-casein (48) and β-galactosidase (68), which is consistent with the presence of hydrophobic interactions. The dodecamer or 19 S species is not enhanced in D_2O; however, since large amounts of aggregates larger than the dodecamer have been prepared (16), studies of the effect of D_2O on still greater aggregation were pursued.

In studying the effect of D_2O on very large phycocyanin aggregates, it was possible to produce samples with as much as 10% of the protein sedimenting as a 67 S peak (Figs. 16–18). Large quantities of species sedimenting faster than 24 S were usually present. Sucrose density gradient studies in D_2O were successful in separating 25 and 35 S fractions in which there was a high visible-to-UV optical density ratio and very high relative fluorescence efficiency (Table VII). The presence of such large aggregates is important in considering the phycobilisomes reported by Gantt and Conti (13) and the possible role of phycocyanin as a structural protein.

VI. PROSPECTS

Much of the work discussed here has been very informative from a structural point of view. It has been possible to elucidate to a limited extent the types of interactions probably involved in determining secondary, tertiary, and quaternary structure of phycocyanin. It is quite likely that many of the observations have applicability to the properties of other proteins. Additional studies may pursue two distinct directions. A significant effort may be invested in the detailed determination of phycocyanin amino acid sequences and also in specific secondary and higher-order structure. This would be the expected sequence of events if a purely physical approach were dominant. It should be a fruitful pursuit. A particular advantage in the use of phycocyanin in this kind of study is the availability of phycocyanin from widely divergent environmental sources (thermophile, psychrophile, halophile, and so on) along with the very good immunogenic properties that permit immunochemical experiments to be executed with ease.

A second approach is to investigate the structural aspects of this protein system as related to its accepted physiological role as an energy transfer pigment in photosystem II. In this regard it is interesting to examine energy transfer from phycocyanin to chlorophyll. A chlorophyll–phycocyanin complex for efficient energy transfer has been postulated (20), and there are good indications that the natural complex can be isolated and synthetic ones may be prepared. An additional physiological role of phycocyanin as a structural component has been postulated (*11,16,33*) and the electron micrographs of phycobilisomes reported by Gantt and Conti (*13*) and confirmed by Edwards et al. (*15*) are good evidence for such a role. If larger aggregates of phycocyanin can be prepared, these may be useful in learning about the basic protein assembly and membrane evolution. In the presence of such aggregates it may be possible to introduce lipids to stimulate larger aggregates and to incorporate large parts of the photosynthetic system. It offers an appealing concept for assembly of a photosynthetic unit on a matrix of phycocyanin.

There is also the very interesting analogy of phycocyanin with phytochrome. The spectrum of the short-wavelength form of phytochrome is similar to that of allophycocyanin. Recent studies (*33,35*) give sufficient information to support the suggestion that allophycocyanin results from irreversible disaggregation of phycocyanin. The possibility of phytochrome being related to phycocyanin is intriguing. No photoreversible behavior of phycocyanin has been reported. Nevertheless, the similarities in aggregation and chromophore require a thorough investigation.

Some aspects of the physiological problems are currently under active investigation and stimulating results have begun to appear. Abeliovitz and Shilo (*72*) have made an interesting report on the behavior of phycocyanin extracted from several different blue-green algae. They have noted that crude extracts of phycocyanin from log phase cells are light-sensitive. Extensive bleaching of the phycocyanin occurs if the buffered protein solution is exposed to light at room temperature for several hours. In related studies they also noted that the same algal cells, when allowed to proceed into late log phase, underwent a loss in the phycocyanin content as assayed in vivo or in vitro by spectrophotometric methods. These studies are of great interest since most studies with phycocyanin have been performed with the protein extracted and purified from laboratory cultures in the stationary phase or from algal blooms. In the log phase cultures, the system is being investigated under conditions in which the most extensive phycocyanin production and utilization are present. The algal physiology is necessarily the most dynamic at this stage. Perhaps under these conditions phycocyanin participates in a photosensitive reaction absent in later growth stages. These observations certainly point to a new and potentially very important area of research involving structure and physiological function of phycocyanin. Additional studies in M. Shilo's laboratory by Padan and associates (*73*) have been directed toward the physiology of the infection of blue-green algae with phage. After the phage infection latent period, there is no carbon dioxide fixation. However, determination of the phycocyanin concentration in various stages of phage infection demonstrates no detectable change in phycocyanin structure or content. This observation is interesting since it has been reported (74) that at later stages of phage infection the lamellae (thalykoids) of the blue-green algae become severely disrupted. If this observation is correct, it might be expected that sites containing phycocyanin structure would reflect these changes. Electron micrographs of the plaques formed by phage lysis exhibit semicrystalline structures resembling phycocyanin. Additional studies to characterize the nature of these structures are in progress. These structures may well represent the kind of very large aggregates being produced in recent studies by Lee and Berns (16,33).

ACKNOWLEDGMENT

This work was supported in part by Grant GB 7519 from the National Science Foundation. I wish to express my appreciation for the collaboration of my co-workers, Drs. E. Scott, J. J. Lee, O. Kao, and R. MacColl and the technical assistance of J. Vaughn. I am grateful to Dr. M. Edwards for her cooperation in the electron microscopy aspects of the work. All micrographs were kindly supplied by Dr. Edwards.

REFERENCES

1. C. Ó hEocha, in *Physiology and Biochemistry of Algae* (R. A. Lewin, ed.), Academic Press, New York, 1962, p. 421.
2. T. Svedberg and I. -B. Eriksson, *J. Am. Chem. Soc.*, **54**, 3998 (1932).
3. T. Svedberg and T. Katsurai, *J. Am. Chem. Soc.*, **51**, 3573 (1929).
4. T. Svedberg and N. B. J. Lewis, *J. Am. Chem. Soc.*, **50**, 525 (1928).
5. A. Tiselius, S. Hjerten, and O. Levin, *Arch. Biochem. Biophys.*, **65**, 132 (1956).
6. J. Porath, *Biochim. Biophys. Acta*, **39**, 193 (1960).
7. H. W. Siegelman, D. J. Chapman, and W. J. Cole, *Biochem. Soc. Symp. No. 28*, 107 (1968).
8. C. Ó hEocha, *Biochem. Soc. Symp. No. 28*, 91 (1968).
9. W. Rüdiger and P. Ó Carra, *European J. Biochem.*, **7**, 509 (1969).
10. J. Meyers and W. G. Kratz, *J. Gen. Physiol.*, **39**, 11 (1955).
11. D. S. Berns, *Plant Physiol.*, **42**, 1569 (1967).
12. D. S. Berns, H. L. Crespi, and J. J. Katz, *J. Am. Chem. Soc.*, **85**, 8 (1963).
13. E. Gantt and S. F. Conti, *Brookhaven Symp. Biol.*, **19**, 393 (1967).
14. E. Gantt, M. R. Edwards, and S. F. Conti, *J. Phycol.*, **4**, 65 (1968).
15. M. R. Edwards, D. S. Berns, W. C. Ghiorse, and S. C. Holt, *J. Phycol.*, **4**, 283 (1968).
16. J. J. Lee and D. S. Berns, *Biochem. J.*, **110**, 465 (1968).
17. A. Hattori and Y. Fujita, *J. Biochem. (Tokyo)*, **46**, 633 (1959).
18. F. T. Haxo, C. Ó hEocha, and P. S. Norris, *Arch. Biochem. Biophys.*, **54**, 162 (1955).
19. A. Hattori, H. L. Crespi, and J. J. Katz, *Biochemistry*, **4**, 1213 (1965).
20. E. Scott and D. S. Berns, *Biochemistry*, **4**, 2597 (1965).
21. G. Neufeld, Ph.D. Thesis, University of Texas, Austin, Texas, 1966.
22. R. Troxler and R. Lester, *Plant Physiol.*, **43**, 1737 (1968).
23. D. S. Berns, E. Scott, and K. T. O'Reilly, *Science*, **145**, 1054 (1964).
24. O. Kao and D. S. Berns, *Biochem. Biophys. Res. Commun.*, **33**, 457 (1968).
25. G. Neufeld and A. Riggs, *Biochim. Biophys. Acta*, **181**, 234 (1969).
26. D. S. Berns, *J. Am. Chem. Soc.*, **85**, 1676 (1963).

27. E. Scott and D. S. Berns, *Biochemistry*, **6**, 1327 (1967).
28. L. J. Boucher, H. L. Crespi, and J. J. Katz, *Biochemistry*, **5**, 3796 (1966).
29. E. Fujimori and J. Pecci, *Biochemistry*, **5**, 3500 (1966).
30. S. Lapanje and C. Tanford, *J. Am. Chem. Soc.*, **89**, 5030 (1967).
31. D. S. Berns and M. R. Edwards, *Arch. Biochem. Biophys.*, **110**, 511 (1965).
32. D. S. Berns, *Biochem. Biophys. Res. Commun.*, **38**, 65 (1970).
33. J. J. Lee and D. S. Berns, *Biochem. J.*, **110**, 457 (1968).
34. D. S. Berns and A. Morgenstern, *Biochemistry*, **5**, 2985 (1966).
35. J. J. Lee, R. MacColl, and D. S. Berns, *Biochem. J.*, in press.
36. D. S. Berns and E. Scott, *Biochemistry*, **5**, 1528 (1966).
37. D. S. Berns and A. Morgenstern, *Arch. Biochem. Biophys.*, **123**, 640 (1968).
38. O. Kao and D. S. Berns, unpublished observations.
39. D. R. Grassetti and J. F. Murray, Jr., *Arch. Biochem. Biophys.*, **119**, 41 (1967).
40. W. Kauzmann, *Advan. Protein Chem.*, **14**, 1 (1959).
41. H. A. Scheraga, *Proteins*, **1**, 477 (1963).
42. P. L. Whitney and C. Tanford, *J. Biol. Chem.*, **237**, PC 1735 (1962).
43. R. Luftig and R. Haselkorn, *J. Virol.*, **1**, 344 (1967).
44. E. Padan, M. Shilo, and N. Kislev, *Virology*, **32**, 234 (1967).
45. J. C. Goedheer, Ph.D. Dissertation, Rijksuniversiteit, Utrecht, 1957.
46. A. K. Ghosh and Govindjee, *Biophysics J.*, **6**, 611 (1966).
47. J. A. Bergeron, *Photosynthetic Mechanisms of Green Plants*, National Academy of Sciences, National Research Council, Washington, D.C., 1963, p. 527.
48. D. S. Berns, J. J. Lee, and E. Scott, *Advan. Chem.*, **84**, 21 (1968).
49. D. S. Berns, *Biochemistry*, **2**, 1377 (1963).
50. J. Lavorel and C. Moniot, *J. Chim. Phys.*, **59**, 1007 (1962).
51. P. Appel and J. T. Yang, *Biochemistry*, **4**, 1244 (1965).
52. G. C. Krescheck, H. Schneider, and H. A. Scheraga, *J. Phys. Chem.*, **69**, 3132 (1965).
53. M. F. Emerson and A. Holtzer, *J. Phys. Chem.*, **71**, 3320 (1967).
54. S. -I. Mizushima, *Advan. Protein Chem.*, **9**, 299 (1954).
55. C. Tanford, *J. Am. Chem. Soc.*, **84**, 4240 (1962).
56. A. Hattori, H. L. Crespi, and J. J. Katz, *Biochemistry*, **4**, 1225 (1965).
57. S. Olsson, *Arkiv Kemi*, **15**, 259 (1960).
58. H. Erlenmeyer, H. Lobeck, and A. Epprecht, *Helv. Chim. Acta*, **19**, 793 (1936).
59. E. A. Halevei, *Progr. Phys. Org. Chem.*, **1**, 109 (1963).
60. A. Streitwieser, Jr., and H. S. Klein, *J. Am. Chem. Soc.*, **85**, 2579 (1963).
61. M. Paabo, R. G. Bates, and R. A. Robinson, *J. Phys. Chem.*, **70**, 540 (1966).
62. L. S. Bartell and R. R. Roskos, *J. Chem. Phys.*, **44**, 457 (1966).
63. M. J. Stern and M. Wolfsberg, *J. Chem. Phys.*, **45**, 2618 (1966).
64. L. Salem, *J. Chem. Phys.*, **37**, 2100 (1962).
65. W. H. Stockmayer, in *Biophysical Science—A Study Program* (J. L. Oncley, ed.), Wiley, New York, 1959, p. 103.
66. K. B. Wiberg, *Physical Organic Chemistry*, Wiley, New York, 1964, p. 131.
67. S. Paglini and M. Lauffer, *Biochemistry*, **7**, 1827 (1968).
68. R. P. Erikson and E. Steers, Jr., *Federation Proc.*, **28**, 468 (1969).
69. S. Inoué, H. Sato, R. Kane, and R. E. Stephens, *J. Cell Biol.*, **27**, 115A (1965).
70. M. A. Raftery and C. Ó hEocha, *Biochem. J.*, **94**, 166 (1965).
71. J. R. Kimmel and E. L. Smith, *Bull. Soc. Chim. Biol.*, **40**, 2049 (1958).
72. A. Abeliovitz and M. Shilo, private communication, 1969.
73. E. Padan, D. Ginzberg, M. Shilo and D. S. Berns, private communication, 1969.
74. K. M. Smith, R. M. Brown, Jr., D. A. Goldstein, and P. L. Walne, *Virology*, **28**, 580 (1966).

CHAPTER 4

TOBACCO MOSAIC VIRUS AND ITS PROTEIN†

Max A. Lauffer

DEPARTMENT OF BIOPHYSICS AND MICROBIOLOGY
UNIVERSITY OF PITTSBURGH
PITTSBURGH, PENNSYLVANIA

I. INTRODUCTION

Tobacco mosaic virus (TMV) is a rod-shaped particle with a molecular weight of about 40,000,000, composed of 95% protein and 5% ribonucleic acid (RNA). When the virus is subjected to dilute alkali (1) or to 67% acetic acid (2), the protein part can be isolated from the RNA in pure form. It is called A-protein because it was first obtained by alkaline degradation in Schramm's laboratory (1). The chemical subunit has a molecular

† This is Publication No. 166 of the Department of Biophysics and Microbiology, University of Pittsburgh. Work was supported by U.S. Public Health Service grant GM 10403.

149

weight of 17.53×10^3 and the amino acid sequence shown in Fig. 1 (3).
At pH values near 7 and at concentrations near 1 mg/ml, A-protein seems
to be a trimer of the ultimate protein subunit (4).

Lauffer et al. (5) showed that TMV A-protein polymerizes endothermi-
cally and reversibly in 0.1 ionic strength (μ) phosphate buffer at pH 6.5.
It was concluded that since heat is absorbed in the polymerization process
the negative standard free energy necessary for the reaction to occur must
derive from an increase in entropy. It was postulated in this early publication that the increase in entropy resulted from the release of water molecules
during the polymerization process. Subsequent research has supplied many
details and has confirmed that water is released upon polymerization.

```
                     5              10              15              20
Acetyl-Ser-Tyr-Ser-Ile- Thr-Thr-Pro-Ser-GluN- Phe-Val-Phe-Leu-Ser-Ser-Ala-Try-Ala-Asp- Pro-Ile-Glu-Leu-Ile-
25                      30              35              40
AspN-Leu-CysH-Thr-AspN- Ala-Leu-Gly-AspN-GluN- Phe-GluN-Thr-GluN-GluN- Ala-Arg-Thr-Val-Val-
45              50              55              60
GluN-Arg-GluN-Phe-Ser- GluN-Val-Try-Lys-Pro- Ser-Pro-GluN-Val-Thr- Val-Arg-Phe-Pro-Asp-
65              70              75              80
Ser-Asp-Phe-Lys-Val Tyr-Arg-Tyr-AspN-Ala- Val-Leu-Asp-Pro-Leu- Val-Thr-Ala-Leu-Leu-
85              90              95              100
Gly-Ala-Phe-Asp-Thr- Arg-AspN-Arg-Ile-Ile-Glu-Val-Glu-AspN-GluN- Ala-AspN-Pro-Thr-Thr-
105             110             115             120
Ala-Glu-Thr-Leu-Asp- Ala-Thr-Arg-Arg-Val- Asp-Asp-Ala-Thr-Val- Ala-Ile-Arg-Ser-Ala-
125             130             135             140
Ile-AspN-AspN-Leu-Ile- Val-Glu-Leu-Ile-Arg- Gly-Thr-Gly-Ser-Tyr- AspN-Arg-Ser-Ser-Phe-
145             150             155
Glu-Ser-Ser-Ser-Gly-Leu-Val-Try-Thr-Ser- Gly-Pro-Ala-Thr.
```

Fig. 1. Amino acid sequence of TMV [Anderer et al. (3).] (Reprinted by permission of
copyright owner, *Zeitschrift für Naturforschung.*)

Thus the polymerization of TMV A-protein is an entropy-driven reaction
in which the required free-energy decrease is derived largely if not entirely
from the changes during polymerization in the interaction between the
protein and water. The stability of the products obtained, whether intermediates with varying degrees of transience or the ultimate cylindrical rod,
results largely if not entirely from protein–solvent interaction. Therefore
any conclusions drawn from such studies can be valid only for aqueous
solutions, a fact that must be remembered constantly when results of
experiments of a thermodynamic nature on aqueous solutions are compared with those obtained by methods such as electron microscopy, in
which the structures observed are thoroughly dry. Whatever contributes
to the stability of structures observed in the electron microscope, the
dominant factor of protein–solvent interaction investigated by thermodynamic methods must be entirely absent. There is therefore a very real

possibility that structures observed by means of electron microscopy do not have their counterpart in aqueous solutions. The situation with respect to structures observed by means of X-ray diffraction studies on crystals is perhaps intermediate between the extremes represented by aqueous solutions and thoroughly dried electron microscope specimens because most analyses are made on crystals or solids that contain at least a small amount of moisture.

The RNA from TMV can be obtained in purified form. The most commonly used method is to extract the virus with phenol. Fraenkel-Conrat and Williams (6) and Lippincott and Commoner (7) showed that copolymerization of TMV A-protein and RNA leads to the formation of rodlike particles, apparently identical with TMV, possessing considerable biological activity. Recent evidence obtained in our laboratory indicates that this process is also endothermic.

Many of the details of the polymerization of TMV A-protein have been described in a recent review (8) which includes numerous historical references. Earlier reviews had been published by Lauffer (9,10) and by Caspar (11). This chapter attempts to bring up to date the experimental aspects of the polymerization of TMV A-protein and of the copolymerization of protein and RNA and to present a consistent and detailed theoretical interpretation of the process.

An understanding of the polymerization of TMV A-protein is important not only because of its significance in relation to the structure of a virus, but also because comparable phenomena are frequently observed in other macromolecular systems (10). Examples are formation of gels by methyl cellulose, formation of micelles and gels by synthetic detergents and soaps, aggregation of copolymers of amino acids, formation of gels by myosin, polymerization of sickle-cell hemoglobin and of G-ADP actin, formation of fibrils by collagen, certain antigen–antibody reactions, formation of gels by nitrocellulose, and binding of ions by proteins. Furthermore, many biological transformations, some highly important, are endothermic processes resembling polymerization of TMV A-protein. Among these are contraction of actomyosin, division of fertilized eggs of several species, formation of pseudopodia by amebas, protoplasmic streaming, rapidly reversible color changes in the skin of the killifish and, finally, formation of mitotic spindles (10).

Since endothermic polymerization of TMV A-protein is a reaction that takes place primarily because water is set free from subunits, it seems likely that these other endothermic polymerization processes, many of great importance in biology, can also be explained on a similar basis. Thus water seems to play a formerly unsuspected role in living systems. In a sense, much of our current understanding was anticipated in 1930 by

Gortner (*12*), who wrote, "It is my belief that many of the reactions characteristic of living processes have to do more with the water relationships of the organism than with any other single factor".

II. STABLE TMV PROTEIN PARTICLES

A. The Protein Monomer

At a TMV protein concentration of 0.1 mg/ml in 0.1 μ borate buffer, pH 9.0, and in 0.05 μ phosphate buffer, pH 6.5, at temperatures between 0 and 6°C, the protein monomer is the stable form. The evidence for this is the observation by Ansevin and Lauffer (*13*) that the sedimentation coefficient is 1.9 S. By the method of Lauffer and Szent-Györgyi (*14*), they made calculations indicating that a sedimentation coefficient of 1.9 S corresponds to a molecular weight of approximately 18,000. Osmotic pressure measurements of Banerjee and Lauffer (*4*) on TMV protein in 67% acetic acid yield an extrapolated value for the number-average molecular weight of 18,200, demonstrating that the monomeric form predominates under these conditions. Also, at pH 13, TMV protein is monomeric (*15,16*). Frist et al. (*17*) obtained a product by treating the protein of cucumber virus 4, a strain of TMV, with succinic anhydride at 5°C. Sedimentation and diffusion measurements extrapolated to zero concentration yielded values that correspond to a molecular weight of 19,000, close to the expected value for the protein monomer.

The protein subunit of the common strain of TMV has one α-carboxyl group, 6 aspartic acid carboxyl groups, 8 glutamic acid carboxyl groups, no histidine groups, no α-amino groups, 2 lysine amino groups, 11 arginine groups, 4 tyrosine hydroxyl groups, and 1 cysteine sulfhydryl group (*3*). If all ionized normally, near pH 7 the particle would have 15 negative charges and 13 positive charges, giving a net charge of -2. No acid-base titration results are available for conditions under which the protein monomer is the dominant particle. Thus this value is not confirmed experimentally.

B. A-Protein

The sedimentation velocity experiments of Ansevin and Lauffer (*13*) show that at TMV protein concentrations greater than 1 mg/ml the sedimentation coefficient is between 4.0 and 4.9 S for protein dissolved in 0.01 μ borate buffer at pH 9.0 and in 0.10 μ phosphate buffer at pH 6.5 at temperatures between 0 and 6°C. As shown by Caspar (*11*), using the method of Lauffer and Szent-Györgyi (*14*), a cyclical trimer of the TMV protein subunit should have a sedimentation coefficient between 4.2 and 4.6 S. Osmotic pressure results of Banerjee and Lauffer (*4*) indicate that at a protein concentration of 2 mg/ml the number-average molecular weight is approximately 50,000 in 0.1 μ phosphate buffer at 5°C and at pH values ranging between 6.5 and 8.0. A similar result was obtained for protein in 0.067 μ phosphate buffer at pH 7.5 for temperatures ranging between 4.6 and 15.8°C. Confirmation was obtained by Paglini and Lauffer (18) at 4°C for protein in 0.1 μ phosphate buffer at pH 6.5 and 7.5. Since a number-average molecular weight of approximately 50,000 could correspond to a solution in which trimers predominate, it is probable that at low temperatures over the pH range from 6.5 to 8 the trimer is the stable configuration. We therefore consider TMV A-protein as a stable particle composed of three protein subunits which can dissociate under some conditions into the ultimate protein particle and which can polymerize under other conditions to form larger structures.

The isoionic point of both TMV and TMV protein, determined as the pH of a protein solution after exhaustive deionization with a mixed-bed ion-exchange resin, was found to be between pH 4.3 and 4.6 (19). The acid-base titration curve obtained by Scheele and Lauffer (20) and shown in Fig. 2 demonstrates that at 4°C, where A-protein seems to exist essentially as a trimer, 4 hydrogen ions are bound per protein monomer, or 12 per trimer between the pH region 6.5–9 and pH 4.65, the highest pH con-

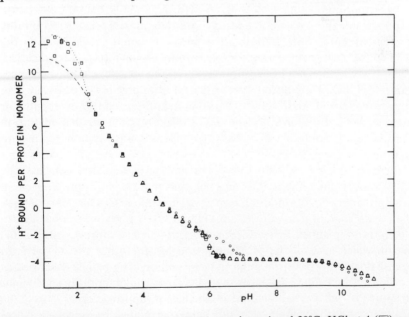

Fig. 2. Acid-base titration curves of TMV protein at 4 and 20°C. HCl at 4 (□) and 20°C (○); KOH at 4 (△) and 20°C (●). [Scheele and Lauffer (20).] (Reprinted by permission of the copyright owner, The American Chemical Society.)

sistent with the experimentally determined isoionic point. This must mean that in the region between pH 6.5 and 9 the protein trimer has 12 negative charges. If all available groups that should be ionized were ionized, then the trimer would have six negative charges at pH 7. It is impossible for the trimer to have 12 negative charges from hydrogen ion dissociation unless six positively ionizing groups are un-ionized or six negatively ionizing groups are ionized at pH values far removed from their normal pK values. The titration curve in Fig. 2 shows further that between pH 7 and pH 1.1, 17 hydrogen ions are bound per protein monomer. This is two more than the total number of carboxyl groups present, which confirms the idea that two groups per protein monomer or six per trimer, which normally titrate at pH values in the alkaline range, are being titrated in the acid range.

Two theories have been advanced to account for this. Conceptually, the simpler hypothesis is that two positive groups, assumed to be lysine amino groups, are "buried" (20). By "buried" it is meant that they are in an environment in which ionization does not take place, such as organic residues with a low dielectric constant. In support of this view is the fact that there is considerable evidence in the literature for restricted reactivity of the ε-amino groups of TMV protein. A more complex hypothesis advanced by Scheele and Lauffer (21) is also in agreement with the major facts. According to this hypothesis, a hydrogen ion is shared between an amino group and a phenolic hydroxyl group in close proximity within the protein structure. Such a shared hydrogen ion would partially satisfy the tendency of each group to capture its own hydrogen ion. A second hydrogen ion would then not be captured by the amine–phenol complex until pH values well below the normal pK values of both groups were reached. As shown by Scheele and Lauffer (21), this hypothesis can also account for many of the known facts. In support of it can be cited the demonstration that in mixed solutions of substituted phenols and arylamines proton sharing does occur (22,23).

Because of the way it fits into the structure of the TMV particle, the TMV protein monomeric unit must be anisometric, with one dimension of approximately 80 Å and the other two much less. Caspar (11) assumed that the trimer, which we call unpolymerized A-protein, is a particle in which the long dimensions of the protein subunits are more or less parallel and in which each of the three units touches the other two to form an aggregate more or less triangular in cross section. This can be described as a cyclical trimer. The only experimental evidence in favor of this particular structure is that the sedimentation coefficient, approximately 4 S, of A-protein corresponds well with the value calculated for such a cyclical trimer. However, thermodynamics provides a more convincing argument in favor of the cyclical arrangement because it involves three "bonds" per trimer. Furthermore, if the aggregates were in some sense linear, there is no readily apparent reason why trimers should be more stable than dimers or tetramers, and so on.

C. Double Discs

In 1963, Markham et al. (24) obtained evidence by means of electron microscopy that TMV protein can under some circumstances polymerize in the form of a double disc with a hole in the center, each member of the pair of discs being made up of 16 protein monomers. Actually, X-ray diffraction studies by Franklin and Commoner (25) on the protein found in plants infected with TMV had indicated the existence of such discs as early as 1955. More recent studies by Finch et al. (26), using X-ray diffraction and electron microscopy, have indicated similar double discs but with 17 protein monomers in each of the two component rings. The discrepancy in the number of protein subunits per double disc, 32 or 34, could possibly mean that both forms actually exist, one form being obtained under one set of conditions and the other under another. Double discs probably exist in solution as well as in the dry state necessary for electron microcsopy and X-ray diffraction experiments. The principal evidence is that components with sedimentation coefficients in the neighborhood of 20 S have often been observed. Caspar (11) has shown that this is approximately the correct sedimentation coefficient for a double disc composed of 32 monomeric units.

It is now known how to produce double discs at will. Figure 3 is an electron micrograph of TMV protein made by drying a solution of 6.5 mg/ml of protein in 0.1 μ phosphate–ammonium acetate buffer at pH 7.5 and 20°C. It shows predominantly disc-shaped particles. When electron micrographs are made of protein at a concentration of 0.2 mg/ml in 0.1 μ phosphate–ammonium acetate buffer at pH 5.25 and 20°C, rodlike particles predominate, as shown in Fig. 4. Thus one obtains double discs when polymerization is carried out at pH value above 7, and rods when it is carried out at lower pH values. Both types of polymerization are favored by high ionic strength.

An important question concerning the double disc is the arrangement of the monomeric units. The two layers must be arranged either face to face or face to back. Lauffer (27,28) concluded that if they are face to face there is a compelling thermodynamic reason for double discs being stable structures in aqueous solution. The X-ray diffraction and the electron microscope studies of Finch et al. (26) show at low resolution strong indications that there are diad axes in the plane of the disc. The simplest interpretation of this is that the two pairs of the discs are arranged face to face

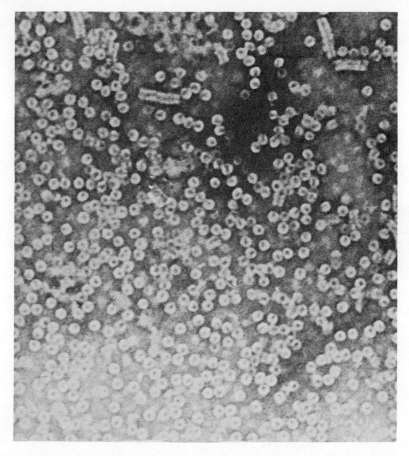

Fig. 3. TMV protein, 6.5 mg/ml, 0.1 μ phosphate–ammonium acetate, pH 7.5, 20°C. Approximately × 165,000. Micrograph prepared by Mrs. Alice P. Estes. [Lauffer and Stevens (8).] (Reprinted by permission of the copyright owner, Academic Press, Inc.)

in the dry or nearly dry state. However, it would be possible to reconcile the data with a face to back arrangement. More recent considerations of the dry structures are presented in Section D.

Double discs can form three-dimensional crystals [Macleod et al. (29); Finch et al. (26)]. Large crystals can be grown from protein solutions at a concentration of 8 mg/ml in tris buffer of 0.1 μ at pH 8.0 by the addition of ammonium sulfate in the same buffer to a final concentration of 0.5 M.

D. Stacked Double Discs

Sometimes TMV protein exhibits in the highly polymerized form a rodlike structure composed of stacked double discs. The first indication of this came from some minor differences in the X-ray diffraction pattern of some preparations of polymerized TMV protein—differences from other preparations and from the arrangement in TMV itself. This was reported by Franklin and Commoner (25). The electron micrographs of Markham et al. (30), as illustrated in Fig. 5, provide conclusive evidence of stacking.

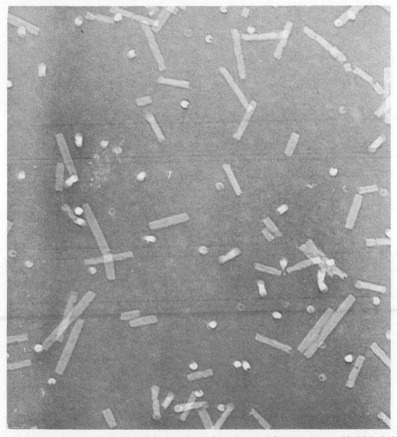

Fig. 4. TMV protein, 0.2 mg/ml, 0.1 μ phosphate–ammonium acetate, pH 5.25, 20°C. Approximately ×165,000. Micrograph prepared by Mrs. Alice P. Estes. [Lauffer and Stevens (8).] (Reprinted by permission of the copyright owner, Academic Press, Inc.)

It is clear from an inspection of Fig. 5, which was made by an image integration procedure, that the separation between rings in a double disc differs from that between double discs. Lest it be thought that this might be an artifact introduced by the integration procedure, it must be noted that the original unaltered photographs of Markham et al. (30), as well as earlier ones published by Dickson and Woods (31), also show a clear differentiation of this sort. It is therefore hard to believe that the bonding between rings in a disc is identical with the bonding between discs. Sometimes rods with the helical arrangement of subunits typical of TMV and stacked discs are found in the same preparation.

Finch (32) reported on optical diffraction studies of electron microscope images of stacked double discs. The three-dimensional particle structure was reconstructed by computation from the images using the Fourier

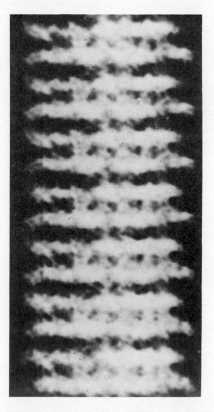

Fig. 5. Integrated photograph of stacked discs of TMV protein. The original was moved through a distance equivalent to 45.4 Å. [Markham et al. (30).] (Reprinted by permission of the copyright owner, Academic Press, Inc.)

inversion method of DeRosier and Klug (33). Finch (32) reported that there are diads perpendicular to the axis of the rods located between the individual members of the double disc and concluded that this requires that the two rings face in opposite directions. More recently, Klug (34) reported further studies from the same laboratory, presumably of a similar nature, which led to the conclusion that protein monomers all face in the same direction in stacked double discs. If this more recent interpretation ultimately prevails, one or the other of the following two conclusions will be necessary concerning the structure of double discs in solution. (1) Protein monomers in the two rings of double disc in solution are arranged face to back. (2) The arrangement of protein monomers in double discs in solution is different from that of double discs in the thoroughly dried condition required for electron microscopy. The first of these conclusions would be difficult, although not necessary impossible, to rationalize thermodynamically. The second of these alternative conclusions presents no difficulty from a thermodynamic point of view because the free energy from solvent interaction for the rotation of a single protein monomer about its long axis at some time after double discs have stacked would be approximately zero.

E. Open Helices

The PM2 strain of TMV produces a nonfunctional viral coat protein which does not combine with RNA to form virus rods. This protein differs from that of the common strain by the presence of isoleucine instead of threonine at position 28 and glutamic acid instead of aspartic acid at position 95 (35–37). The PM2 protein resembles the protein of the normal strain in that it can be reversibly polymerized. However, Siegel and associates (38) showed that the PM2 protein polymerizes into single, double, and double-double open helices with a pitch of 280–290 Å and a diameter of 120 Å. Figure 6 shows some of the open helices.

There is some evidence that the normal strain of TMV also polymerizes in the form of open helices, at least as a transient structure. Lauffer et al. (39) found that the enthalpy and entropy of polymerization of the protein from the PM2 strain were comparable to the values for the protein from the normal strain.* Furthermore, electron micrographs obtained after spraying at 20°C from TMV protein solutions at a concentration of 0.2 mg/ml and 0.1 μ phosphate–ammonium acetate buffer at pH 5.25 showed fully formed TMV-like rods and also some definitely thinner rods reminiscent of the helices obtained by Siegel et al. for the PM2 protein; see Fig. 4. Unfortunately, thus far it has been possible to obtain this result only one time. Because of this and of the possibility of staining artifacts, the electron microscope evidence is less conclusive than the thermodynamic evidence.

F. Helical Rods

The normal product of polymerization of TMV protein is a rod with a helical arrangement of the protein subunits resembling that found in the

*Srinivasan and Lauffer showed recently that in 0.1 μ phosphate buffer, $\triangle H$ and $\triangle S°$ for PM2 polymerization are about the same for TMV–P at pH 6.75 (65).

TMV particle itself. Figure 7 is a diagrammatic representation of the structure of the TMV particle as deduced by Franklin and Holmes (40). Franklin (41) showed that the X-ray diffraction pattern of polymerized TMV protein was quite similar to that of TMV itself. The major difference

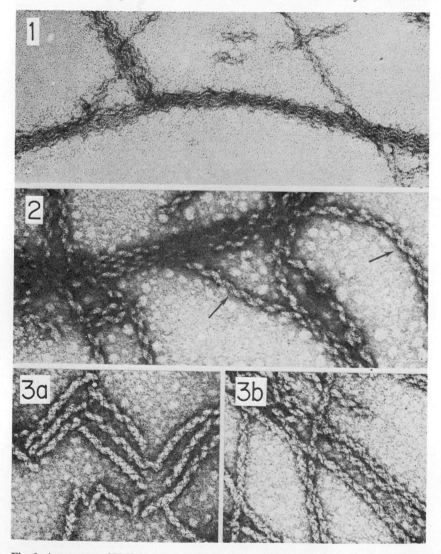

Fig. 6. Aggregates of PM2 protein formed in vitro. × 200,000. (1) Closely packed groups of single helices. Double helices are also seen. (2) Double helices. Arrows point to regions of distinct doubleness. (3a) and (b) Double-double helices shown under two different staining conditions. Regions of double helix are also visible. [Siegel et al. (38).] (Reprinted by permission of the copyright owner, Academic Press, Inc.)

TMV

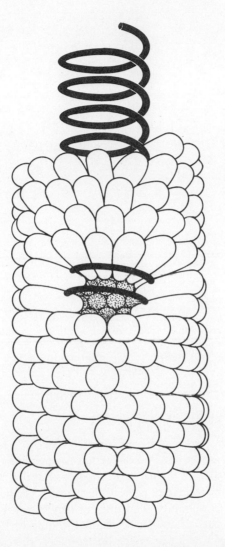

Fig. 7. The structure of TMV. Inner helix represents RNA, and helically assembled ellipsoids of revolution represent protein subunits. Sketch based on model originally constructed by Dr. Rosalind Franklin, but reversed in pitch to conform to the current view that TMV is a left-handed helix. [Lauffer and Stevens (*8*).] (Reprinted by permission of the copyright owner, Academic Press, Inc.)

can be shown by comparing the radial density distribution of TMV, first measured by Caspar (*42*), with that of polymerized TMV protein measured by Franklin (*41*). This difference is shown in Fig. 8. This is the basis for locating the RNA strand 40 Å from the axis of the rod in the virus particle.

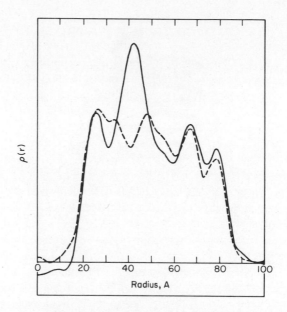

Fig. 8. Electron density distribution of TMV. The solid line shows the radial density distribution of TMV after refinement of the contribution of the central diffraction peak. The broken line, the radial density of the repolymerized, RNA-free virus protein. [Franklin and Holmes (40).] (Reprinted by permission of the copyright owner, Ejnar Munksgaard, Ltd.)

G. Other Possible Stable Aggregates

In addition to the particle types already discussed, Caspar (11) predicted two other stable aggregates. One of them is a particle composed of seven monomers arranged with their long dimensions parallel and in hexagonal close packing with respect to cross section. Another is a helical rod of three turns containing 49 subunits. The only evidence in favor of the actual existence of such particles is the occasional finding of boundaries in the ultracentrifuge which sediment with approximately the correct sedimentation coefficients. Because of the well-known complications associated with sedimentation of aggregating particles in equilibrium, such evidence taken alone is not conclusive. Thus these suggested stable aggregates might or might not exist. It is true that we have used sedimenting boundaries as evidence for the existence of an aggregate in solution when there was conclusive evidence from other sources that such aggregates existed in the dried state as shown by electron microscopy or in the semidried state as shown by X-ray diffraction. However, we are reluctant to base the existence of an aggregate as a stable particle solely on sedimentation behavior.

III. DISSOCIATION OF A-PROTEIN

As shown above, A-protein, as normally encountered in dilute solutions having pH values near and slightly below neutrality, exists primarily in the form of a stable trimer of the protein subunit. Ansevin and Lauffer (13) first observed that A-protein, which has a sedimentation coefficient at temperatures between 4 and 6°C of approximately 4 S in dilute buffers near neutrality at concentrations of 1 mg/ml or more, can be dissociated into particles with a sedimentation coefficient of 1.9 S at a concentration of approximately 0.1 mg/ml, corresponding to the protein monomer. Such dissociated protein repolymerizes when it is concentrated by evaporation and is capable of reconstituting with RNA to form nucleoprotein. Lauffer (28) and Caspar (11) analyzed these results in terms of an equilibrium between three protein subunits and one A-particle and obtained the result that the free energy of the association of the monomer into trimer is approximately-4000 cal/mole of monomer. Westover and Stevens (43) carried out a more detailed analysis by means of sedimentation equilibrium. Their results are consistent with the view that the main components of the solution are monomers in equilibrium with trimers, and that the standard free energy of trimer formation is − 4170 cal/mole of monomer.

IV. THE POLYMERIZATION PROCESS

A. Preliminary Analysis

An analysis of the results obtained by light scattering and osmometry over a period of many years in our laboratory indicates clearly that two distinctly different reactions have been studied. These can be illustrated by results obtained at a TMV protein concentration of 1 mg/ml in 0.1 μ phosphate buffer at pH 6.5. Under these conditions one of these processes takes place at temperatures between approximately 4 and 14°C and the second at temperatures between approximately 14 and 20°C or higher. Thus for historical reasons the first is referred to as "low-temperature" polymerization and the second as "high-temperature" polymerization, but it must be realized that under different conditions of protein concentration, pH, ionic strength, and chemical composition of the medium low-temperature polymerization can occur at temperatures well above 14°C, and under still other conditions high-temperature polymerization can occur below 14°C. Low-temperature polymerization, under the conditions specified above, is characterized by a positive enthalpy of about 30,000 cal/mole, a positive entropy of about 126 eu (4), by small volume changes (8), and by no change in charge (20). In contrast, high-temperature polymerization is characterized by a positive enthalpy of 206,000 cal/mole, a positive entropy of 739 eu, a large volume change, and a charge change of about one ion per chemical subunit, under the conditions specified above (8,44,20).

Two distinctly different points of view have been employed in attempts to understand the polymerization process. Caspar (11) treated the polymerization primarily as a cooperative process; Lauffer (8–10) used the ideas inherent in Flory's theory of condensation polymerization (45).

B. Cooperativity

Caspar (11) describes the polymerization of TMV protein as a cooperative reaction. Many considerations commend this point of view when attention is focused on the protein monomer as the smallest component and the fully formed helical rod with well-defined "crystal" structure on the other extreme. The total distinction between the protein monomer and a helical rod has many of the features of a phase transition, the classic cooperative process. In the final helical rod, each protein unit is surrounded by six neighbors and it is "bonded" to them. Multiple bonding is a characteristic of cooperativity. Finally, under some conditions the transition from an unpolymerized to a highly polymerized state takes place over a narrow temperature range. This type of behavior is characteristic of cooperative processes. The explanation is that it is much more difficult to form a "bond" between the first two particles than between subsequent ones. On the average, this must be true of the TMV protein helical rod.

While all of this is true and reasonable, nevertheless, separate stages of the polymerization process need not have all of the characteristics of a cooperative reaction. As a matter of fact, two of the stages of the polymerization process identified and studied in great detail in our laboratory can be described very satisfactorily in terms of Flory's mathematical equations for condensation polymerization. The crucial assumption involved in deriving these equations is that the free energy for the formation of a bond is independent of the number of bonds previously formed. This assumption is directly opposite that involved in cooperative processes. It must be emphasized, however, that Flory's mathematics applies only to discrete stages of the polymerization reaction, not to the full transition from isolated protein monomers to fully formed helical rods. It is thus quite proper to think of the ultimate TMV protein polymers as involving cooperativity and at the same time to view discrete stages of the polymerization process as being noncooperative. It is interesting to observe that the stage in the polymerization process that involves the steep temperature dependence does not involve a cooperative process of the type being discussed. In this case, therefore, one of the "obvious" reasons for thinking the process of cooperative is no evidence at all for cooperativity.

The word, "cooperative," is sometimes used in a different sense to describe the situation in which a "bond" between two protein units involves more than interaction between one atom on one with one atom on the other. Examples would be a number of hydrogen bonds or entropic interactions with a number of solvent molecules for each "bond" between reactive sites on two adjacent units. Since the "bonds" between TMV protein units in solution are at least predominantly entropic interactions, cooperativity in this sense is certainly involved. This type of cooperativity is not at all inconsistent with the independence of the reactivity of different combining sites on a single protein unit, and it is consistent with the high values of ΔH and $\Delta S°$ observed.

C. Condensation Polymerization

Flory (45) derived a theory of linear condensation polymerization based on the assumption that the chemical reactivity of a combining site was completely independent of the length of the chain of which it is the terminus. If any stage of the polymerization of TMV A-protein involves linear aggregation (not necessarily in a straight line), Flory's mathematical treatment should be applicable because the units are sufficiently large so that even on the monomer a combining site can be thought to be independent of combining sites on diametrically opposite sides of the particle. There is no reason to believe that polymerization of TMV A-protein bears any other resemblance to linear condensation polymerization but, as a matter of fact, equations derived from Flory's treatment of linear condensation polymerization have proved to be very useful in interpreting this reaction.

Consider a unit volume of a solution containing N particles (including all polymers and residual monomers), N_x particles composed of x monomeric units, and N_0 monomeric units, including free units and those involved in the polymer particles. Assume that each monomeric unit has one combining site of a particular kind. On an uncombined monomer and on one end of a polymer chain, this site is free. On all other monomeric units, the site is combined. Define p as the probability that such a site is combined. In the solution therefore the number of combined sites is $N_0 p$ and the number of free sites is $N_0(1-p)$; this is also the number of particles, N. The probability of finding a polymer consisting of x monomeric units is N_x/N and is the product of the probability of finding $x-1$ combined sites, p^{x-1}, and the probability of finding one uncombined site, $1-p$. Thus Eqs. (1) and (2) can be written

$$N_x/N = (1-p)p^{x-1} \qquad (1)$$

$$N_x/N_0 = (1-p)^2 p^{x-1} \qquad (2)$$

Consider the reaction, $A_1 + A_1 = A_2$, with equilibrium constant K and all other similar reactions, $A_2 + A_1 = A_3$, $A_2 + A_2 = A_4$, and so on. Since the basic assumption requires equal reactivity regardless of the value of x, all of these reactions have the same equilibrium constant K. From the mass action equation, the first reaction can be written

$$[A_2]/[A_1] = K[A_1] \tag{3}$$

By a simple extension of this argument, the general equation is obtained

$$[A_x]/[A_1] = (K[A_1])^{x-1} \tag{4}$$

From Eq. (2) it follows that N_x/N_1 is p^{x-1}. Therefore

$$p = K[A_1] = K[A]_0(1-p)^2 \tag{5}$$

$$K = p/[A]_0(1-p)^2 \tag{6}$$

It must be emphasized that the K of Eq. (6) is the simple second-order equilibrium constant for the reaction $A_1 + A_1 = A_2$ and all similar reactions involving the polymers of A.

For a solution ideal except for the tendency of A to polymerize, osmotic pressure depends on the number of particles per unit volume. It follows directly from Eq. (1) and (2) that

$$\pi/\pi_0 = 1-p \tag{7}$$

In this equation π is the osmotic pressure of a solution of polymers and π_0 is the osmotic pressure the same solution would have if all the material were in the unpolymerized state.

The equation for the turbidity of a solution ideal except for the tendency of the material to polymerize is derived as follows. From elementary light-scattering theory

$$\tau_0 = HcM_0 \tag{8}$$

In this equation τ_0 is the turbidity of a solution of pure monomer of molecular weight M_0 at a concentration of c grams per milliliter. H is a proportionality constant. This simple equation should be valid for such a solution if the particles are much smaller than the wavelength of light. The contribution of an x-mer to the turbidity τ_x is $Hc_x M_x$. Since c_x is $N_x(xM_0)/Av$, where Av is Avogadro's number, and c is $N_0 M_0/Av$

$$\tau_x = HcM_0(1-p)^2 x^2 p^{x-1} \tag{9}$$

$$\tau = \sum_1^\infty \tau_x = HcM_0(1-p)^2 \sum_1^\infty x^2 p^{x-1} \tag{10}$$

$$\sum_0^\infty p^x = 1/(1-p).$$

Differentiate with respect to p:

$$\sum_1^\infty xp^{x-1} = 1/(1-p)^2.$$

Differentiate a second time with respect to p:

$$\sum_2^\infty x(x-1)p^{x-2} = 2/(1-p)^3.$$

Let $x' = x-1$:

$$\sum_1^\infty (x'+1)\, x'p^{x-1} = 2/(1-p)^3 = \sum_1^\infty x'^2 p^{x-1} + \sum_1^\infty x'p^{x-1}.$$

The primes are not relevant, therefore:

$$\sum_1^\infty x^2 p^{x-1} = 2/(1-p)^3 - \sum_1^\infty xp^{x-1} = 2/(1-p)^3 - 1/(1-p)^2 =$$

$$(1+p)/(1-p)^3.$$

Substitute in Eq. (10) to obtain Eq. (11).

$$\tau = HcM_0\,(1+p)/(1-p) = \tau_0\,(1+p)/(1-p) \tag{11}$$

Solving for p gives

$$p = \frac{(\tau/\tau_0) - 1}{(\tau/\tau_0) + 1} \tag{12}$$

In the light-scattering and osmotic pressure studies, monomer, symbolized by subscript 0, refers to the smallest polymerizing unit which we believe to be A-protein, a trimer of the ultimate protein subunit.

V. LIGHT-SCATTERING STUDIES

Polymerization of TMV protein has been followed by light-scattering or turbidity measurements and results have been reported by Smith and Lauffer (44), Khalil and Lauffer (46), and Shalaby and Lauffer (47). In general, the method used consisted of measuring with a Beckman DU spectrophotometer the optical density at a wavelength of 320 μ. Since TMV protein does not absorb at this wavelength, the optical density is a measure of scattered light or turbidity. In most experiments optical density was measured as a function of temperature. The compartment in the spectrophotometer into which the sample cuvet was placed was equipped with a temperature-controlling mechanism. Stevens and Paglini (48) showed that this method yields turbidity measurements for TMV approximately equal to those calculated by integrating the intensity of light scattered in a Brice-Phoenix light-scattering apparatus over all directions. Typical results are shown in Fig. 9, in which open circles represent increasing temperatures and closed circles decreasing temperatures.

Most of the light-scattering studies were carried out on solutions with a total protein concentration of about 1 mg/ml. It is reasonable to assume that the simplest form of the light-scattering equation [Eq. (8)] applies for

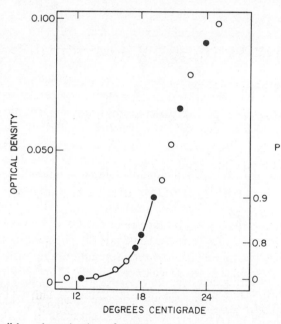

Fig. 9. Reversible polymerization of TMV protein in 0.1 μ phosphate buffer at pH 6.5. ○ corresponds to rising temperatures and ● to falling temperatures. Left ordinate is $OD - OD_0$ and right ordinate is fraction of linkages formed. [From Lauffer (9); data of Smith and Lauffer (44).] (Reprinted by permission of the copyright owner, The University of Texas Press.)

such solutions and that the solutions can be considered ideal except for the tendency to polymerize. If Eq. (12) is substituted into Eq. (6) and the result is substituted into

$$-RT \ln K = \Delta F° = \Delta H - T\Delta S°$$ (13)

one obtains

$$\ln (\tau^2 - \tau_0{}^2) = \Delta S°/R + \ln 4 [A_0] \tau_0{}^2 - \Delta H/RT$$ (14)

[Smith and Lauffer (44)]. Since $(\tau - \tau_0) = 2.303 (OD - OD_0)$, where $(OD - OD_0)$ is the difference between the observed optical density attributable to light scattering at a given temperature and the minimum observable value before polymerization (at the lowest temperature) Eq. (14) can be transformed into Eq. (15)

$$2 \ln 2.3 \, (OD - OD_0) + \ln \left(1 + \frac{2\tau_0}{2.3 \, (OD - OD_0)} \right) =$$

$$\frac{\Delta S^\circ}{R} + \ln 4 \, [A_0] \, \tau_0{}^2 - \frac{\Delta H}{RT} \qquad (15)$$

The measurements of Smith and Lauffer (44) permit the evaluation of $OD - OD_0$ but not the calculation of τ_0 from OD_0 because of the possibility of having a small constant amount of dirt or denatured protein in each sample. The value of τ_0 must be calculated from the protein concentration and an assumed value for M_0, the molecular weight of the polymerizing unit, using Eq. (8). It is known from the results of Banerjee and Lauffer (4) that at temperatures between 4.6 and 11°C the molecular weight of the polymerizing unit of TMV protein dissolved in 0.1 μ phosphate buffer at pH 6.5 is near 52,500. However, the possibility does exist that in the temperature range 15–20°C, the region in which the data of Smith and Lauffer (44) were obtained, the polymerizing unit might be larger, possibly with a molecular weight of 105,000. H has a value of 4.23×10^{-5} (cm/g)^2 (44). At a protein concentration of 0.001 g/ml, $\tau_0 = 0.0022$ and $[A_0] = 1.9 \times 10^{-5}$ moles/liter if M_0 is taken to be 52,500. If, however, M_0 is taken to be 105,000, τ_0 is 0.0044 and $[A_0]$ 0.95 $\times 10^{-5}$. The minimum usable value of $OD - OD_0$ is about 0.005, and the minimum usable value of $2.3(OD - OD_0)$ is thus 0.0115. Therefore whether τ_0 is 0.0022 or 0.0044, the minimum usable values of $2.3(OD - OD_0)$ are greater than $2\tau_0$. Inspection of Eq. (15) shows that when $2.3(OD - OD_0)$ is greater than $4\tau_0$ the change upon further increase of $(OD - OD_0)$ of the second term on the left is a small fraction of the change in the first term. Since $-\Delta H/R$ is the derivative with respect to $1/T$ of the left side of Eq. (15), the value of ΔH is nearly independent of the assumed value of τ_0 if only data for which $2.3(OD - OD_0)$ is greater than $4\tau_0$ are used. These are the most accurate measurements.

Errors in the value assigned to $\Delta S^\circ/R$ when data are fitted to Eq. (15) resulting from the wrong assumed value of M_0 can be estimated as follows. If M_0 is 52,500, $\ln 4[A_0] \, \tau_0{}^2$ is -21.7; if 105,000, it is -21.0. The difference between them is the error in the assigned value of $\Delta S^\circ/R$ if the wrong value of M_0 is chosen and is trivial compared to the value of $\Delta S^\circ/R$ of $+739/2$ evaluated by Smith and Lauffer (44) on the assumption that M_0 is 52,500. Thus for all practical purposes the values of ΔS° and ΔH obtained by fitting data to Eq. (15) are independent of whether the polymerizing unit in these experiments is a cyclical trimer of molecular weight 52,500 or a particle of double that weight.

Equation (15) fits the data obtained by Smith and Lauffer (44), Khalil and Lauffer (46), and Shalaby and Lauffer (47) at low and intermediate values of $OD - OD_0$, but definitely not at higher values. Equation (16) fits all their data [Smith and Lauffer (44)]

$$\ln [2\tau_0 + 2.3(OD - OD_0)] + \ln 2.3(OD - OD_0) - 2 \ln 2.3 (OD_m - OD)$$

$$= \frac{\Delta S^\circ}{R} + \ln 4[A]_0\tau_0{}^2 - 2 \ln [2\tau_0 + 2.3(OD_m - OD_0)] - \frac{\Delta H}{RT} \quad (16)$$

The first two terms of the left side of Eq. (16) are an alternative way of writing the terms on the left side of Eq. (15). OD_m is a hypothetical maximum OD. Equation (16) can be derived from a number of different assumptions. Since data are not available to permit a decision between them, it must be regarded as an empirical equation. It can be fitted by means of a computer. OD_m should be proportional to $[A]_0$ because for a given molecular-weight distribution light scattering is proportional to total concentration. Results obtained by Banerjee and Lauffer (4) seem to indicate that this parameter is essentially independent of concentration, but their measurements were not carried to sufficiently high degrees of polymerization to afford a critical determination of this.

Inspection of the results of Khalil and Lauffer (46) shows clearly that OD_m varies greatly with pH; it has high values at pH 6 or 6.2 and low values at pH 6.8. Since Scheele and Lauffer (20) found that hydrogen ions are dissociated when the pH of polymerized TMV protein is raised from 6.0 to 6.8, it seems reasonable to assume that the decrease in the maximum turbidity, probably related in some way to maximum particle size, is associated with electrical charge on the polymerized particles. One way of approaching this problem is to inquire how electrical work is related to particle size. Since the end result of what we call high-temperature polymerization seems to be a cylinder with helical structure resembling that of TMV, regardless of the nature of the initial stages of the polymerization, one possible approach to this question is in terms of the change in electrical work associated with the polymerization of charged cylinders.

It is therefore necessary to calculate the electrical work W_{el} for charging a cylinder in order to determine whether the reaction $2P \rightarrow P_2$ is affected by electrostatic considerations. P is a cylinder of radius R, length L, and net charge $Z\varepsilon = Q$. Hill (49) gives the following equation

$$W_{el} = \frac{Z^2\varepsilon^2}{DL} \left[\frac{K_0(\kappa a)}{\kappa a K_1(\kappa a)} + \ln \frac{a}{R} \right] \quad (17)$$

D is the dielectric constant, a is the radius of closest approach $= R + r_i$, where r_i is the mean ionic radius, κ is the Debye-Hückel constant, and

$K_0(x)$ and $K_1(x)$ are modified Bessel functions of x of the second kind. Since

$$W_{el} = \int_0^Q \xi \, dQ,$$

where ξ is the electrokinetic potential and $K_0(\kappa a)/\kappa a K_1(\kappa a) \simeq \ln[1+ 1/\kappa a]$, when $R >> r_i$ the above equation reduces to

$$\xi = \frac{2Q}{DL} \ln\left(1 + \frac{1}{\kappa a}\right) \tag{18}$$

Gorin (50) gives the following equation for ξ

$$\xi = \frac{2Q}{D(L+2R)} \left[\frac{K_0(\kappa a)}{\kappa a K_1(\kappa a)} + \ln\frac{a}{R} \right] \tag{19}$$

This reduces as above to

$$\xi = \frac{2Q \ln\left(1 + \dfrac{1}{\kappa a}\right)}{D(L+2R)} \tag{20}$$

It approaches the Hill equation [Eq. (18)] when $L >> R$, which in turn is the same as the equation for the potential, neglecting end effects, between the plates of a cylindrical condenser when $L >> R$ and $1/\kappa$ is the distance between the plates.

An alternative equation for ξ and therefore for W_{el}, can be derived for a model that is probably reasonable for a TMV or TMV protein rod. The model is a cylinder of length L and radius R with a *uniform* charge Q on its curved surface. This differs from a charged cylindrical conductor, for which the charge per unit length is greater at the ends than in the middle. This equation is

$$\bar{\xi} = \frac{2Q \ln\left(1 + \dfrac{d}{R}\right)}{DL^2} \left[\left(R + \frac{d}{2}\right)^2 + L^2 - \left(R + \frac{d}{2}\right) \right]^{1/2} \tag{21}$$

This can be derived as follows. Symmetry permits this to be treated as a problem of a *uniform* line charge Q/L, located at the axis of the cylinder. The component in the radial direction $E_{A\,dl}$ of the electric intensity, given by Coulomb's law, at the end R of a cylindrical radius originating at A on the line arising from an element $Q\,dl/L$ of charge located l centimeters from A, is

$$E_{A\,dl} = \frac{Q\,dl}{DL(R^2 + l^2)} \frac{R}{(R^2 + l^2)^{1/2}}$$

Consider the point A as the origin. The total electric intensity in the radial direction at the end of R, E_A, is obtained by integrating between $-A$ and $L-A$

$$E_A = \frac{QR}{DL} \int_{-A}^{L-A} \frac{dl}{(R^2+l^2)^{3/2}} = \frac{Q}{DRL} \left\{ \frac{L-A}{[R^2+(L-A)^2]^{1/2}} + \frac{A}{(R^2+A^2)^{1/2}} \right\}$$

If ξ_A is the potential difference between two points at $R+d$ and R on the radius originating at A

$$\xi_A = \int_R^{R+d} E_A \, dR = \int_R^{R+d} \frac{Q}{DRL} \left\{ \frac{L-A}{[R^2+(L-A)^2]^{1/2}} + \frac{A}{(R^2+A^2)^{1/2}} \right\} dR$$

While this can be integrated exactly, the result is cumbersome. A highly precise approximation can be obtained by integrating the following, where the quantity outside the integral sign is treated as a constant

$$\xi_A = \frac{Q}{DL} \left\{ \frac{L-A}{\left[\left(R+\frac{d}{2}\right)^2+(L-A)^2\right]^{1/2}} + \frac{A}{\left[\left(R+\frac{d}{2}\right)^2+A^2\right]^{1/2}} \right\} \int_R^{R+d} \frac{dR}{R}$$

$$\xi_A = \frac{Q}{DL} \left\{ \frac{L-A}{\left[\left(R+\frac{d}{2}\right)^2+(L-A)^2\right]^{1/2}} + \frac{A}{\left[\left(R+\frac{d}{2}\right)^2+A^2\right]^{1/2}} \right\} \ln\left(1+\frac{d}{R}\right)$$

The average value of ξ, $\bar{\xi}$, is

$$\int_0^L \xi_A \, dA \Big/ \int_0^L dA$$

$$\bar{\xi} = \frac{Q \ln\left(1+\frac{d}{R}\right)}{DL^2} \left[\int_0^L \frac{A \, dA}{\left[\left(R+\frac{d}{2}\right)^2+A^2\right]^{1/2}} - \int_0^L \frac{(A-L)\,d(A-L)}{\left[\left(R+\frac{d}{2}\right)^2+(A-L)^2\right]^{1/2}} \right]$$

When the integration is performed, Eq. (21) results.

Equation (21) reduces to Eq. (18) when $L >> [R+(d/2)]$ and when $1/\kappa a$ is substituted for d/R. When $L \geqq 2[R+(d/2)]$, it reduces to an equation intermediate between Eqs. (18) and (20) in which the denominator is $D\{L+[R+(d/2)]\}$. Furthermore, it is accurate not only for very long rods, but even when $[R+(d/2)] > L$, especially when $R >> d$. When $1/\kappa a$ is substituted for d/R and $1/\kappa$ for d and $\bar{\xi}\ dQ$ is integrated from 0 to Q, one obtains

$$
\begin{aligned}
W_{el} &= \frac{Q^2 \ln\left(1+\dfrac{1}{\kappa a}\right)}{DL^2}\left\{\left[\left(R+\frac{1}{2\kappa}\right)^2 + L^2\right]^{1/2} - \left(R+\frac{1}{2\kappa}\right)\right\} \\
&= \frac{Q^2 \ln\left(1+\dfrac{1}{\kappa a}\right)\left(R+\dfrac{1}{2\kappa}\right)}{DL^2}\left\{\left[1+\left(\frac{L}{R+\dfrac{1}{2\kappa}}\right)\right]^{1/2} - 1\right\}
\end{aligned}
\tag{22}
$$

Substituting $Z\varepsilon$ for Q, where Z is the number of ionic charges and ε is the electronic charge, 4.8×10^{-10} absolute electrostatic units, multiplying by Avogadro's number and converting from ergs to calories by dividing by 4.185×10^7 gives

$$
\begin{aligned}
W_{el} = &\frac{3.32 \times 10^{-3}\,Z^2 \ln\left(1+\dfrac{1}{\kappa a}\right)}{DL^2}\left(R+\frac{1}{2\kappa}\right) \\
&\left\{\left[1+\left(\frac{L}{R}+\frac{1}{2\kappa}\right)\right]^{1/2} - 1\right\}\ \text{cal/mole}
\end{aligned}
$$

Assume that for a slightly hydrated polymerized TMV protein rod $R = 9.3 \times 10^{-7}$ cm and $a = 9.5 \times 10^{-7}$ cm. Since there are 16 1/3 protein units per 23 Å or 0.71 units per 10^{-8} cm, Z/L is $0.71z \times 10^8$ ionic charges per centimeter, where z is the charge on a unit. At 25°C and 0.1 μ, $D = 78.54$ and $\kappa = 1.04 \times 10^7$ cm^{-1}. For such values $\ln\left[1+(1/\kappa a)\right] = 1/\kappa a$ or $1/9.9$, $[R+(1/2\kappa)]$ is $[9.3+(1/2 \times 1.04)] \times 10^{-7}$ or 9.78×10^{-7} cm. $D\kappa$ is practically constant between 10 and 25°C, and therefore W_{el} is almost independent of the temperature in the range involved in these studies.

With the numbers shown above, it is possible to calculate ΔW_{el} for the reaction $2P \rightarrow P_2$, for various values of z and of the ratio $L/[R+(1/2\kappa)]$ for the rod P.

ΔW_{el} is W_{el} for P_2 minus twice W_{el} for P. The results of such a calculation are shown in Table I.

It is difficult or perhaps impossible to compare the calculations shown in Table I directly with the thermodynamic parameters obtained in the studies of Smith and Lauffer (44), Khalil and Lauffer (46), and Shalaby and Lauffer (47). The reason is that ΔW_{el} corresponds to the reaction between short cylinders, but the thermodynamic parameters are valid only for the earliest stages for polymerization before cylinders are formed. As shown previously, the early stages of high-temperature polymerization probably involve formation of double-threaded open helices similar to

TABLE I

$+\Delta W_{el}$ for Polymerization of TMV Protein Rods

L, Å	$\dfrac{L}{R+1/2\kappa}$	$+\Delta W_{el}$, k cal/mole			
		$z = \pm 1$ (pH 5.2)	$z = \pm 2$ (pH 5.9)	$z = \pm 3$ (pH 6.6)	$z = \pm 4$ (pH 7.5)
49	1/2	3.5	14	32	57
98	1	8.9	36	80	143
196	2	13.5	54	122	216
392	54	17.6	70	158	282
784	108	19.3	78	174	308
∞	∞	21.2	85	191	340

those of the protein from the PM2 strain of TMV. It is reasonable to assume that after the open helices become long enough they either condense or polymerize laterally with other open helices to form short, closed helices or cylinders (39). Further polymerization would then involve end-to-end joining of short cylinders. If this view is correct, then $-\Delta F°$ for further polymerization of short cylinders should be many times that for the initial reaction. From the model of Lauffer et al. (39), it can be estimated that the appropriate factor is approximately 8. Khalil and Lauffer (46) obtained values of $\Delta S°$ and ΔH of 755 eu and 212,000 cal/mole for high-temperature polymerization at pH 6.6 in 0.1 μ phosphate buffer. At 18°C, $\Delta F°$ is -7700 cal/mole. Eight times this is $-61,600$ cal/mole. From the results of Scheele and Lauffer (20), it can be inferred that at pH 6.6 the polymerized protein has a charge z of -3 per protein unit. From the turbidities at the point where the data of Smith and Lauffer (44) for 0.1 μ phosphate buffer at pH 6.5 begin to depart from the simple theory, it can be estimated that the number-average molecular weight is slightly more than 500,000. This corresponds to a cylinder for which $L/[R+(1/2\kappa)]$ is about 1/2. Table I shows that ΔW_{el} for combination of two such short rods is $+32,000$ cal/mole. This number is about half the value of $-\Delta F°$, neglecting electrical effects, for polymerization of such particles. Therefore electrical work is great enough to inhibit polymerization. Table I also shows that for longer rods of the same charge, or for rods with higher charge, the electrical work term can be much greater, while for lower charges the electrical work term can be lower even for infinitely long rods. In a general way, these calculations correlate with the experimental findings that the maximum degree of polymerization of TMV protein decreases as pH and negative charge increase. Thus the dependence of the electrical work term on cylinder length and charge per unit length provides a plausible explanation of the effect of pH upon the departure of polymerization data from the simple mathematical formulation for condensation polymerization.

The idea that high-temperature polymerization involves three steps, formation of double-stranded threads, condensation or polymerization of threads into very short cylinders, and further polymerization of cylinders, is partially supported by temperature-jump studies of Scheele and Schuster (62). They subjected TMV protein in 0.1 M salt solutions at 19°C to temperature jumps of 1°C. They found three distinct relaxation processes during the first 2 sec. Since their studies are still in a preliminary stage, it is not yet possible to identify the nature of each step. Nevertheless, the number of steps found does coincide with the three here postulated.

VI. OSMOTIC PRESSURE STUDIES

The mathematical equation appropriate for condensation polymerization can also be used to interpret data obtained by osmotic pressure measurements on the polymerization of TMV A-protein. In this case it is necessary to solve Eq. (7) for p and substitute this into Eq. (6). The result is

$$\pi = \frac{\pi_0}{K[A]_0} \left(\frac{\pi_0}{\pi} - 1 \right)$$

Substitution of

$$\pi_0 = RT[A]_0 \quad \text{and} \quad [A]_0 = \frac{1000c}{M_0}$$

into this gives Eq. (23).)

$$\pi = \frac{1000(RT)^2}{KM_0} \frac{c}{\pi} - \frac{RT}{K} \tag{23}$$

If the mathematics of condensation polymerization applies, Eq. (23) predicts that if osmotic pressure is plotted against concentration divided by osmotic pressure a straight line should be obtained. The intercept should be RT/K, which permits an evaluation of K and the slope $1000(RT)^2 KM_0$ from which M_0 can be calculated. As shown in Fig. 10, data obtained by Banerjee and Lauffer (4) on TMV A-protein dissolved in 0.1 μ phosphate buffer at pH 6.5 fits this relationship at temperatures of 4.6, 8.7, and 11.0°C. The maximum concentration involved in these experiments was 9 mg/ml. Paglini and Lauffer (18) found that at 4°C the same equation fits the data obtained in 0.1 μ phosphate and barbital buffers at both pH 6.5 and 7.5.

In this case the data fit the equation at concentrations up to approximately 20 mg/ml; marked deviations occurred at higher concentrations. In all cases M_0 was in the neighborhood of 50,000, indicating that the cyclical trimer is the polymerizing unit.

From such curves, as mentioned previously, the equilibrium constant K at each temperature can also be evaluated. Figure 11 shows a plot of log K vs. $1/T$ for data obtained with protein dissolved in 0.1 μ phosphate buffer at pH 6.5. From this type of graph, ΔH was evaluated to be $+30,000$

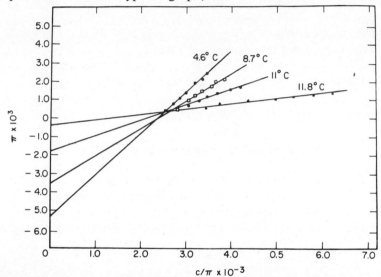

Fig. 10. Plot of π vs. c/π for TMV protein in phosphate buffer, pH 6.5, 0.1 μ, as a function of temperature; π is osmotic pressure in centimeters of water and c protein concentrations in milligrams per milliliter [Lauffer and Stevens (8). Data of Banerjee and Lauffer (4).] (Reprinted by permission of the copyright owner, Academic Press, Inc.)

cal/mole and $\Delta S°$ to be $+124$ eu. The electron micrograph shown in Fig. 3 demonstrates that the end product of polymerization in 0.1 μ phosphate buffer at pH 7.5 is the double disc.

Because the mathematical formulation for condensation polymerization is based on the assumption that the polymerization process is linear and open-ended, it is surprising that Eq. (23) fits data in a process leading to the formation of double discs. The assumption involved in the condensation polymerization mechanism requires $-\Delta F°$ to be the same for adding to a chain regardless of its length. If $-\Delta F°$ is twice the normal value when the final unit that closes a ring is added, then discs should form in sufficient number to distort significantly the number average molecular-weight distribution even at relatively low values of π/π_0. Since the data do fit the equation, it can be concluded that ring closing is inhibited, probably because it is accompanied by strain.

The theoretical justification for this conclusion follows. As discussed above, there are conflicting reports on whether double discs contain 32 or 34 chemical subunits. For the sake of discussion, remembering that one A_1 is made up of three chemical subunits, one can think of double discs

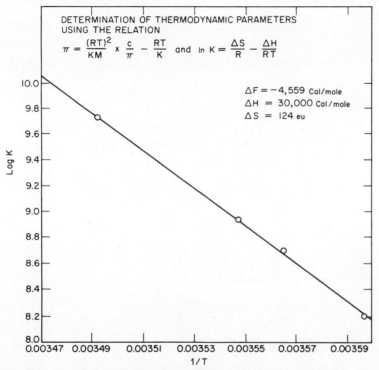

Fig. 11. Plot of log K vs. $1/T$, where K is equilibrium constant and T absolute temperature. [Banerjee and Lauffer (4).] (Reprinted by permission of the copyright owner, The American Chemical Society.)

being formed, on the average, by adding a final ring-closing A_1 to an A_{10} molecule to give the double disc A_{11}. Since two bonds are formed, ΔF° for this final step is expected to be $2\Delta F^\circ + \Delta Fs$ or $-RT \ln K^2 + RT \ln \sigma$, or $-RT \ln K^2/\sigma$, where K is the bimolecular equilibrium constant for condensation polymerization and $RT \ln \sigma$ is defined as ΔFs, the free energy change associated with strain. Thus the equilibrium constant for this final step is K^2/σ. Therefore one can write:

$$\frac{[A_{11}]}{[A_1]} = \frac{K^2}{\sigma} [A_{10}] \tag{24}$$

Substitution of Eq. (4) into Eq. (24) gives

$$\frac{[A_{11}]}{[A_1]} = \frac{K}{\sigma}(K[A_1])^{10} \tag{25}$$

A solution in which the linear polymers A_2 through A_{10} and also the closed ring A_{11} are in equilibrium with A_1 can now be considered. For such a solution

$$\frac{\pi}{\pi_0} = \frac{[A]}{[A]_0} = \frac{\{1-(K[A_1])^{10}\}(1-K[A_1]) + \dfrac{K}{\sigma}(K[A_1])^{10}(1-K[A_1])^2}{1-11(K[A_1])^{10} + 10K[A_1]^{11} + 11\,\dfrac{K}{\sigma}\,K[A_1]^{10}(1-K[A_1])^2} \tag{26}$$

Equation (26) is derived as follows

$$\frac{[A]}{[A_1]} = \sum_1^{10}\frac{[A_x]}{[A_1]} + \frac{[A_{11}]}{[A_1]}$$

$$\frac{[A]_0}{[A_1]} = \sum_1^{10}\frac{x[A_x]}{[A_1]} + \frac{11[A_{11}]}{[A_1]}$$

$$\sum_1^{10} y = \sum_1^{\infty} y - \sum_{11}^{\infty} y$$

From Eq. (4)

$$\sum_1^{10}\frac{[A_x]}{[A_1]} = \sum_1^{\infty}(K[A_1])^{x-1} - \sum_{11}^{\infty}(K[A_1])^{x-1}$$

$$= \sum_1^{\infty}(K[A_1])^{x-1} - (K[A_1])^{10}\sum_1^{\infty}(K[A_1])^{x-1}$$

$$= \frac{1}{1-K[A_1]} - \frac{(K[A_1])^{10}}{1-K[A_1]}$$

The expression for $[A_{11}]/[A_1]$ is given by Eq. (25). Therefore

$$\frac{[A]}{[A_1]} = \frac{1-(K[A_1])^{10} + \dfrac{K}{\sigma}(K[A_1])^{10}\,(1-K[A_1])}{1-K[A_1]}$$

Similarly

$$\sum_1^{10}\frac{x[A_x]}{[A_1]} = \sum_1^{\infty}x(K[A_1])^{x-1} - \sum_{11}^{\infty}x(K[A_1])^{x-1}$$

$$= \sum_1^{\infty}x(K[A_1])^{x-1} - 10(K[A_1])^{10}\sum_1^{\infty}(K[A_1])^{x-1} - (K[A_1])^{10}\sum_1^{\infty}x(K[A_1])^{x-1}$$

$$= \frac{1}{(1-K[A_1])^2} - \frac{10(K[A_1])^{10}}{(1-K[A_1])} - \frac{(K[A_1])^{10}}{(1-K[A_1])^2}$$

$[A]_0/[A_1]$ is obtained by adding $11(K/\sigma)(K[A_1])^{10}$ to this. Finally, Eq. (26) is obtained by dividing $[A]/[A_1]$ by $[A]_0/[A_1]$.

In Eq. (26) $[A]$ is the total number of moles per liter of particles A_1 through A_{11} and $[A]_0$ is the total number of moles per liter of polymerizing unit (trimers of the chemical subunit) in them. It must be remembered that in simple linear polymerization, $K[A_1]$ is equal to p, the probability that a site is combined, which can vary only between 0 and 1. When $K[A_1]$ is considerably less than unity, for example $1/2$, the equation can be simplified to Eq. (27)

$$\frac{\pi}{\pi_0} = \frac{(1 - K[A_1]) + \dfrac{K}{\sigma}(K[A_1])^{10}(1 - K[A_1])^2}{1 + 11\dfrac{K}{\sigma}(K[A_1])^{10}(1 - K[A_1])^2} \tag{27}$$

When the denominator approaches unity, the numerator approaches $(1 - K[A_1])$. Since $K[A_1]$ equals p, this is Eq. (7) for linear condensation polymerization.

Now consider a solution of 10 mg/ml of TMV A-protein that obeys Eq. (7) up to a value of $p = K[A_1] = 1/2$. The only way Eq. (7) can be obeyed is for the denominator of Eq. (27) to approach to within a few percent, for example, 5%, of unity. From Eq. (2) it is evident that for such a system, with $K[A_1] = p = 1/2$, $[A_1]$ must be equal to $(1 - 1/2)^2[A]_0$. $[A]_0 = 10/52,500 = 0.00019$ and $[A_1]$ is about 0.00005. Thus K is about 10^4, $(K/\sigma)(K[A_1])^{10}$ is $(10^4/\sigma)(1/2)^{10} \simeq 10/\sigma$ and $11(10/\sigma)(1 - 1/2)^2 = 0.05$. Therefore $\sigma = 550$. $\Delta Fs = RT \ln \sigma = 3.5$ kcal/mole at 280 Å. Such a value for ΔFs would permit condensation polymerization mathematics to apply up to $p = 1/2$.

It cannot be expected that Eq. (23) will be applicable to solutions too concentrated to be considered ideal. As a first approximation

$$\pi_0 = \pi_{0i}(1 + M_0 Bc) \tag{28}$$

where π_0 is the osmotic pressure of unpolymerized A-protein (trimer) and subscript i refers to an ideal solution. B is the second virial coefficient defined in the manner of Tanford (51). For a hydrated protein, according to Eq. (24) of Lauffer (52), B is $1000\xi_0/M_0^2 + (1000\beta_{22}^0/2M_0^2) + 1000$ $x^2/4M_0^2\ m_3'$. In this formulation ξ_0 is a hydration factor and is equal to the weight of solvent bound by 1 mole of protein, divided by 1000. Since it is usually small compared to the other terms, it will be neglected. $\beta_{22}^0/2$ for spherical particles with no interaction is solely an excluded volume term equal to the hydrated volume of 1 mole of protein divided by 250 times the specific volume of the solvent. For a specific volume of solvent of 1 ml/g, a molecular weight of protein of 52,500 and a hydrated specific volume of protein of 0.775, $\beta_{22}^0/2$ has a value of 163 and $1000M_0\beta_{22}^0/2M_0^2$ has a value of 3.1. The value 0.775 for the specific volume of protein was derived from the fact that TMV A-protein at about 10°C has a partial specific volume of 0.73 and a hydration of 0.20 g water per gram of protein [Jaenicke and Lauffer (53)]. It is appropriate here to relate β_{22}^0 to the excluded volume term only because this treatment deals with interaction in terms of extent of polymerization. The final term is the Donnan contribution. If x is 12 [Scheele and Lauffer (20)] and m_3' is 0.1, $1000M_0\,x^2/4m_3'M_0^2$ has a value of 6.87. Thus M_0B is 10 for unpolymerized protein with a molecular weight of 52,500. When protein polymerizes, the contribution to B of the excluded volume term, being more-or-less inversely proportional to M, should change. The contribution of the Donnan term should not change because there is no change in charge per unit mass associated with a low-temperature polymerization (20). Since the Donnan term is more than twice that for excluded volume, the value of $1000MB$ for polymers should be greater than that for monomers, but π depends most heavily on the smallest particles in a distribution. Therefore, the effective value of $1000\,MB$ should not vary greatly when polymerization takes place. Therefore the factor $1+10c$ is taken to be appropriate for partially polymerized protein as well as for the unpolymerized material.

Equation (29) is the equilibrium expression for condensation polymerization involving nonideal solute species

$$(\gamma_{x-1}/\gamma_x)\,\gamma_1 K = p/(1-p)^2\,[\text{A}]_0 \tag{29}$$

Consider the reaction, $\text{A}_{x-n}+\text{A}_n = \text{A}_x$. By analogy to the derivation of Eq. (4)

$$[\text{A}_x]/[\text{A}_{x-n}] = (\gamma_{x-n}/\gamma_x)\,K\gamma_n\,[\text{A}_n] = (\gamma_{x-n}/\gamma_x)\,(K\gamma_1[\text{A}_1])^n$$

Consistent with the assumptions of condensation polymerization, K is a constant independent of x and n. From Eq. (1), $[\text{A}_x]/[\text{A}_{x-n}] = p^n$. Therefore $p = (\gamma_{x-n}/\gamma_x)^{1/n}K\gamma_1[\text{A}_1]$. If γ_{x-1}/γ_x is independent of x, $(\gamma_{x-n}/\gamma_x)^{1/n} = \gamma x - 1/\gamma$. Furthermore, from Eq. (2), $[\text{A}_1]/[\text{A}_0] = (1-p)^2$. Therefore $p = (\gamma_{x-1}/\gamma_x)\,\gamma_1 K[\text{A}_1] = (\gamma_{x-1}/\gamma_x)\gamma_1 K[\text{A}]_0(1-p)^2$. This can be rearranged to give Eq. (29).)

Equation (6) is the same except that all activity coefficients γ are equal to one. Thus in Eq. (23) K must be replaced by $(\gamma_{x-1}/\gamma_x)\gamma_1 K$. Probably γ_{x-1}/γ_x does not differ greatly from 1; therefore it is assumed to be 1. The activity coefficient of the monomeric species before any polymerization takes place can be calculated from the excluded volume effect. While its value probably changes somewhat after polymerization occurs, it is difficult to calculate this change. It is assumed that the change is so small that γ_1

remains effectively constant. $\gamma_1 = 1+\beta_{22}^0 [A]_0 = 1+\beta_{22}^0\ c/M_0 = 1+$ $2x1000\ 163c/52,500 = 1+6.2c$. Therefore the K in Eq. (23) must be multiplied by $1+6.2c$. Thus the total correction of Eq. (23) for nonideality is to divide π on the left by $1+3.8c$ and to multiply c/π on the right by $1+10c$.

When the data of Paglini and Lauffer (18) for the polymerization in 0.1 μ barbital buffer at pH 7.5 and at 4°C are corrected in this manner, the results are as shown in Fig. 12. The same data are shown in the uncorrected form in Fig. 13. The straight line was fitted by the method of least squares.

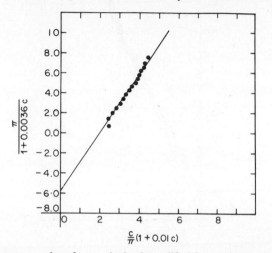

Fig. 12. A least-squares plot of π vs. c/π, both modified for nonideal solutions. TMV A protein in 0.1 μ barbital buffer, pH 7.5, 4°C. [Paglini and Lauffer (18).] (Reprinted by permission of the copyright owner, The American Chemical Society.)

The value of K corresponding to this analysis is 4.11×10^3, and the value of M_0 is 46,000. More serious than the small deviation of the data from a straight line is the fact that the calculated molecular weight for the polymerizing unit is somewhat too low. As shown above, it should be some integral multiple of 17,500, most likely 3. The detailed treatment of osmotic pressure data just outlined permits osmotic pressure data obtained over a wide range of concentrations to be interpreted. Paglini and Lauffer (18) evaluated K at different temperatures for polymerization for TMV protein in 0.1 μ phosphate buffer at pH 6.5 and for polymerization in 0.1 μ phosphate buffer at pH 7.5. From these data values of ΔH and $\Delta S°$ were evaluated. At pH 6.5 values of $+34,000$ cal/mole and $+139$ eu were obtained for the enthalpy and the entropy, respectively, in good agreement with previous values of $+30,000$ cal/mole and $+124$ eu obtained by Banerjee and Lauffer (4). At pH 7.5 the values were $+19,000$ cal/mole and $+86$ eu.

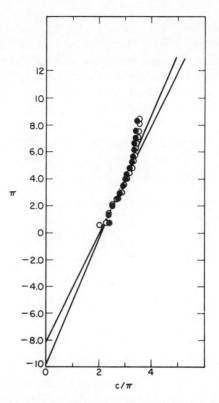

Fig. 13. π (osmotic pressure in centimeters of water) vs. c/π (c is concentration in milligrams per milliliter) for TMV A-protein in 0.1 μ barbital buffer, pH 6.5 ($\bigcirc$) and pH 7.5 ($\bullet$), at 4°C. [Data of Paglini and Lauffer (*18*).]

VII. CALORIMETRIC STUDIES

Direct measurements of the heat absorbed when TMV protein polymerizes and when TMV is reconstituted from protein and RNA have been made by Stauffer et al. (*54*) and by Srinivasan and Lauffer (*55*) with a microcalorimeter. Since all measurements had to be made at room temperature, polymerization could be brought about only by changing pH, ionic strength, or concentration or by adding RNA to protein.

Heat measured in the calorimeter divided by the mass of reactant yields the heat absorbed per unit weight of material involved in the reaction. This must be converted to a molar quantity if results from the calorimeter are to be compared with results from equilibrium studies. If $\bar{M}_{n1}$ and $\bar{M}_{n2}$ are the number-average molecular weights of the material in the unpolymerized and polymerized states, respectively, the number of molecules per gram in the unpolymerized state is Avogadro's number Av divided by $\bar{M}_{n1}$, and in the polymerized state, Av divided by $\bar{M}_{n2}$. The change in the number of molecules per gram during polymerization is then $Av(\bar{M}_{n2} - \bar{M}_{n1})/\bar{M}_{n1}\bar{M}_{n2}$. Since one bond is broken for each molecule formed, the heat per bond is q, the heat per gram, divided by the change in number of molecules, and the molar enthalpy ΔH is Av multiplied by this value

$$\Delta H = q(\bar{M}_{n1}\ \bar{M}_{n2})/(\bar{M}_{n2} - \bar{M}_{n1}) \tag{30}$$

When $\bar{M}_{n2}$ is much greater than $\bar{M}_{n1}$, this reduces to the simple formula

$$\Delta H = q\bar{M}_{n1} \tag{31}$$

If the polymerized state corresponds to a closed ring or disc, the change in number of molecules per gram during polymerization is still given by the formula mentioned above, but the change in the number of free ends of one kind per gram is $[N(1-r)/\bar{M}_{n1}]-0$, where r is the ratio of the number of discs to the total number of molecules. This is true because there are no free ends in the disc. The formula above applies for a case in which the final state is composed solely of discs. By reasoning identical with that in the preceding paragraph

$$\Delta H = q\bar{M}_{n1}/(1-r) \tag{32}$$

Several experiments were carried out by Stauffer et al. (54) in which TMV protein was depolymerized by dilution. In all cases the initial state was protein at a concentration 50 mg/ml in pH 7.5, 0.1 μ phosphate buffer. Considerable evidence indicates that under these conditions the protein exists in the form of double discs. The final state corresponded to protein concentrations between 2 and 4 mg/ml in phosphate buffer at pH 7.5, μ 0.1; at pH 8.0, μ 0.1; or at pH 7.5, μ 0.01. The results are shown in Fig. 14. They should be interpreted in accordance with Eq. (32). However, it is not easy to determine r, but its value should approach one in a very concentrated solution and zero in infinitely dilute solutions. Thus the difficulties can be avoided by calculating ΔH from $q\bar{M}_{n1}$ and extrapolating to zero concentration, where r should vanish. In this instance $\bar{M}_{n1}$ is the final number-average molecular weight since the experiment involves depolymerization rather than polymerization.

As shown in Fig. 14, when the data are thus treated, the extrapolated values of ΔH fall within the range 25,000–30,000 cal/mole. This is close to the range of values obtained by Banerjee and Lauffer (4) and by Paglini and Lauffer (18) from equilibrium measurements, 18,000–34,000 cal/mole. It should be noted, however, that the equilibrium measurements refer to ΔH values for the initial stage of polymerization in which the fraction of material in the double-disc state is negligibly small. The value obtained in the measurements under discussion, even when extrapolated to zero concentration, are average quantities per mole of bond for all bonds broken when double discs are dissociated. While it is a reasonable assumption that the thermodynamic parameters are the same for all successive stages during polymerization short of the final closing of the ring, it is necessary to assume that the enthalpy for the ring-closing step is higher because of strain in order to explain the fact that condensation polymerization mathematics works at all for a process in which double discs are the result.

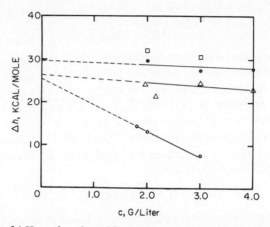

Fig. 14. Graph of ΔH as a function of final TMV A-protein concentration. Initial state: 50 mg protein per milliliter in 0.1 μ phosphate buffer, pH 7.5. $\bigcirc$, Final state, 0.1 μ, pH 7.5; $\triangle$, final state, 0.1 μ, pH 8.0; $\bullet$, final state, 0.01 μ pH 7.5. [Stauffer et al. (54).] (Reprinted by permission of the copyright owner, The American Chemical Society.)

Measurements were also made of the heats associated with polymerization or depolymerization of TMV protein as a result of drastic changes in pH (54). Two types of experiments were performed. In one, unpolymerized protein at a concentration of 4 mg/ml in 0.04 μ phosphate buffer at a pH value above 7.2 was polymerized by diluting with buffer which produced a final pH of 6.2, a final ionic strength of 0.12, and a final protein concentration of 2 mg/ml. In the other, polymerized protein at a concentration of 4 mg/ml in phosphate buffer of 0.04 μ at pH 6.2 was depolymerized by diluting with a buffer, which brought the final pH to some value above 7.3, the final ionic strength to 0.12, and the final protein concentration to 2 mg/ml. To interpret these data it would have been desirable to multiply the heat per gram by the appropriate number-average molecular weight at the pH, ionic strength, and protein concentration of the unpolymerized material. However, it proved to be extremely difficult to obtain accurate results for the number-average molecular weight under the conditions of these experiments. The reason is that at room temperature even unpolymerized protein has a number-average molecular weight in the neighborhood of 100,000 at low protein concentrations, in contrast to values of approximately half as great as 4°C. Thus because of the high molecular weight and low protein concentration the osmotic pressure measurements are not as accurate as one might wish. Therefore, in this study, for the experiments involving polymerization, all of the number-average molecular weights obtained at ionic strength 0.04 at concentration 4 mg/ml at pH values between 7.2 and 8 were averaged to obtain 98,500. The heats of polymerization, as measured in the calorimeter, were also averaged to obtain 0.788 cal/g. This value was multiplied by 98,500 to give a mean value of 77.6 kcal/mole for ΔH. Similarly, the number-average molecular weights at protein concentrations of 2 mg/ml in 0.12 μ phosphate buffers at pH values between 7.3 and 8 were averaged to yield 107,000. The heats of depolymerization were averaged to give 0.808 cal/g. From this an average value for ΔH of 86.4 kcal/mole was obtained, which was averaged with the figure for polymerization to yield an overall average of 82 kcal/mole. This number is large and it is positive for polymerization. From equilibrium studies at constant pH and ionic strength, enthalpies ranging between approximately 100,000 and 300,000 cal/mole were obtained, depending on pH and ionic strength. The equilibrium values and the calorimetric values can not be compared directly because in the equilibrium measurements the initial and final states differed only in respect to temperature, whereas in the calorimetric experiments the initial and final states were at the same temperature but differed with respect to pH, ionic strength, and protein concentration.

It is possible, however, to perform an indirect experiment which comes close to yielding the heat of polymerization at constant pH and constant temperature. Srinivasan and Lauffer (55) obtained the data shown in Fig. 15 for heat changes resulting from a change in pH for both TMV and TMV protein at a concentration of 2.5 mg/ml in 0.01 μ phosphate buffer. Auxiliary experiments showed that the heat absorbed in changing the pH of protein from 7.5 to 5.7 is the same for both 0.1 μ and 0.01 μ buffers, except that the sharp rise takes place at a slightly higher pH value in the more concentrated buffer. Auxiliary experiments also showed that changing the ionic strength and composition of buffer without changing pH results in no measurably thermal effects for TMV. Figure 15 shows that when the pH values of TMV solutions are reduced from 7.5 to approximately 3.5 there is an evolution of heat. The amounts involved are approximately those expected for the negative heat of hydrogen ion binding. Since the pH titration curves of polymerized TMV protein and TMV are very nearly the same over this pH range, [Scheele and Lauffer (20)], it is reasonable to assume that the results obtained for TMV are those that would be obtained for TMV protein if it were polymerized over the whole pH range and if the structure were identical with the arrangement of protein subunits in the virus over the whole pH range. On examining the protein a gradual absorption between pH 7.5 and 6.2 and a sharp increase in absorption near pH 6.2 followed by a gradual increase in heat absorbed as the pH is lowered from 6.2 to 5.0 are observed. The rise above pH 6.2 can be attributed to the major polymerization reaction and the changes between pH 6.2 and

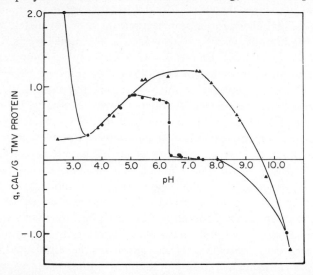

Fig. 15. Graph of Q (calories per gram of protein) as a function of pH. ▲, TMV in 0.01 μ phosphate buffer; ●, TMV A-protein in 0.01 μ phosphate buffer. [Srinivasan and Lauffer (55).] (Reprinted by permission of the copyright owner, The American Chemical Society.)

5.0 can be interpreted as a tightening of the structure of the polymerized protein rod. Between pH 5 and 3.5 the polymerized protein and the virus give off heat and the amounts are the same. This region of the protein curve is therefore attributed to hydrogen ion binding just as for the virus. Therefore from Fig. 15, without making any highly questionable assumption, it is possible to read off the heat that would be absorbed per gram of TMV protein if it could be polymerized in either 0.01 or 0.1 μ phosphate buffer at pH 7.5 to form a structure in which the packing of the proteins subunits is the same as in TMV. The value is 1.2 cal/g. Actually, X-ray diffraction evidence referred to earlier indicates that the packing of protein is the same in virus and in polymerized protein at approximately pH 5. If

TABLE II

Heat of Interaction of TMV Protein and RNA

Reactant		Product	Value
TMV protein; (phos); pH 5.0; 0.01 μ	$\xrightarrow{Q_1}$	TMVP; (phos); pH 7.5; 0.01 μ	−0.880
TMV protein; (phos); pH 7.5; 0.01 μ	$\xrightarrow{Q_2}$	TMVP; (pyrophos); pH 7.3; 0.1 M	+0.785
TMV RNA; (phos); pH 5.0; 0.01 μ	$\xrightarrow{Q_3}$	TMV RNA; (phos); pH 7.5; 0.01 μ	0
TMV RNA; (phos); pH 7.5; 0.01 μ	$\xrightarrow{Q_4}$	TMV RNA; (pyrophos); pH 7.3; 0.1 M	−0.258
TMV protein; (pyrophos); pH 7.3; 0.1 M $\left.\rule{0pt}{14pt}\right\}$ TMV RNA; (pyrophos); pH 7.3; 0.1 M	$\xrightarrow{Q_5}$	TMV; (pyrophos); pH 7.3; 0.1 M	+0.198
TMV; (pyrophos); pH 7.3; 0.1 M	$\xrightarrow{Q_6}$	TMV; (phos); pH 7.5; 0.01 μ	0
TMV; (phos); pH 7.5; 0.01 μ	$\xrightarrow{Q_7}$	TMV; (phos); pH 5.0; 0.01 μ	−0.330
TMV protein; (phos); pH 5.0; 0.01 μ + TMV RNA; (phos); pH 5.0; 0.01 μ $\left.\rule{0pt}{14pt}\right\}$	$\xrightarrow{Q}$	TMV; (phos); pH 5.0; 0.01 μ	−0.485 cal/g TMV protein

one assumes that the number-average molecular weight at a concentration of 2.5 mg/ml in 0.10 μ phosphate buffer at pH 7.5 is the same as the average value at a concentration of 2 mg/ml in 0.12 μ phosphate buffer, 107,000, one can calculate a value of 129,000 cal/mole for polymerization at constant pH and constant temperature. This is considerably less than the maximum values of ΔH obtained from equilibrium studies. Deviations of this sort are not uncommon; they are usually explained in terms of the difference between the heat capacities of reactants and products.

The heat involved in the interaction between TMV protein and RNA must be obtained by a roundabout procedure involving an application of Hess's law. The results are shown in Table II, which shows that in a hypothetical reaction in which 1 g of polymerized TMV protein combines with the appropriate amount of RNA at pH 5.0 in 0.01 μ phosphate buffer −0.485 calories are absorbed. When this value is multiplied by the molecular weight of protein monomer, it corresponds to −8.5 kcal/mole. Since three nucleotides are associated with each protein monomer, this amounts to −2.8 kcal/mole of nucleotide, which is more than adequate to account for the enhanced stability of TMV over that of the polymerized protein. This point is made clear if it is considered that when protein alone is polymerized in 0.1 μ phosphate buffer at pH 6.5, ΔH is +206 kcal/mole and $\Delta S°$ is +739 eu. From these values it can be calculated that at 23°C $\Delta F°$ is −12.5 kcal/mole. At pH 6.5 and 23°C, polymerized protein is very stable. If the polymerizing unit of the protein is considered to be cyclical trimer, this would be associated with nine nucleotides in TMV which should attribute nine times (−2.8) or −25 kcal/mole. This enthalpy contribution to the free energy could lead to an enormous increase in stability.

VIII. ION BINDING

In order to explain the increase in entropy associated with polymerization of TMV A-protein, Lauffer et al. (5) assumed that water molecules were released during the polymerization process. As shown in the next section, this actually does happen. However, it is theoretically possible that an entropy change could be brought about by a change in ion binding during the polymerization. Equilibrium dialysis and sedimentation studies carried out by Jaenicke and Lauffer (56) using ^{32}P showed no appreciable binding of phosphate by TMV A-protein in the neighborhood of pH 6.5. Shalaby et al. (19) measured ion binding by TMV and TMV A-protein by using cation- and anion-exchange membrane electrodes. All measurements were made at 22°C. TMV bound small amounts of K^+ and Cl^- and larger amounts of Ca^{2+} at pH values above 7 but none of these ions at the isoionic point, pH 4.6. The binding of Ca^{2+} confirmed numerous earlier reports of divalent cation binding. The protein did not bind Cl^-, K^+, or Ca^{2+}, either at the isoionic point, pH 4.6, where the protein is polymerized, or at pH values between 7.5 and 9, where the protein is unpolymerized. This indicates that when polymerization is brought about by change in pH at room temperature there is probably no binding of ions other than H^+.

The experiments just described, however, do not prove conclusively that there is no change in ion binding when protein is polymerized by raising the temperature at pH 6.5. Accordingly, Banerjee and Lauffer (57) investigated Cl^- and K^+ binding by TMV and by TMV A-protein at pH 6.5, both at low temperatures at which the protein is unpolymerized and at room temperature at which it is polymerized. Commercial electrodes specific for K^+ and Cl^- were used. The data showed conclusively that neither ion is bound by the protein at pH 6.5 whether in the unpolymerized or in the polymerized state. Thus all attempts to demonstrate change in ion binding associated with polymerization have failed. The experiments also showed that TMV does not bind K^+ or Cl^- under any of the conditions mentioned at pH 6.5.

IX. WATER AND ENTROPIC REACTIONS

Even though the idea is old that water is released in endothermic or entropic structure-forming reactions, direct experimental evidence dates to the experiment of Stevens and Lauffer (58). In their experiment TMV A-protein dissolved in aqueous solution of glycerol buffered to pH 7.5 was placed in a cellophane sac suspended from a delicate quartz helix in a medium containing the same low-molecular-weight components. The protein was then polymerized by changing the pH to 5.5. If water is released on polymerization, this changes the buoyant weight of the protein, which is measured by a change in the equilibrium position of the previously calibrated stretched helix. The apparatus is illustrated in Fig. 16. Stated simply, the idea is that free water, electrolyte, and glycerol are free to come to equilibrium on both sides of the membrane. Thus the weight of each is canceled by buoyancy. Bound water and protein, however, are retained inside the sac and thus are weighed after correction for buoyancy. If, upon polymerization, bound water is released, it will come to equilibrium and will no longer be weighed. Because the density of a glycerol solution is greater than that of water, bound water will have a negative weight. When it is released, the net buoyant weight of the contents of the sac increases. More recently, Jaenicke and Lauffer (56) carried out similar experiments in which polymerization was brought about by raising the temperature. The results are shown in Fig. 17. It can be seen that there is a definite change in the spring extension when the temperature is increased, that this change is much greater for TMV A-protein which does polymerize than for TMV which does not, and that the change is completely reversible.

An exact interpretation of these experiments is much more involved than that implied by the simple explanation given above. The reason is

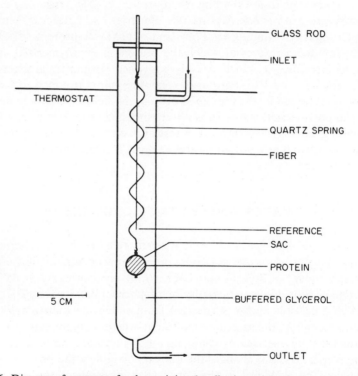

GLASS ROD

INLET

THERMOSTAT

QUARTZ SPRING

FIBER

REFERENCE

SAC

PROTEIN

5 CM

BUFFERED GLYCEROL

OUTLET

Fig. 16. Diagram of apparatus for determining the effective weight change accompanying polymerization of TMV protein. [Stevens and Lauffer (58).] (Reprinted by permission of the copyright owner, The American Chemical Society.)

that contributions to the buoyant weight are also made by redistribution of electrolyte in accordance with Donnan equilibrium, by the binding of electrolyte, and by response of the glycerol concentration to the osmotic pressure of the system. Furthermore, there is an increase in the partial specific volume of protein when it is polymerized and there are changes in partial specific volume resulting from temperature changes (59). The exact theory follows.

The buoyant weight W of γ grams of material inside the dialysis sac is given by Eq. (33)

$$W = \gamma_2 G(1 - V_1\rho) + \gamma_2 G'(1 - V_4\rho)$$

$$+ \gamma_2(1 - V_2\rho) + \frac{M_3\gamma_1}{1000} m_3''(H_3 - 1)(1 - V_3\rho)$$

$$\frac{M_4\gamma_1}{1000} m_4'' \left\{ \left[H_4 + \left(\frac{q}{2m_4''} \right)^2 \right]^{1/2} - \frac{q}{2m_4''} - 1 \right\}(1 - V_4\rho) \quad (33)$$

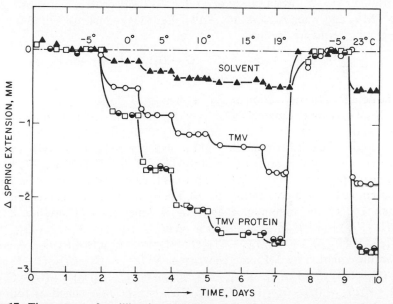

Fig. 17. Time course of equilibration of TMV and TMV A-protein in a spring balance. Total amount of material in the sac: 209 mg of TMV (○), 200 mg of TMV protein (□), 140 mg of TMV protein (◑) normalized for 200 mg of material and 100 mg of spring (spring coefficient, 0.526 mg/mm). Solvent: 25% (weight/volume) glycerol in phosphate buffer, pH 6.77, $\mu = 0.05 + 3 \times 10^{-3}$ M EDTA. [Jaenicke and Lauffer (56).] (Reprinted by permission of the copyright owner, The American Chemical Society.)

where ρ is the density of the external solution, V the average partial specific volume, G the bound water in grams per gram of component 2, G' the bound salt in grams per gram of component 2, M the molecular weight, m'' the molality in external solution, and q the molality of positive ion inside sac resulting from ionization of protein salt [Lauffer (59)]. $H_3 = f_1' f_3''/f_3' f_1'' \exp(h_1 - h_3)$ and $H_4 = [(f_1')^2 (f_{4+}'')(f_{4-}'')/[f_{4+}')(f_{4-}')(f_1'')^2] \exp(2h_1 - h_4)$, where f is the activity coefficient and $h_i = (P'\bar{V}_i' - P''\bar{V}_i'')/RT$, when P is pressure and $\bar{V}_i$ is partial molar volume. Subscripts 1, 2, 3, and 4 refer to water, protein, glycerol, and salt, respectively; single prime refers to "free" components inside the sac and double prime to components in the external solution.

When γ_2/γ_1 is sufficiently small, when $q = xm'_2 << m''_4$, when $W' = W/\gamma_2$, when

$$G'' = G' + \frac{M_4 m''_4}{2M_2 1000}\left(2\bar{V}_1 - \bar{V}_4 - \frac{1000x}{m''_4}\right) \simeq G' - \frac{M_4 x}{2M_2}$$

when $\Delta\Phi \equiv \Phi_{T_2} - \Phi_{T_1}$, where Φ represents any temperature-dependent variable $T_2 > T_1$, when subscript a refers to TMV protein and subscript b to TMV, and when $\Delta\Delta\Phi \equiv \Delta\Phi_a - \Delta\Phi_b$, Eq. (33) reduces to Eq. (34) after negligible terms are dropped

$$\Delta\Delta W' = (1 - V_1\rho)_{T1} \Delta\Delta G - (G_a - G_b)_{T2} \Delta(V_1\rho) - \Delta\Delta V_2\rho_{T2} \\ + (1 - V_4\rho)_{T1}\Delta\Delta G'' \tag{34}$$

The details of the derivation of Eq. 34 from Eq. 33 follow.

Since h_i is approximately equal to $m'_2 \bar{V}_i/1000$ [Stevens and Lauffer (58)], $W' = G(1 - V_1\rho) + (1 - V_2\rho) + (M_3 m''_3/M_2 1000)(\bar{V}_1 - \bar{V}_3)(1 - V_3\rho) + G''$ $(1 - V_4\rho)$. The third term on the right is almost zero for polymerized TMV protein and TMV because of the high value of M_2. Even when M_2 has a value of approximately 600,000, the probable value of the "unpolymerized" protein at the high concentrations used here, the value of this term is less than 1% of the minimum change in W' measured. Therefore $W' = G(1 - V_1\rho) + (1 - V_2\rho) + G''(1 - V_4\rho)$. $\Delta W'_a = G_{aT2}(1 - V_1\rho)_{T2} - G_{aT1}(1 - V_1\rho)_{T1} - (V_{2a}\rho)_{T2} + (V_{2a}\rho)_{T1} + G''_{aT2}(1 - V_4\rho)_{T2} - G''_{aT1}(1 - V_4\rho)_{T1}$. A comparable equation for ΔW_b, can be written: $\Delta\Delta W' = (G_a - G_b)_{T2}(1 - V_1\rho)_{T2} - (G_a - G_b)_{T1}(1 - V_1\rho)_{T1} - [(V_{2a} - V_{2b})\rho]_{T2} + [(V_{2a} - V_{2b})\rho]_{T1} + (G''_a - G''_b)_{T2}$ $(1 - V_4\rho)_{T2} - (G''_a - G''_b)_{T1}(1 - V_4\rho)_{T1}$. This can be rearranged as follows. To the first two terms on the right, subtract and add $(G_a - G_b)_{T2}(1 - V_1\rho)_{T1}$. This gives $(G_a - G_b)_{T2}[1 - (V_1\rho)_{T2} - 1 + (V_1\rho)_{T1}] + \Delta\Delta G(1 - V_1\rho)_{T1}$ or $-(G_a - G_b)_{T2}\Delta(V_1\rho) + \Delta\Delta G(1 - V_1\rho)_{T1}$. The other term can be treated in a comparable manner to give, $\Delta\Delta W' = (1 - V_1\rho)_{T1}\Delta\Delta G - (G_a - G_b)_{T2}\Delta(V_1\rho)$ $- \Delta\Delta V_2\rho_{T2} - (V_{2a} - V_{2b})_{T1}\Delta\rho + (1 - V_4\rho)_{T1}\Delta\Delta G'' - (G''_a - G''_b)_{T2}\Delta(V_4\rho)$. Since salt binding by polymerized TMV protein at T_2 and by TMV at T_2 are both zero and since the charge on the two are about the same, the final term on the right is zero. The fourth term on the right was shown to be negligible. Therefore Eq. (34) is appropriate for interpreting the data under discussion.

The actual method of experimentation involves measurement of the change in buoyant weight with temperature for TMV protein and for TMV. The difference between these two changes is $\Delta\Delta W$. This experiment is repeated with a number of solutions in glycerol–water mixtures of different compositions and therefore different densities (ρ). The results are then interpreted by using Eq. (34). If the solvent contains no glycerol but only water and a very dilute buffer, $1 - V_1\rho$ and $\Delta(V_1\rho)$ are practically zero and $\Delta\Delta W'$ is practically equal to $\Delta\Delta V_2\rho_{T2}(1 - V_4\rho)_{T1}$. Since $\Delta\Delta V_2$ and ρ can be determined independently, the other term can be evaluated. This represents the difference between the changes in salt binding by TMV protein and TMV. It turns out to be a very small number corresponding to a fraction of a mole of salt per mole of protein monomer. As discussed in the preceding section, the change in salt binding upon polymerization is probably zero. With these evaluated, another experiment can be performed in a glycerol–water solvent and the first two terms in Eq. (34) determined. Values of ρ and of V_1 at a high temperature T_2 and a low temperature T_1 must be evaluated from independent experiments. Also, $G_a - G_b$ at temperature T_2, where the protein is polymerized, must be measured independently. One is then in a position to calculate $\Delta\Delta G$. In this manner Jaenicke and Lauffer (56) obtained as an average of several experiments $\Delta\Delta G = -0.033 \pm 0.00035$ g per gram of protein. This value corresponds to a water release of 32 moles of water per mole of protein monomer or 96 moles per mole of protein trimer or A-protein. Very similar results were obtained by Stevens and Lauffer (58) when protein polymerization was brought about by changing pH at constant temperature. These experiments demonstrate conclusively that water is actually released during the endothermic polymerization process, and thus the theory that the free energy decrease needed to drive this reaction comes from an increase in entropy associated with water release is valid.

The equilibrium studies of Smith and Lauffer (44) provide a different method for estimating water release upon polymerization of TMV protein. From the results of Shalaby and Lauffer (47) and of Khalil and Lauffer (46), it can be estimated that $\Delta H°$ and $\Delta S°$ for polymerization at pH 6.7 and 0.05 μ are not much different from the values for polymerization at pH 6.5 and 0.1 μ. The entropy thus obtained, $+738.7 \pm 22.5$ eu, can be taken as representative of the value for the experimental conditions used in this study. This is a net entropy increase for the polymerization process. When correction is made for the entropy changes associated with binding of hydrogen ions and with joining of protein molecules, the entropy increase attributable to the release of water is 865 eu. If the entropy change on freeing a water molecule from protein is approximately the same as that of releasing it from ice, 5.26 eu, 865 eu correspond to the release of 164 water molecules per bond. Since there is reason for believing that the polymerizing unit is a trimer, this value can be compared with 96 moles of water released per trimer obtained in the present study.

The values of G_a and G_b in Eq. (34) were determined by means of independent studies in a spring balance [Jaenicke and Lauffer (53)]. When

Eq. (34) is differentiated with respect to ρ and all terms shown (53) to be negligible are dropped, Eq. (35) is obtained

$$\frac{\partial W}{\partial \rho} = \gamma_2 G V_1 + \gamma_2 V_2 \tag{35}$$

The result of experiments in which buoyant weights were measured as a function of density are shown in Fig. 18. When independent values for V_2 are employed, G for TMV and TMV protein were calculated from the slopes to be 0.16 and 0.20 g water per gram protein, respectively.

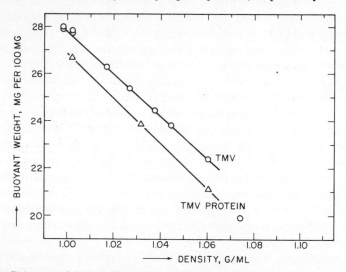

Fig. 18. Buoyant weight W (milligrams per 100 mg) as a function of density from spring balance experiments in aqueous buffers and glycerol buffer mixtures, 20°C. TMV in water and phosphate buffer, pH 6.77, $\mu = 0.05$. TMV A-protein in phosphate buffer, pH 6.7, $\mu = 0.05$. [Data of Jaenicke and Lauffer (53).]

The value of $\Delta\Delta V_2$ in Eq. (34) can be obtained from measurements with a dilatometer. Stevens and Lauffer (58) found that when TMV protein was titrated at 4°C from pH 7.5 to 5.5, there is an increase in volume of 0.00741 ± 0.0003 ml/g. Jaenicke and Lauffer (56) found that when polymerized TMV protein at pH 6.8 is titrated to pH 5.5 there is a volume increase of 0.00247 ml/g at 25°C. Since there is no reason to believe that the volume change observed by Stevens and Lauffer at 4°C would have been different if the experiment had been carried out at 25°C, the difference between these two values, 0.0049 ± 0.00003 ml/g, represents the volume increase when unpolymerized protein at pH 7.5 is titrated to polymerized protein at pH 6.8. Since Scheele and Lauffer (20) have found that unpolymerized protein does not bind H^+ between pH 7.5 and 6.8, there is no need to make a correction for electrostriction. This value is thus the volume change resulting from polymerization. Jaenicke and Lauffer (56) showed that the change in partial specific volume with temperature at pH 6.8 was the same, within experimental error, for TMV and for polymerized TMV protein both in water and in 25% glycerol. Thus the value 0.0049 represents $\Delta\Delta V_2$.

The fact that there is an increase in partial specific volume when TMV protein polymerizes is of interest in its own right. From this fact is can be inferred with thermodynamic certainty that an increase in pressure shifts an equilibrium mixture of polymerized and unpolymerized TMV protein toward the side of depolymerization. Furthermore, the increase in partial specific volume can be interpreted on the assumption that "bound" water has a higher density or lower specific volume than "free" water.

The increase in partial specific volume of polymerized TMV protein when the pH is dropped from 6.8 to 5.5 is also of considerable intrinsic interest. This is attributable in part to change in electrostriction. Since, according to Scheele and Lauffer (20), TMV binds approximately the same amount of hydrogen ion between pH 6.8 and 5.5, the portion attributable to electrostriction can be estimated from the volume increase for TMV over this pH range, 0.0009 ml/g. Thus the difference between volume increases for polymerized TMV protein and for TMV in this pH range, 0.0016 ml/g, must stem from something other than electrostriction. The most probable interpretation is that as pH is reduced with polymerized protein, with the resultant decrease in net negative charge, the lowered electrostatic repulsion permits tighter packing of the protein subunits in the polymer structure, thereby allowing for additional release of water and parallel increase in partial specific volume. The results obtained with the calorimeter by Srinivasan and Lauffer (55) over the same pH range are in complete accord with this view.

Dilatometric experiments can be analyzed further to show the difference between the change in partial specific volume of TMV protein associated with low-temperature and with high-temperature polymerization. Stevens and Lauffer (58) observed a change in partial specific volume of -0.00046 ml/g when TMV protein at 4°C in 0.1 μ buffer at pH 7.6 was diluted from 22 to 4 mg/ml. With Eq. (6) and the value of K for low-temperature polymerization at pH 7.5, 0.1 μ of 4.14×10^3 found by Paglini and Lauffer (18), taking the molecular weight of the polymerizing unit to be 52,500, it can be calculated that p is 0.2 at 4 mg/ml and 0.475 at 22 mg/ml. Thus the change in p is 0.275. Therefore if the change in p had been from 1, where all possible bonds are formed, to 0, where no bonds are present, the change in partial specific volume would have been $-0.00046/0.275$ or -0.0017 for depolymerization or $+0.0017$ ml/g for low-temperature polymerization.

In contrast with this low value, Stevens and Lauffer (58) found a change in partial specific volume of TMV protein for high-temperature polymerization of $+0.0074$ when protein at pH 7.6 and 22 mg/ml was changed to pH 5.5 and 4 mg/ml. To obtain a better value, to compare with the value for low-temperature polymerization, two corrections must be made. First, 0.475×0.0017 or 0.0008 must be added to obtain the value $+0.0082$ that would have been obtained if polymerization had started with unpolymerized A-protein of molecular weight 52,500. Then 0.0014, the value obtained with a TMV control, must be subtracted to eliminate the effect of hydrogen ion binding which has no direct relevance to polymerization. The result is $+0.0068$ ml/g. The ratio of ΔV_2 for high-temperature polymerization to ΔV_2 for low-temperature polymerization is $4 : 1$. It is interesting to observe that the ratio of the calculated calorimetric value of ΔH for high-temperature polymerization at constant pH and temperature, $+129,000$ cal/mole, to the calorimetric value of ΔH for low-temperature polymerization, 25,000–30,000 cal/mole falls in the range 4.3–$5.2 : 1$.

X. OPTICAL ROTATORY DISPERSION

The optical rotatory dispersion (ORD) of TMV, TMV protein, in both the unpolymerized and polymerized states, and TMV RNA was investigated by Simmons and Blout (60). The rotatory dispersion of the unpolymerized protein showed an incipient Cotton effect with an inflection around 293 μ and a large negative Cotton effect with a trough at 232 μ. The dispersion data for the unpolymerized protein were interpreted to indicate that it contains between 25 and 35% α-helix. When the protein subunits were polymerized into helical rods, the optical rotation changed to substantially more positive values and there was essentially no rotation above 270 μ. Simmons and Blout stated that this change in dispersion resulting from the formation of helical rods could not be interpreted precisely. Lauffer et al. (39) reported unpublished experiments of Khalil which showed that at pH values of 6 and 6.2 there is a marked change in dispersion correlated with polymerization when the temperature is raised from 7 to 25°C, but that at pH 6.8 there is no substantial change between 12 and 34°C.

Jaenicke and Lauffer (56) obtained the data shown in Fig. 19. They confirmed the deep trough with a minimum at 232 μ for the protein in the unpolymerized state. At pH 7 in 0.01 μ phosphate buffer, the depth of the trough changes very little with temperature. However, in a phosphate–acetate buffer of the same ionic strength at pH 5.9, the depth of the trough decreased remarkably when the temperature was raised from 3°C, where the protein is unpolymerized in this buffer, to 27°C, where the protein is polymerized. Changes in circular dichroism, CD, during polymerization of TMV protein have also been reported (63).

If, as is conventionally assumed, changes in ORD indicate changes in conformation, then these data would indicate that glycerol does not affect the conformation of TMV protein. However, does the large reversible change in the depth of the trough associated with polymerization brought about by changes in temperature indicate a remarkable change in conformation when TMV protein polymerizes? Certainly, the possibility of change

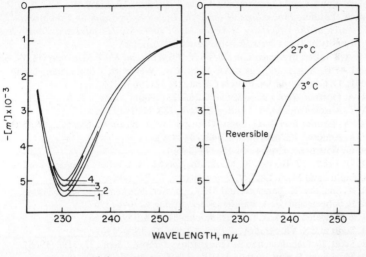

(a)   (b)

Fig. 19. ORD of TMV A-protein. Influence of temperature. (a) Phosphate, pH 7.0, $\mu = 0.01$, $c_2 = 1.6$ mg/ml, 1-mm path length. (1) 3°C unchanged after temperature cycle; (2) 15°C; (3) 20°C; and (4) 28°C. (b) Phosphate–acetate, pH 5.9, $\mu = 0.01$, $c_2 = 0.2$ mg/ml, 5-mm path length. [Jaenicke and Lauffer (56).] (Reprinted by permission of the copyright owner, The American Chemical Society.)

in conformation when protein is transferred from an essentially aqueous environment to an essentially organic environment, which happens when protein polymerizes, cannot be overlooked. However, recent experiments of Jaenicke (61) indicate that this shift might be an artifact. Jaenicke showed that the change in dispersion that accompanies polymerization can be duplicated by contaminating unpolymerized TMV protein with aluminum oxide, thus producing an artificially turbid solution. When these turbid systems were clarified by centrifugation or filtration, the normal dispersion of unpolymerized protein was restored. Thus the change observed when TMV A-protein polymerizes might be the result of the increase in turbidity; scatter distorts ORD and CD curves (64).

REFERENCES

1. G. Schramm, *Naturwissenschaften*, **31**, 94 (1943).
2. H. Fraenkel-Conrat, *Virology*, **4**, 1 (1957).
3. F. A. Anderer, B. Wittmann-Liebold, and H. G. Wittmann, *Z. Naturforsch.*, **20b**, 1203 (1965).
4. K. Banerjee and M. A. Lauffer, *Biochemistry*, **5**, 1957 (1966).
5. M. A. Lauffer, A. T. Ansevin, T. E. Cartwright, and C. C. Brinton, Jr., *Nature*, **181**, 1338 (1958).
6. H. Fraenkel-Conrat and R. C. Williams, *Proc. Natl. Acad. Sci. U.S.*, **41**, 690 (1955).
7. J. A. Lippincott and B. Commoner, *Biochim. Biophys. Acta*, **19**, 198 (1956).
8. M. A. Lauffer and C. L. Stevens, *Advan. Virus Res.*, **13**, 1 (1968).
9. M. A. Lauffer, *Proceedings of the 15th Annual Symposium on Fundamental Cancer Research at the University of Texas M.D. Anderson Hospital and Tumor Institute*, University of Texas Press, Austin, Texas, 1962, p. 180.
10. M. A. Lauffer, in *Symposium on Foods: Proteins and Their Reactions* (H. W. Schultz and A. F. Anglemier, eds.), Avi Publ. Co., Westport, Connecticut, 1964b, p. 87.
11. D. L. D. Caspar, *Advan. Protein Chem.*, **18**, 37 (1963).
12. R. A. Gortner, *Trans. Faraday Soc.*, **26**, 678 (1930).
13. A. T. Ansevin and M. A. Lauffer, *Nature*, **183**, 1601 (1959).
14. M. A. Lauffer and A. G. Szent-Györgyi, *Arch. Biochem. Biophys.*, **56**, 540 (1955).
15. F. A. Anderer, *Z. Naturforsch.*, **14b**, 24 (1959).
16. H. G. Wittmann, *Experienta*, **15**, 174 (1959).
17. R. H. Frist, I. J. Bendet, K. M. Smith, and M. A. Lauffer, *Virology*, **26**, 558 (1965).
18. S. Paglini and M. A. Lauffer, *Biochemistry*, **7**, 1827 (1968).
19. R. A. Shalaby, K. Banerjee, and M. A. Lauffer, *Biochemistry*, **7**, 955 (1968).
20. R. B. Scheele and M. A. Lauffer, *Biochemistry*, **6**, 3076 (1967).
21. R. B. Scheele and M. A. Lauffer, *Biochemistry*, **8**, 3597 (1969).
22. R. Scott and S. Vinogradov, *J. Phys. Chem.*, **73**, 1890 (1969).
23. R. Scott, D. Depalma, and S. Vinogradov, *J. Phys. Chem.*, **72**, 3192 (1968).
24. R. Markham, S. Frey, and G. J. Hills, *Virology*, **20**, 88 (1963).
25. R. E. Franklin and B. Commoner, *Nature*, **175**, 1074 (1955).
26. J. T. Finch, R. Leberman, C. Yu-Shang, and A. Klug, *Nature*, **212**, 349 (1966).
27. M. A. Lauffer, *Chimia (Aarau)*, **20**, 89 (1966).
28. M. A. Lauffer, *Biochemistry*, **5**, 2440 (1966).
29. R. Macleod, G. J. Hills, and R. Markham, *Nature*, **200**, 932 (1963).
30. R. Markham, J. H. Hitchborn, G. J. Hills, and S. Frey, *Virology*, **22**, 343 (1964).
31. J. L. Dickson and R. E. Woods, *Virology*, **10**, 157 (1960).
32. T. J. Finch, paper presented at a conference on Interaction Between Subunits of Biological Macromolecules, June 24–27, 1968, Cambridge University, England.
33. D. J. DeRosier and A. Klug, *Nature*, **217**, 130 (1968).
34. A. Klug, paper presented at a symposium on Assembly of Large Structures, Third International Biophysics Congress, August 29–Sept. 23, 1969, Cambridge, Massachusetts.
35. H. G. Wittmann, *Z. Vererbungslehre*, **97**, 297 (1965).
36. M. Zaitlin and W. F. MacCaughey, *Virology*, **26**, 500 (1965).
37. M. Zaitlin and W. Ferris, *Science*, **143**, 1451 (1964).
38. A. Siegel, G. J. Hills, and R. Markham, *J. Mol. Biol.*, **19**, 140 (1966).
39. M. A. Lauffer, R. A. Shalaby, and M. T. M. Khalil, *Chimia (Aarau)*, **21**, 450 (1967).
40. R. E. Franklin and K. C. Holmes, *Acta Cryst.*, **11**, 213 (1958).
41. R. E. Franklin, *Nature*, **177**, 928 (1956).

42. D. L. D. Caspar, *Nature*, 177, 928 (1956).
43. C. Westover and C. J. Stevens, private communication, 1968.
44. C. E. Smith and M. A. Lauffer, *Biochemistry*, 6, 2457 (1967).
45. P. J. Flory, *J. Am. Chem. Soc.*, 58, 1877 (1936).
46. M. T. M. Khalil and M. A. Lauffer, *Biochemistry*, 6, 2474 (1967).
47. R. A. Shalaby and M. A. Lauffer, *Biochemistry*, 6, 2465 (1967).
48. C. L. Stevens and S. Paglini, private communication, 1966.
49. T. L. Hill, *Arch. Biochem. Biophys.*, 57, 229 (1955).
50. M. H. Gorin, in *Electrophoresis of Proteins and the Chemistry of Cell Surface* (H. A. Abramson, L. S. Moyer, and M. H. Gorin, eds.), Van Nostrand Reinhold, New York, 1942, p. 126.
51. C. Tanford, *Physical Chemistry of Macromolecules*, Wiley, New York, 1961, p. 126.
52. M. A. Lauffer, *Biochemistry*, 5, 1952 (1966).
53. R. Jaenicke and M. A. Lauffer, *Biochemistry*, 8, 3077 (1969).
54. H. Stauffer, S. Srinivasan, and M. A. Lauffer, *Biochemistry*, 9, 193 (1970).
55. S. Srinivasan and M. A. Lauffer, *Biochemistry*, 9, 2173 (1970).
56. R. Jaenicke and M. A. Lauffer, *Biochemistry*, 8, 3083 (1969).
57. K. Banerjee and M. A. Lauffer, *Biochemistry*, 10, 1100 (1971).
58. C. L. Stevens and M. A. Lauffer, *Biochemistry*, 4, 31 (1965).
59. M. A. Lauffer, *Biochemistry*, 3, 731 (1964).
60. N. S. Simmons and E. R. Blout, *Biophys. J.*, 1, 55 (1960).
61. R. Jaenicke, private communication, 1969.
62. R. B. Scheele and T. M. Schuster, *Biophys. Soc. Abstr.*, 236a (1970).
63. D. Schubert and B. Krafczyk, *Biochim. Biophys. Acta*, 188, 155 (1969).
64. D. W. Urry, T. A. Hinners and L. Mesotti, *Arch. Biophys. Biochem.*, 137, 214 (1970).
65. S. Srinivasan and M. A. Lauffer, unpublished results, 1971.

CHAPTER 5

MYOSIN: MOLECULE AND FILAMENT

Susan Lowey

THE CHILDREN'S CANCER RESEARCH FOUNDATION

AND

THE DEPARTMENT OF BIOLOGICAL CHEMISTRY
HARVARD MEDICAL SCHOOL, BOSTON, MASSACHUSETTS

INTRODUCTION

Muscle is an excellent cellular system to study from a molecular point of view. The contractile proteins are few in number and those present form highly organized polymeric structures known as the thick and thin filaments: the thick filament is composed of several hundred large, complex molecules called myosin, while the thin filament consists primarily of a small, globular protein called actin. The arrangement of these filaments in regular lattices possessing considerable three-dimensional order has made possible detailed morphological studies of the muscle cell. These studies have established that contraction proceeds by a relative sliding force generated between the two sets of interdigitating filaments. Tension is most likely maintained by direct linkage between the proteins comprising these filaments (*1*).

Myosin and actin are the two most abundant proteins in the myofibril. Secondary in concentration if not importance are tropomyosin and troponin. This complex is associated with actin in the thin filament, where it serves to regulate the myosin–actin interaction (*2*). These proteins can be dispersed in solution and the properties of the individual monomers studied or, alternatively, in vitro aggregates of these proteins can be used to elucidate the nature of the protein–protein interactions in the native filaments. Thus the muscle cell offers a unique opportunity for examining the structural and functional aspects of a cell at all levels of organization: it is possible to begin with a description of the monomeric proteins; proceed to the assembly of these proteins into filamentous structures; and, ultimately, understand how the interactions within a given filament and between different filaments regulate the specialized functions of the cell.

This brief chapter is limited to what is known about the substructure and assembly of the myosin molecule. This discussion is not intended as a review of the myosin literature. Several comprehensive reviews have been published in recent years (*3–5*). Rather, we plan to sketch a portrait of what we consider some of the principal features and areas of interest pertaining to this complex molecule.

I. THE THICK FILAMENT

The thick filaments of vertebrate skeletal muscle are spindle-shaped objects about 1.6 μ long and 150 Å thick (*6*). They show a smooth "bare zone" in the middle of the filament with a rough surface extending away from the center toward either end (Fig. 1). The unusual appearance of the myosin filament reflects the equally unusual form of the myosin molecule; a protein consisting of a globular head about 100–200 Å in diameter attached to a rodlike tail about 1400 Å long (Fig. 2). Since myosin normally exists as a filamentous structure in vivo, high concentrations of salt are needed to disperse it to a monomeric form. Usually, myosin is dissolved in 0.5–1 M KCl buffered with phosphate, but in the case of preparations used for shadow casting, a volatile solvent such as 1 M ammonium acetate can be used. When the ionic strength of myosin solutions is gradually lowered from > 0.5 to < 0.1 M, the native filamentous structure is regained. At intermediate ionic strengths, small aggregates of myosin are

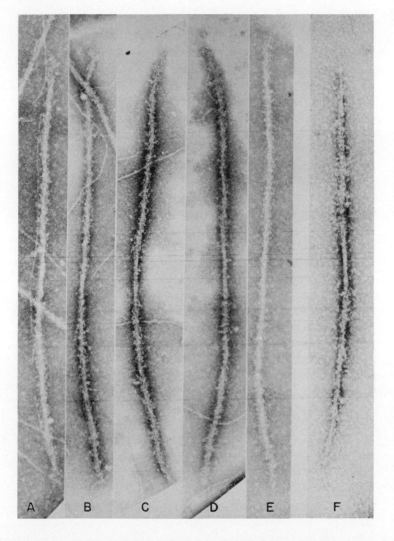

Fig. 1. (A) to (E) Thick filaments from muscle. (F) Synthetic filament made from myosin. (Reprinted from H. E. Huxley (6) by courtesy of Academic Press, Inc.)

seen which have a very clearly defined bare region about 2000 Å long with globular appendages on either side (Fig. 3). It is apparent from the shape of the myosin molecule and the filament that the latter is built by an anti-parallel aggregation of the myosin monomers with the rod portion of the molecule forming the bare region (Fig. 4) (6). The assembly process and the three-dimensional packing of the molecules in the filament are only partially understood at present. However, certain properties of myosin

Fig. 2. Myosin molecules shadow-cast unidirectionally with platinum. (Reprinted from Reference *21* by courtesy of The National Academy of Sciences.)

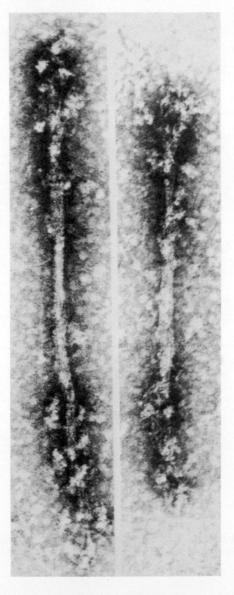

Fig. 3. Short synthetic myosin filaments. (Reprinted from H. E. Huxley (*134*) by courtesy of W. H. Freeman and Co.)

and its subfragments are relevant to a more complete understanding of the structure of the thick filament and the aggregation phenomenon. We shall therefore describe the myosin molecule in some detail before returning to a fuller discussion of the role of myosin in the thick filament.

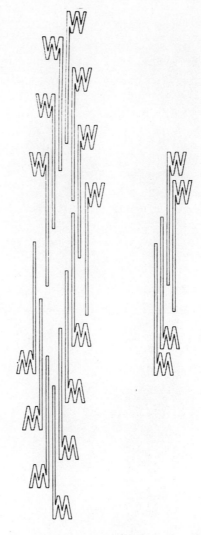

Fig. 4. Schematic diagram of myosin molecules illustrating antiparallel growth to form filaments of increasing length. (Reprinted from H. E. Huxley (*134*) by courtesy of W. H. Freeman and Co.)

II. THE MYOSIN MOLECULE

Over 100 years ago it was discovered that a viscous protein could be extracted from minced muscle with strong salt solutions. The significance of this new protein was quickly recognized; when a salt solution of the protein was squirted into water the protein gelled in the form of threads which "contracted" upon the addition of ATP (7). Soon afterward, it was shown that this protein also had ATPase activity, thereby linking the contractile properties with an energy source. What was called "myosin" in the early days proved later to be a complex of primarily two proteins, actin and myosin. However, even pure myosin remains a highly unusual macromolecule. It is one of the few proteins known at present that has two distinct domains which are covalently linked: one a typical globular conformation with active sites for binding nucleotide and actin, and the other a highly α-helical conformation serving a purely structural role. By means of proteolytic degradation of the molecule, it has been possible to study these domains separately, a procedure that has greatly simplified the problem of determining the substructure of myosin. Another approach, which has proved of great value in our understanding of myosin, has been to study the molecule under a variety of denaturing conditions, such as alkaline pH, chemical modification of the molecule, or complete denaturation in solvents containing guanidine hydrochloride or high concentrations of urea. After a brief description of some of the principal features of the native molecule, the properties of the proteolytic subfragments of myosin and the composition of myosin in terms of the noncovalently bound polypeptide chains are discussed.

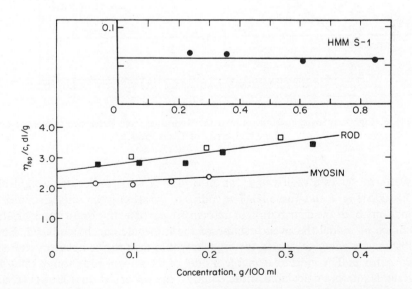

Fig. 5. Concentration dependence of the reduced specific viscosity for myosin and its principal papain subfragments. (Reprinted from Reference *36* by courtesy of Academic Press, Inc.)

A. Myosin in Solution

All the hydrodynamic properties of myosin indicate a long, thin molecule resembling a rigid rod (8,9). The viscosity of 2.1 dl/g is two orders of magnitude higher than that usually observed for the more typical globular proteins such as G-actin (0.04 dl/g) (Fig. 5). The small sedimentation rate (6.4 S) for a molecule of such large size and the hypersharp character of the sedimentating peak are both indicative of a highly asymmetric molecule (Table I). Further evidence for this picture of the solvated molecule is provided by light-scattering data (9,10); the angular dependence of the

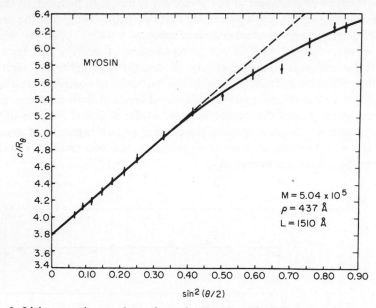

Fig. 6. Light-scattering envelope of myosin. (Reprinted from Reference *10* by courtesy of the University of Texas Press.)

scattering gives a radius of gyration of 437 Å for a molecular weight of 500,000 (Fig. 6 and Table II). The radius of gyration ρ can only give information about the distribution of mass in the particle, but by an appropriate choice of model the overall shape of the molecule can be deduced. For instance, a spherical model for myosin would lead to the absurd value of 40 Å for ρ. The more reasonable model of a uniform rod, about 1500 Å long, leads to a calculated value equal to the observed one. However, as early as 1957, it was recognized that myosin was not a uniform rod; the curvature in the light-scattering envelope at high angles was best accounted for by assuming a nonuniform particle (*11*); the conformation and hydro-

TABLE I

Physicochemical Parameters for Myosin and its Subfragments[a]

Parameter	Myosin	HMM	LMM	HMM S-1	HMM S-2	Rod
Molecular weight $\times 10^{-3}$	510 ± 10	340 ± 10	140 ± 5	115 ± 5	62 ± 2	220 ± 10
Intrinsic viscosity, dl/g	2.1 ± 0.1	0.49 ± 0.02	1.2 ± 0.1	0.064 ± 0.01	0.4 ± 0.1	2.4 ± 0.1
Intrinsic sedimentation coefficient, $\times 10^{13}$	6.4 ± 0.1	7.2 ± 0.1	2.9 ± 0.1	5.8 ± 0.1	2.7 ± 0.1	3.4 ± 0.1
Rotatory dispersion constant, (b_0)	400 ± 10	320 ± 10	630 ± 10	230 ± 10	610 ± 10	660 ± 10
α-Helix, %	57	46	90	33	87	94

[a] Reprinted from Reference 36 by courtesy of Academic Press, Inc.

TABLE II

Light-Scattering Data from Myosin and Meromyosins[a,b]

	Myosin		
Concentration, %	$M \times 10^{-3}$	ρ, Å	L, Å (rod)
0.0181	504	437	1510
0.0193	499	433	1500
0.0203	532	426	1475
0.0248	523	440	1520
0.0286	541	430	1490
0.0367	526	439	1520
Average	525	434	1503

	HMM			
Concentration, %	$M \times 10^{-3}$	ρ, Å	L, Å (rod)	L, Å (ellipse)
0.0505	347	183	634	820
0.0302	330	176	608	788
0.0522	340	160	554	716
0.0292	322	156	541	699
0.0265	346	178	615	796
0.0348	344	162	562	725
0.0279	350	147	510	658
Average	340	166	575	743

	LMM		
Concentration, %	$M \times 10^{-3}$	ρ, Å	L, Å (rod)
0.0263	138	221	766
0.0334	136	226	783
0.0376	132	215	745
0.0424	133	225	780
Average	135	222	769

[a] Reprinted from Reference *10* by courtesy of the University of Texas Press.

[b] *L*, Length; *M*, molecular weight; ρ, radius of gyration.

dynamic properties of the tryptic fragments of myosin suggested a rodlike region attached to more globular entities (*12*); and the early electron micrographs of myosin showed a rodlike molecule with an increase in mass at one end (*6,13,14*). By using the most recent structural data for myosin (see Section II.B), the radius of gyration can be calculated for a molecule whose mass (about 500,000) is distributed equally between a rod

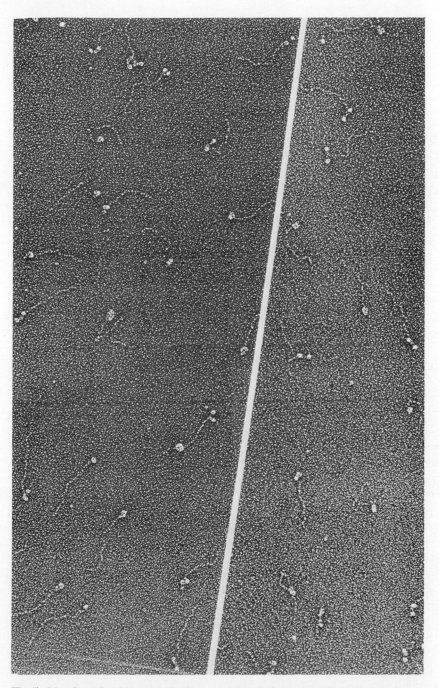

Fig. 7. Myosin molecules rotary shadow-cast with platinum. (Reprinted from Reference *36* by courtesy of Academic Press, Inc.)

1400 Å long and a sphere. Despite the obvious oversimplifications inherent in such a model, the calculated ρ closely approximates the measured value. We can conclude that myosin exists in solution in a relatively extended state with a high degree of rigidity in the rod portion of its structure.

B. Electron Microscopy of Myosin

The unusual shape of myosin lends itself to direct visualization by the electron microscope (Fig. 2). By evaporating platinum from a heated tungsten filament under high vacuum and at an angle to the specimen, metal is deposited everywhere on the specimen except where the molecule throws a shadow. This technique provides sufficient contrast between the macromolecule and the supporting grid to enable one to "see" the molecule. The image one does see clearly establishes the asymmetric nature of the myosin molecule inferred from hydrodynamic studies but, more important, localizes the globular regions at one end of the rod (6,13,14). A subject of considerable controversy in the myosin literature has been the question how many subunits are contained in this globular region (15–19). Physicochemical studies have supported anywhere from one to three globular units in myosin, but no final agreement seemed possible until the number of subunits could finally be "seen". Some substructure had been observed in the globular head of myosin by electron microscopy, but sufficient detail was not present to allow any firm conclusions (14,20).

The solution to the problem was found by shadow casting the molecules while rotating the specimen (21). An advantage of rotary shadowing over standard shadow casting is that less metal accumulates on any one side of the particle, resulting in more structural detail. By this procedure two globular subunits attached to a flexible, rodlike tail are readily observed in myosin (Fig. 7). The technique of rotary shadow casting has been successfully applied in the past to the DNA molecule where it was desired to preserve the structural continuity of the long, winding DNA strand (22). A similar problem occurs with myosin; if the rod portion of myosin lies in an unfavorable position relative to the beam of platinum metal, portions of its length or the entire rod might be obscured. Furthermore, one globule could easily overshadow the other and give the erroneous appearance of a single head (Fig. 2). By rotary shadowing the myosin molecule, both the flexible nature of the rod and the double-headed character of the globular region are readily visualized. Measurements on several hundred myosin molecules give an average contour length for the rod region of 1340 Å and an average diameter of 90 Å for each of the two globules; see Fig. 8 (top) and Table III.

TABLE III

Average Lengths for Myosin and Its Subfragments[a,b]

Preparation	Number counted	Peak length, Å	Number-average length, Å	Weight-average length, Å
Rod in myosin	493	1370	1340	1382
Isolated rod	306	1370	1360	1386
Rod in HMM	400	540	528	571
HMM S-2	549	460	474	500
LMM+rod	530	820	1045	1200
LMM	628	730	785	870
Single globules in myosin	216	114	117	121
HMM S-1	210	92	95	99

[a] Reprinted from Reference *36* by courtesy of Academic Press, Inc.
[b] The averages given in this table were all calculated from the data represented in the histograms. Approximately 25 Å should be subtracted from these dimensions in order to correct for the accumulation of metal on the molecules during the replication process.

Before going into any further detail about the substructure of the myosin molecule by electron microscopy, we summarize briefly the preparation and properties of its proteolytic subfragments.

C. Proteolytic Subfragments of Myosin

The large size and unusual shape of myosin make it a particularly difficult molecule to study by the conventional techniques of physical and protein chemistry, and X-ray crystallography. Some of these difficulties have been overcome by degrading myosin to smaller units with proteolytic enzymes. A classic example of the value of enzymes in delineating the functional regions of a macromolecule is γ-globulin. Papain cleaves γ-globulin into two major kinds of fragments; one fragment retains all the antibody specificity of the parent molecule, while the other is involved in complement fixation and, unlike the native molecule, is readily crystallizable (*23*). Although myosin bears, at best, a very superficial resemblance

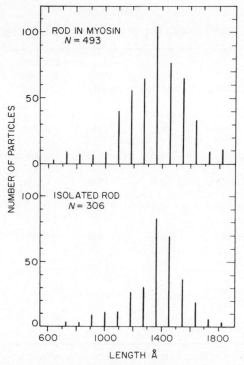

Fig. 8. Distribution of lengths for the rod portion of the myosin molecule (top) and for the isolated papain fragment (bottom). (Reprinted from Reference *36* by courtesy of Academic Press, Inc.)

to γ-globulin, the molecules do have certain features in common which lend themselves to analysis by enzymes. Both proteins have several biological functions which are separated from one another by a considerable distance in the native molecule. These specialized regions are covalently attached, unlike other multifunctional, multisubunit proteins such as the enzyme aspartate transcarbamylase (*24*), or fatty acid synthetase (*25*).

The first efforts to digest myosin with proteolytic enzymes date back to 1950 (*26–29*). It was discovered that trypsin (and chymotrypsin) were able to hydrolyze myosin into two types of large fragments called light meromyosin (LMM) and heavy meromyosin (HMM), in reference to their respective molecular weights (Table I and Fig. 9A). The difference in solubility of the fragments provides a simple means of fractionation. By lowering the ionic strength of digested myosin, the LMM precipitates while the HMM remains in solution. The LMM precipitate can be dispersed in salt and further purified by alcohol treatment (*30*). (The unusually high α-helix content of LMM protects it against denaturation from high con-

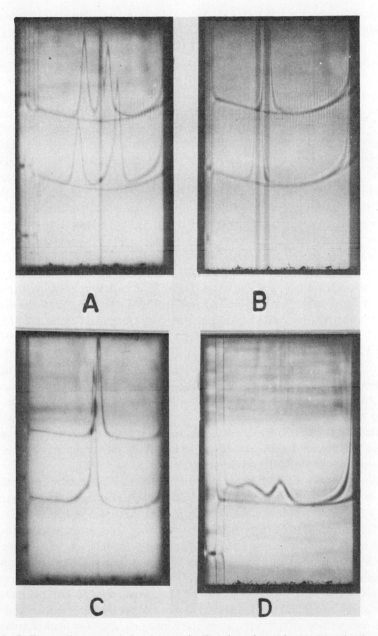

Fig. 9. Sedimentation patterns for the tryptic digestion of myosin. (A) Myosin digested for 5 min (top) and 40 min (bottom). (B) LMM prepared from digests in (A). (C) HMM prepared from digests in (A). (D) HMM digested further with excess trypsin. (Reprinted from Reference 36 by courtesy of Academic Press, Inc.)

centrations of alcohol.) HMM contains the active sites of myosin and, similar to many enzymes, is readily denatured by exposure to organic solvents, heat, freezing, metal ions, and most nonphysiological conditions.

Apart from demonstrating that myosin has distinct functional regions, what insight into the substructure of myosin have the meromyosins provided? Optical rotatory dispersion (ORD) studies of the meromyosins have shown that LMM has greater than 90% α-helix as compared to about 45% α-helix for HMM (31). The excellence of the wide-angle X-ray diagram for LMM, compared to the rather poor quality of the HMM pattern, implies that LMM consists largely of coiled-coil α-helices, whereas HMM has a preponderance of nonhelical regions which presumably contain the nucleotide- and actin-binding sites (32). The slow migrating, hypersharp sedimentation peak and high viscosity both suggest that LMM is an extremely asymmetric molecule (Fig. 9B and Table I). On the basis of these observations it seemed appropriate to adopt a rigid rod model for the calculation of the molecular length from the light-scattering radius of gyration (Fig. 10). Simultaneous measurements of the weight and length of several LMM preparations, and the assumption of an α-helical conformation for the polypeptide chains, resulted in the conclusion that LMM consists of approximately two polypeptide chains per molecule (Table II) (10,16). For example, given a molecular weight of 135,000 and a length of 770 Å for LMM, a mean residue weight of 115 and a 1.5 Å rise per residue for the α-helix, there would be $[(135,000/115) \times 1.5 \text{ Å}]/770 \text{ Å} = 2.3$ chains per molecule. LMM, similar to all α-helical fibrous proteins, is distinguished by an unusually high proportion of polar amino acid residues (Table IV). The role of the two α-helical chains, apart from conferring a degree of structural rigidity to the myosin molecule, is presumably to stabilize the large surface charge by hydrophobic bonding and strong van der Waals interactions between the neighboring chains (33,34).

Because of its lower helix content, HMM is a much more difficult molecule to analyze than LMM. Although HMM resembles a typical globular protein in terms of solubility, amino acid composition (Table IV), and optical rotatory properties, its viscosity is approximately 10-fold higher than that of most enzymes (Table I). These data, in conjunction with a large radius of gyration (Fig. 10) and an α-type X-ray pattern (32) originally led to the proposal that HMM probably contained regions of coiled-coil α-helices similar to LMM (12). Since myosin had been shown to consist of 1 mole of HMM and 1 mole of LMM (35), it followed that the rod portion of myosin probably contained two α-helical chains in a coiled-coil conformation (16,33). The demonstration of double-strandedness by direct observation of two subunits in the globular region of myosin and HMM has amply confirmed this hypothesis (Fig. 11) (21,36).

TABLE IV

Amino Acid Compositions of Myosin and Its Subfragments[a,b]

Amino acid	Myosin	HMM	HMM S-1	LMM	HMM S-2	Rod
Lys	92	86	83	94	121	107
His	16	14	18	21	9.0	15
Arg	43	34	34	60	37	56
Cys	8.8	7.4	11	4.0	6.4	4.6
Asp	85	82	85	83	87	88
Thr	44	44	49	33	43	38
Ser	39	39	41	34	35	39
Glu	157	137	117	210	242	219
Pro	22	32	37	0	0	0
Gly	40	50	61	18	19	20
Ala	78	73	70	81	87	85
Val	43	48	55	38	24	33
Met	23	26	28	19	26	23
Ile	42	44	53	39	34	36
Leu	81	73	75	96	99	99
Tyr	20	21	34	9	2.6	5.9
Phe	29	36	52	4	8.6	7.0

[a] Reprinted from Reference *36* by courtesy of Academic Press, Inc.
[b] Expressed as number of residues per 10^5 g.

The sedimentation profile of HMM frequently indicates small amounts of a slower sedimenting species (Fig. 9C). This heterogeneity implies that tryptic digestion of myosin does not stop with the meromyosins but progressively degrades these proteins to still smaller fragments. By using an excess of trypsin, it was shown that HMM can be completely converted to a less asymmetric molecule, called heavy meromyosin subfragment 1 (HMM S-1), plus large amounts of poorly defined low-molecular-weight components and dialyzable peptides (Fig. 9D) (*37*). LMM undergoes a similar degradation, but the heterogeneity is more difficult to demonstrate since all rodlike molecules, independent of length, sediment at about equal rates (Table I). However, light scattering has shown that the average length of LMM fragments can decrease from about 900 Å to less than 700 Å by increasing the time of tryptic digestion (*38*).

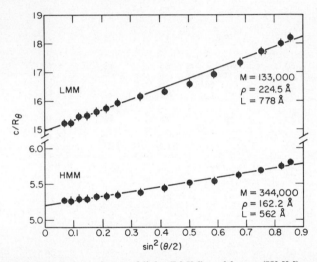

Fig. 10. Light-scattering envelopes of light (LMM) and heavy (HMM) meromyosins. (Reprinted from Reference *10* by courtesy of the University of Texas Press.)

A more promising enzyme for separating the active globular regions (HMM S-1) from the helical portions of myosin proved to be the protease papain (*39,40*). This enzyme, which was so valuable in isolating the active fragments of γ-globulin, has been equally successful in preparing a wide variety of subfragments from myosin (Fig. 11). These studies with papain have been described elsewhere (*36*), but the principal findings and conclusions are briefly summarized here.

The scheme in Fig. 12 outlines the major fragments obtained with trypsin and papain and indicates that there are two general regions in myosin, A and B, which are particularly vulnerable to enzymic attack. Papain has a slight preference for region B over A but attacks both areas readily, whereas trypsin shows great selectivity toward the A region. Only at high concentrations does trypsin cleave site(s) B, and then it liberates HMM S-1 at the expense of HMM S-2 which is to a large extent degraded to smaller peptides (Fig. 9D). In this respect, the diagram is somewhat

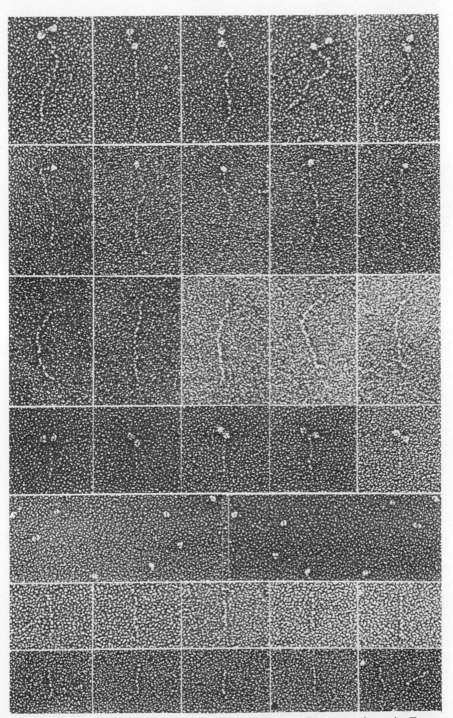

Fig. 11. A composite of selected myosin molecules and subfragments of myosin. From top to bottom are: myosin, single-headed myosin, rods, HMM, HMM S-1, LMM, and HMM S-2. (Reprinted from Reference *36* by courtesy of Academic Press, Inc.)

misleading since HMM S-2 cannot be obtained in any appreciable yield
by tryptic digestion (41). It should also be emphasized that the gross size
and shape of HMM S-1 produced by either trypsin or papain are similar,
but the covalent breaks in the polypeptide chain depend on the specificity
of the enzyme. In regard to the relative merits of papain vs. trypsin in the
digestion of myosin, papain is clearly the enzyme of choice for preparing
a wide variety of subfragments. Only in the case of HMM is trypsin pre-
ferable, since papain cleaves the globules from the tail too readily to

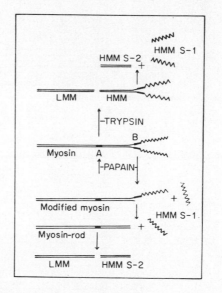

Fig. 12. Schematic representation of the mechanism of action of trypsin and papain on
myosin. (Reprinted from Reference 36 by courtesy of Academic Press, Inc.)

permit the isolation of HMM in appreciable yields. All the fragments
depicted in Fig. 12 have been isolated and purified, including the single-
headed or "modified myosin" (42). In addition, a single-headed version
of HMM has been recently prepared by splitting one globule off the tryptic HMM with low concentrations of papain. This modified HMM can be
purified by elution off DEAE-cellulose under the same conditions as HMM
(Fig. 14B) (42).

A typical papain digestion of myosin is illustrated in Fig. 13A. When
the digest is dialyzed against a low ionic strength, all the components that
contain the LMM regions are precipitated; specifically, undigested myosin,

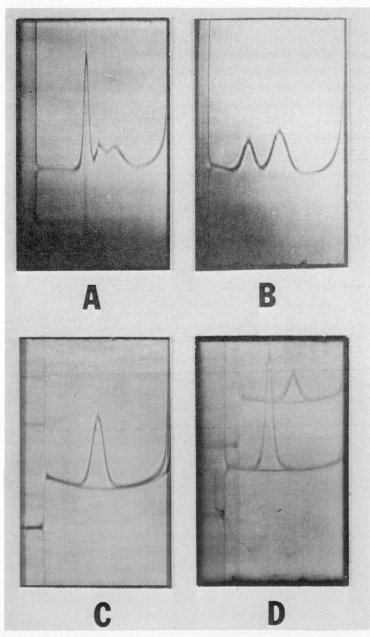

Fig. 13. Sedimentation patterns for the components of a myosin digest separated by ion-exchange chromatography (refer to Fig. 14). (A) Myosin digested with papain. (B) The water-soluble fraction of the digest in (A) consisting of HMM S-1 and HMM S-2. (C) HMM S-1 isolated by DEAE-cellulose chromatography. (D) HMM S-2 isolated by DEAE-cellulose chromatography. (Reprinted from Reference *36* by courtesy of Academic Press, Inc.)

modified myosin, rod, and LMM. The water-soluble subfragments remaining in the supernatant are HMM S-1 and HMM S-2 (Fig. 13B). These "subunits" are readily separated by salt elution from a DEAE-cellulose ion-exchange column (Fig. 14A). HMM S-1, the more positively charged fragment is eluted first, followed by HMM S-2. The latter is best detected

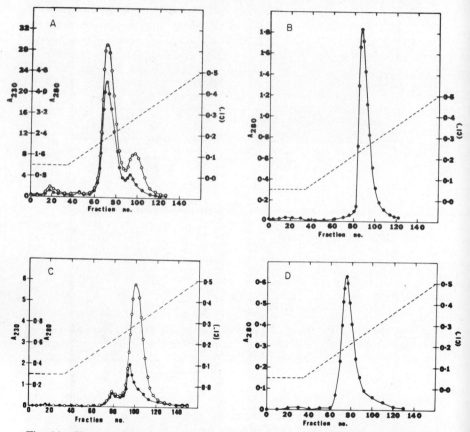

Fig. 14. Chromatography of myosin subfragments on DEAE-cellulose. (A) Water-soluble fraction from a papain digest (refer to Fig. 13B). (B) HMM prepared by brief tryptic digestion. (C) Rechromatography of fractions 90 to 120 (HMM S-2) from (A). (D) Rechromatography of fractions 60 to 80 (HMM S-1) from (A). (Reprinted from Reference 36 by courtesy of Academic Press, Inc.)

at 230 mμ because of its low aromatic residue content (Table IV). The concentrated, separated fractions are relatively homogeneous as judged by a single peak on DEAE-cellulose, and in the ultracentrifuge (Figs. 13C and D and 14C and D). The physico-chemical characteristics of these subfragments are given in Table I, their amino acid compositions in Table IV, and their dimensions in Table III. Despite their apparent homogeneity, the polypeptide chains of these subfragments have clearly undergone some hydrolysis, which becomes evident in strong denaturing solvents. For example, HMM S-1, which has a molecular weight of 115,000 in KCl–tris solvents, shows a range of sizes from 20,000 to 80,000 g/mole in sodium dodecyl sulfate (SDS) (Fig. 15).

In the papain digestion described above, a certain proportion of the water-insoluble fragments consists of rods. It is relatively simple to separate rods from myosin and single-headed myosin since alcohol irreversibly denatures the latter. However, there is no clear way to separate LMM from

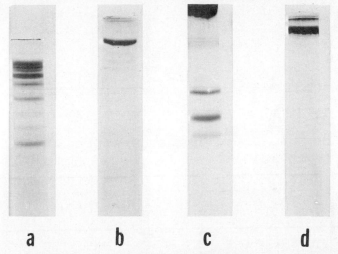

a   b   c   d

Fig. 15. Characterization of myosin and its subfragments by polyacrylamide gel electrophoresis in SDS. (a) HMM S-1 prepared by papain digestion. (b) Rod subfragment prepared by papain digestion. (c) Purified rabbit myosin. (d) Myosin from which the light chains have been removed by alkali treatment.

rods. The contamination by LMM can be minimized by limiting the amount of papain used in the digestion, but we were unable to entirely eliminate the presence of LMM. In order to prepare pure rods from myosin, we took advantage of the fact that aggregates of myosin have the HMM S-1 moieties on the surface of the filament and the rods tightly packed into the core of the filament; thus site(s) A is inaccessible to enzymic attack. By digesting myosin in the precipitated state with papain, it was possible to isolate relatively pure HMM S-1 from the supernatant fraction and to show that the 3 S component in the water-insoluble fraction was the rod portion of myosin. Depending on the degree of papain digestion, myosin and single-headed myosin were present along with rods, but they could easily be removed by alcohol treatment (Fig. 16). The properties of the rod are summarized in Tables I, III, and IV, and a field of rod particles shadowed

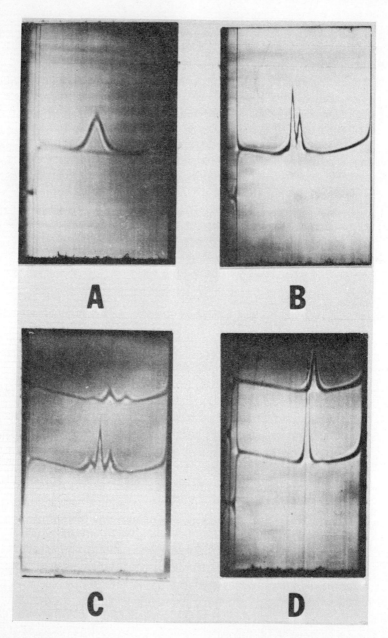

Fig. 16. Sedimentation patterns of subfragments produced from precipitated myosin with increasing concentrations of papain. (A) Supernatant of digestion or HMM S-1. (B) Water-insoluble fraction containing undigested myosin and single-headed myosin. (C) Same as (B) but also containing rods. (D) Rod subfragment of myosin. (Reprinted from Reference *36* by courtesy of Academic Press, Inc.)

Fig. 17. Isolated myosin rods by rotary shadow casting. (Reprinted from Reference *36* by courtesy of Academic Press, Inc.)

by platinum is shown in Fig. 17. The distribution of lengths for the rod gives an average value of about 1400 Å, which agrees well with the measurements of the rod portion of myosin (Fig. 8). Since we know the molecular weight of the rod from sedimentation equilibrium measurements, and the length from electron microscopy, the number of α-helical chains can be deduced by an approach analogous to that used for LMM. The result, not surprisingly, is two α-helical chains. It is interesting to note that the rod is sufficiently protected from papain proteolysis during the digestion to show a subunit molecular weight of 110,000 in SDS gels (Fig. 15).

The many studies on the proteolytic subfragments of myosin described above can best be summarized in a model for myosin (Fig. 18). Myosin appears to consist of two major polypeptide chains; each chain is α-helical

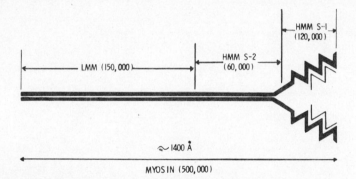

Fig. 18. Schematic representation of the myosin molecule. (Reprinted from Reference *36* by courtesy of Academic Press, Inc.)

in the rod region of the molecule and terminates in a globular conformation. The reasons for introducing an additional chain into each HMM S-1 globule are discussed in the next section.

D. Noncovalent Subunit Structure of Myosin

Physico-chemical and X-ray diffraction studies of LMM and HMM first introduced the idea that the rod portion of myosin contains two subunits (*10,16,33*). Little could be said about the globular region since the conformation was too complex to permit the simple type of analysis possible with α-helices. The problem had to remain unsolved until two globular subunits could be visualized directly by electron microscopy (*21*). Our studies on the subunit structure of myosin concerned themselves primarily with the native conformation of the molecule. At the same time, other investigators were examining the substructure of myosin in strong denaturing solvents, such as guanidine hydrochloride, in which the molecular weight of myosin was shown to drop to about 200,000 (*17,44*). Furthermore, it appeared from ultracentrifugal (*17,43*) and electrophoretic (*15*) analyses that myosin consisted of only a single type of polypeptide chain. This assumption, coupled to the significantly higher molecular weight of 600,000 for native myosin determined by these workers, led to a model of three similar polypeptide chains for the subunit composition of myosin (*17,43,44*).

Apart from the erroneous molecular weight for myosin, these early studies overlooked the presence of a small amount of low-molecular-weight protein in the myosin preparations. This material is easily missed since it accounts for only about 10% of the mass of myosin. It had been recognized for many years that myosin exposed to urea (*45*), heat denaturation (*46*), chemical modification (*47–49*), or alkaline conditions (*50–52*) liberated small amounts of low-molecular-weight protein. Since myosin is known to be difficult to purify, it was generally assumed that this minor component was a contaminant. However, despite the introduction of more elaborate purification procedures (*53,54*), the presence of this component persisted, and it became increasingly difficult to dismiss it as an "impurity." This noncovalently bound, low-molecular-weight fraction of myosin which we refer to as "light chains" can be isolated in many ways; almost all of the methods utilized the difference in solubility between the light and heavy chains. For example, after separating the two chains at pH 11.4, the heavy chains can be precipitated with 0.8 M potassium citrate while the majority of the light chains remain in solution (*52*). The supernatant is then dialyzed exhaustively against water and freeze-dried. Light chains prepared in this manner undoubtedly retain some of their native conformation since they react immunochemically with rabbit antisera to chicken myosin and chicken HMM S-1 (*55*). The isolated heavy chains aggregate too readily to be studied in the native state. However, in a dissociating solvent such as guanidine-hydrochloride they give a molecular weight of about 210,000 (*56,57*).

The light chains are heterogeneous both in charge, composition, and size; chicken and rabbit myosin give three major bands upon disc gel electrophoresis in tris–HCl at pH 9. These electrophoretic components can be separated into three peaks on DEAE-cellulose, with a phosphate gradient at pH 5.9 (Fig. 19). The amino acid analyses of the column-purified light chains show clear differences in composition, more so between the DTNB light chains and the A chains than between the A1 and A2 groups (Table V). "Fingerprints" of the chymotryptic digests of the separated chains provide further evidence that the A1 and A2 light chains are chemically related and differ from the DTNB chain. In particular,

each A-chain has one cysteine peptide per mole which is identical in
sequence for both A1 and A2:

Met-Ala-Gly-Gln-Glu-Asp-Ser-Asn-Gly-CMC-Ile-Asn-Tyr,

where CMC is carboxymethyl cysteine (58,59).

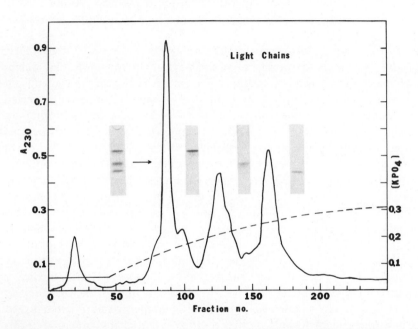

Fig. 19. Ion-exchange chromatography of performic acid-oxidized light chains prepared
by guanidine hydrochloride denaturation of myosin. The inserts are polyacrylamide
disc gel electrophoresis (tris, pH 9.5) patterns of total and separated light chain fractions.

The DTNB light chain has two thiol peptides per mole having the
sequences (58):

1. Thr-Thr-Gln-CMC-Asp-Arg-Phe
2. Lys-Asn-Ile-CMC-Tyr

On the basis of the amino acid composition and the peptide maps, minimum
molecular weights of 21,000, 17,000, and 19,000 can be calculated for the
A1, A2, and DTNB light chains, respectively. These values lie in the range
of sizes determined for light chains by high-speed equilibrium ultracentrifugation (47,56). Another estimate of the molecular weight of the individual
light chains can be obtained from SDS gels (Fig. 15). The lowest molecular

TABLE V

Amino Acid Composition of the
Light Chains of Myosin[a,b]

Amino acid	Al	DTNB	A2
Lys	21.0	15.9	11.6
His	2.0	1.1	1.9
Arg	4.2	6.1	4.3
Cmc	0.96	1.9	0.93
Asp	19.8	23.1	18.3
Thr	8.0	10.3	7.1
Ser	8.8	5.8	8.2
Glu	29.2	23.8	24.4
Pro	12.1	6.1	3.7
Gly	12.0	13.0	12.0
Ala	22.8	14.5	12.5
Val	10.6	8.9	10.2
Met	6.0	6.4	5.6
Ile	9.0	9.8	7.0
Leu	13.8	9.5	12.8
Tyr	3.0	2.0	3.0
Phe	8.1	13.0	8.5

[a] Weeds and Lowey, unpublished data (62).
[b] Expressed as moles per mole.

weight bands are 25,000, 18,000 and 16,000 for A1, DTNB and A2, respectively.

The nomenclature we have adopted for the light chains originates in the observation that myosin reacted with 5,5'-dithiobis(2-nitrobenzoic acid) (DTNB) liberates a light chain fraction which is approximately half of the total light chain material in myosin (57). Furthermore, upon regeneration of the thiol groups in the DTNB-treated myosin, the full ATPase activity of myosin is recovered (*57*). Consequently, the DTNB light chains are not essential for enzymic activity and their functional role in myosin—if any—is not understood at present. The remaining light chains in myosin can only be removed under denaturing conditions (for example, at pH 11.4, from which the name alkali (A) light chains is derived) with concomitant loss of ATPase activity and aggregation of the heavy chains. As described above, the DTNB light chains move as one band electrophoretically and contain two thiol sequences per mole. The alkali light chains show two electrophoretic bands which can be separated on DEAE into two fractions, A1 and A2, each containing a single thiol sequence per mole.

Attempts to establish the stoichiometry of the light chains in myosin have usually relied upon knowing the weight percent of light chains in a given weight of myosin and the molecular weight of the light chains and myosin. These figures were derived from measurements by ultracentrifugation (*51,56*), light scattering, and direct preparative isolation procedures (*60*). The values for the weight percent of light chains have ranged from 5% (*52*) to 15% (*47*) with the general consensus of opinion about 10%. A molecular weight of 480,000 for myosin and an average weight of 20,000 for the light chains lead to the conclusion that myosin contains 2 moles of light chains per mole myosin (*56*). This approach does not attempt to differentiate among the different kinds of light chains in myosin.

The property of a single cysteine residue per mole of A1 or A2, and only two cysteine residues per mole of DTNB light chain, presents a unique opportunity to determine the stoichiometry of the three classes of light chains in myosin. Because the "radioisotope dilution technique" is, in our opinion, the most accurate means of establishing the amounts of the different light chains in myosin, it is described in some detail below (*61,62*).

Purified light chains (either A1, A2, or DTNB) were reacted with radioactive iodoacetic-^{14}C acid to label the thiol group, and a given number of counts was added to a known weight of completely carboxymethylated cold myosin. Both the control of pure hot light chains and the mixture of hot light chains and cold myosin were digested with chymotrypsin and simultaneously subjected to high-voltage paper ionophoresis. The radioactive cysteine peptides were eluted from the paper and the number of counts and micromoles of peptide were determined for both the control and the mixture. From the extent of dilution of the counts in the mixture, the moles of light chains per mole of myosin (p) could be determined:

$$p = nM/yw \ (y/x-1)$$

where n = number of counts from ^{14}C-labeled light chains, w = weight of unlabeled myosin, y = specific activity of peptides from pure ^{14}C-labeled light chains, x = the specific activity of peptides in the mixture, and specific activity is expressed in counts per mole of peptide. The results of several such experiments yielded 1.7 moles DTNB light chains per mole myosin and 2.0 moles of alkali light chains (the values for A1 were essentially the same as for A2). Similar radioisotope dilution experiments on HMM S-1 gave 1 mole of alkali light chains per mole HMM S-1. Less than 0.4 moles of DTNB light-chain peptides could be detected in this subfragment; apparently, papain degrades the DTNB light chains (62). The results on HMM S-1 confirm the conclusion that the DTNB light chains are not necessary for enzymatic activity.

The schematic model in Fig. 18 implies that myosin has a twofold axis of symmetry parallel to the long axis of the molecule. This supposes that the two halves of myosin are either identical or very similar in composition. At present, there is little chemical evidence to support this hypothesis. The heterogeneity of the alkali light chains argues against two identical halves, unless two populations of myosin molecules are assumed in which each molecule has two identical light chains. It is of course conceivable that any given muscle may contain several closely related forms of myosin or isoenzymes. The major evidence in support of two similar heavy chains comes from the sequence studies on the cysteinyl peptides of myosin (63). There are 8–9 moles of cysteic acid per 10^5 g myosin or 40–45 moles cysteine residues per mole myosin. If myosin contains two identical classes of polypeptide chains, one should expect to find no more than 20 to 22 unique thiol sequences. To date, 21 different thiol peptides have been sequenced (63). Of these, three can be assigned to the light chains (58), three to the LMM region of myosin (64), and two to the HMM S-2 portion of the rod (63). The remaining sulfhydryl peptides must be located in the HMM S-1 portion of the myosin molecule.

As the thiol peptides account for no more than 5–10% of the sequences in myosin, it is obvious that a great deal of work remains to be done before much can be said about the primary structure of myosin. Recently, it has been shown that a subfragment resembling LMM can be produced from myosin by a limited cleavage of the methionine residues with cyanogen bromide (CNBr) (65). Since this fragment is essentially intact, it should prove a useful starting point for the sequence determination of myosin.

In principle, it should be possible to determine the number of polypeptide chains in myosin by characterizing its C- and N-terminal amino acid residues. In practice, the detection of 2 moles of an amino acid per 500,000 g/mole is sufficiently difficult to make this experiment far from routine. At present, 2 moles of N-acetyl serine have been identified on the heavy chain in the HMM region (66,67). Therefore, it appears that the N-terminal end of the major polypeptide chain of myosin is initiated in the globular head of the molecule. This finding is confirmed by the CNBr studies of myosin; CNBr cleavage of methionyl peptides results in the formation of C-terminal homoserine; the absence of this residue in the LMM fragment implies that LMM must be the C-terminus of myosin (*65*). In regard to the identification of the C-terminal residues in myosin, the only definitive finding is that the alkali light chains contain a C-terminal isoleucine residue (*51,59*).

Before concluding our description of the primary structure of myosin, it is of interest to note that several unusual amino acid residues have been found in myosin, among them 3-methyl histidine (*68*), ε-*N*-monomethyl-lysine, and ε-*N*-trimethyllysine (*69,70*). Methylated lysines have been previously reported in flagella of some species (*71*), histones (*72*), and cytochrome c (*73*). Whether these methylated residues have any particular functional significance in myosin remains to be shown.

III. ACTIVE SITES IN MYOSIN

Muscle contracts by means of an interaction between the myosin- and actin-containing filaments. In vitro interactions between myosin and actin are in part dependent upon the ionic milieu and the concentration of ATP. In solutions of high ionic strength ($>$ 0.2 M KCl), myosin and F-actin form a soluble complex which can be dissociated by ATP. This dissociation is most effective in the presence of magnesium ion (the most abundant divalent cation in the myofibril), which enhances the binding of ATP and thereby inhibits the ATPase activity of myosin. If the ionic strength is lowered to physiological levels ($<$ 0.15 M), actin is partially recombined with myosin–Mg-ATP, and the enzymic activity is restored. As the supply of ATP is depleted by hydrolysis, actin and myosin become more fully complexed, the ATPase activity rises, and precipitation and superprecipitation occur (*4*). The final ATP-deficient actomyosin gel can be thought of as analogous to the condition of rigor in muscles lacking ATP.

The most likely interpretation of these observations is that each myosin molecule has distinct and separate sites for actin and nucleotide binding (*4,74*). However, the interdependence of these sites is illustrated by the influence actin has on the hydrolytic activity and the ability of ATP to dissociate actin from myosin. In the following section we summarize some experiments aimed at establishing the stoichiometry of these sites.

A. Nucleotide and Actin Binding

The binding of ATP can be measured by inhibiting the hydrolytic activity of the molecule. In a solvent containing 1.5 M NaCl and 1 mM in Mg^{2+}, the rate of hydrolysis is sufficiently reduced to permit such determinations by gel filtration chromatography (*75*). The results of these and related experiments indicate two ATPase sites per mole of myosin (*75,76*). By using the competitive inhibitor, Mg-ADP, we have measured the binding of nucleotides to myosin by the classic technique of equilibrium dialysis (Fig. 20). Within the experimental error, the Scatchard plots appear linear, suggesting the existence of two equivalent binding sites per mole protein, $\bar{\nu}$. The considerable scatter in these data is partly attributable to the incomplete removal of the contaminating AMP deaminase and myokinase in some preparations, and slight irreversible inactivation of the myosin in other preparations (*77*). The addition of the chelating agent, EDTA, or

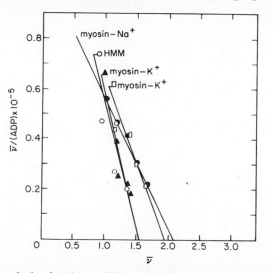

Fig. 20. Scatchard plot for the equilibrium binding of ADP to myosin and HMM. (Reprinted from Reference *77* by courtesy of the American Chemical Society.)

the substitution of calcium for magnesium, or the presence of actin, markedly reduces the binding of nucleotide to myosin (*77*). As these substances all share the property of being activators of myosin ATPase, it is clear that activation is accompanied by a reduction in binding at the hydrolytic site.

It is generally accepted that both HMM S-1 subfragments possess binding sites for actin (*78*). This property has, in fact, been used to isolate the globular fragments from the papain digestion mixture by complexing the HMM S-1 with F-actin (*79*). Whether 2 moles of actin can bind simultaneously to the double-headed HMM is less clear; interpretations of kinetic data involving ATP hydrolysis, have favored either a 1 : 1 complex (*80*) or, in some cases, a 1 : 2 complex of moles HMM to moles actin (*81*). Direct measurement of the binding of F-actin to HMM by centrifuging

the complex and measuring the concentration of free HMM in the supernatant yielded 1 mole of actin monomer per mole HMM (78). The obvious
implication of these data is that only one globule of HMM is available
for attachment to actin monomer at any given time. This type of constraint
could conceivably promote the necessary force to slide the thin filament
relative to the thick filament. However, our own experiments do not
support such a model; we have determined the binding of HMM (and
HMM S-1) to F-actin by methods similar to those used previously (78)
but find, contrary to earlier reports, that 2 moles of actin monomer can
bind to 1 mole of HMM (Fig. 21) (42). Measurement of the maximum

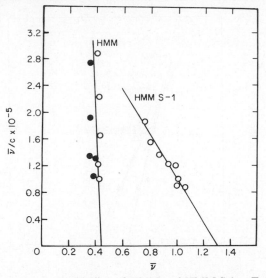

Fig. 21. Scatchard plot for the binding of HMM and HMM S-1 to F-actin, where $\bar{\nu}$ is
the average number of moles of HMM (or HMM S-1) bound per mole of actin monomer.
[S. S. Margossian and S. Lowe, unpublished data (42).]

actin-activated ATPase indicates that HMM splits ATP at twice the rate
of HMM S-1 (Fig. 22); furthermore, the association between HMM and
actin in the presence of ATP is several orders of magnitude weaker than
in the absence of nucleotide (42). Thus it appears that both globules of
HMM participate in the binding to F-actin. The close resemblance between
electron micrographs of negatively stained actin filaments "decorated"
with HMM S-1 and those "decorated" with HMM (82) support this conclusion. The globular subfragments, whether free or joined in HMM,
must bind to the actin units in quite a regular, ordered arrangement to
give the distinctive "arrowhead" appearance of these structures. This
unusual feature signifies the underlying structural polarity of the thin
filaments, a property that would not be evident from the actin filaments

alone (Fig. 23A and B). Thus both the thin and the thick filaments (see Figs. 3 and 4) have the directionality required by the sliding filament mechanism of contraction.

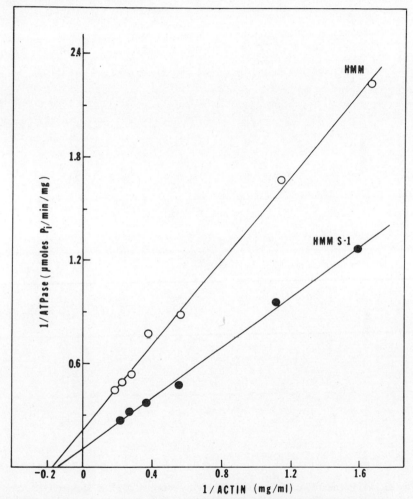

Fig. 22. Actin-activated Mg-ATPase activity of HMM S-1 and HMM. [S. S. Margossian and S. Lowey, unpublished data (42).]

B. Role of Light Chains

Ever since the discovery of a low-molecular-weight, noncovalently bonded protein fraction in myosin, the problem of the relation of the light chains to the major polypeptide chains in myosin has existed. The probability of the light chains being a tightly bound impurity seems remote; all

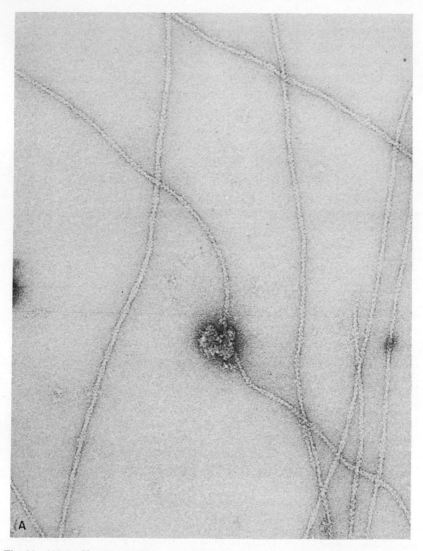

Fig. 23. (A) Purified F-actin filaments. (Photo by courtesy of H. E. Huxley.) (B) Actin filaments "decorated" with HMM S-1. (Reprinted from P. B. Moore, H. E. Huxley, and D. J. De Rosier (*82*) by courtesy of Academic Press, Inc.)

preparations of myosin so far examined contain some light chains, no matter what the source or the purification procedure. Even HMM S-1 that has undergone proteolytic degradation and subsequent DEAE chromatography contains 1 mole of light chains essential to its activity. However, the finding that reaction of myosin with DTNB liberates half of the light-chain material in rabbit skeletal myosin, without any impair-

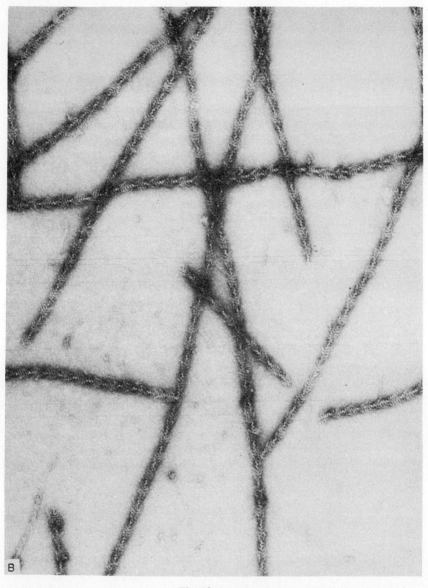

Fig. 23. (*cont.*).

ment of activity, does raise the possibility that some reagent may yet be found that releases the remaining light chains without affecting the ATPase activity. Barring this event, one may consider the origin and location of the 2 moles of "alkali" light chains in myosin.

The liberation of alkali light chains is always accompanied by the simultaneous denaturation of myosin. The addition of ATP simultaneously slows the denaturation process and the release of light chains, leading to the inference that the light chains are perhaps located in the region of the hydrolytic site (*83,138*). This structure would be similar to that found in antibodies, in which the antigen combining site is composed of portions of the light and heavy chains of the γ-globulin molecule (*23*). Stronger evidence for the presence of light chains in the vicinity of the catalytic site comes from the demonstration that an analogue of ATP, which can act as a substrate for myosin, is bound to the light chains under conditions where hydrolysis is inhibited (*135*).

In order to prove unequivocally that a subunit is an integral part of a protein's function, it is necessary to be able to remove and restore the subunit with a parallel loss and gain of activity. Such experiments have been reported for myosin, although sometimes with conflicting results (*52,56,60,136*). Nevertheless, it has been shown that when light chains are separated from heavy chains in 4 M LiCl, either by gel filtration chromatography (*83*) or selective precipitation (*136*), no activity is detectable in the isolated components. However, recombination of the chains prior to desalting results in a significant recovery of activity (*83,136*). Thus present data support the concept that light chains serve some functional role in the active site of myosin.

It is now well established that myosin is synthesized on large polysomes containing about 50 to 60 ribosomes. This is in the size range that would be predicted for monocistronic messenger RNA coding for a 200,000 g/mole polypeptide chain (*84*). The light chains could be synthesized independently on much smaller polyribosomes or, alternatively, they could constitute part of the heavy chain. It is conceivable that myosin exists originally as a single chain, similar to other precursor proteins such as proinsulin (*85*), chymotrypsinogen (*86*), or poliovirus coat protein (*87*), and is then activated through cleavage of a few peptide bonds by an intracellular proteolytic enzyme. Recent reports in the literature tend to support the first alternative, namely, that light chains are synthesized on a different polysome fraction from the heavy chains (*137*). Radioactive myosin was synthesized in a cell-free system containing one of several classes of polysomes of different sizes. After addition of carrier myosin, the product was purified by ion-exchange chromatography and the subunits analyzed by SDS gel electrophoresis. All the radioactivity was in the heavy chain band when the large polysome fraction (50–60 ribosomes) was used in the incubation mixture, whereas counts appeared in the light chain bands when the smaller polysome fractions (5–9 ribosomes) or undissociated polysomes were used. These results suggest that two classes of polysomes synthesize the light and heavy chains which then assemble to form the intact myosin molecule (*137*).

The role of light chains as regulatory proteins is an attractive postulate. It has been shown that the level of ATPase activity can be correlated with the speed of a particular muscle (*88*). Since none of the muscles we use for our myosin preparations is entirely pure (*89*), some of the heterogeneity encountered in the isolated light chains may be attributable to a mixture of different muscles. Inspection of a number of different muscle types and species has shown distinct differences in the light chain composition. Myosin with a relatively high ATPase activity from the muscles of chicken breast and leg, rabbit leg and back and M. Latissimus dorsi posterior (a predominantly white muscle from the wing of the chicken) has three light chains of molecular weight 16,000, 18,000 and 25,000 (see Fig. 15c). Myosin of lower activity from the muscles of rabbit and chicken heart and M. Latissimus dorsi anterior (a predominantly red muscle) shows only two light chain components of molecular weight about 20,000 and 27,000 (*139*). The significance of these findings in relation to the activity of myosin and the speed of the muscle remains to be determined.

IV. ASSEMBLY OF MYOSIN AND ITS SUBFRAGMENTS INTO ORDERED STATES

Thus far we have discussed the myosin molecule as a discrete entity composed of several subunits which interact to form a functional unit. Although this is a necessary step in comprehending how myosin works, at some point myosin must be placed in its native environment, namely, as one molecule among several hundred like molecules, all organized in a highly specific, cooperative manner to form a functional myosin filament.

Myosin is uniquely designed to aggregate into a filament; the variation in solubility among its different subfragments is directly related to the functional role of each region of the myosin molecule. The LMM region is the least soluble, hence is primarily responsible for anchoring myosin in the core of the thick filament. HMM S-2 is much more soluble than LMM and, correspondingly, has weaker interactions with the backbone of the myosin filament. HMM S-1 is highly water-soluble and its interactions are probably entirely limited to the actin-containing thin filament.

Myosin appears to aggregate by a self-assembly process; that is, all the information required to build a thick filament is contained in the myosin molecule, and it is unlikely that another protein(s) participates in the assembly. An appropriate aqueous environment is, however, essential to the assembly process, and factors such as pH and ionic strength are critical in regulating the growth of the thick filament.

Studies relating the aggregation of myosin to its molecular structure and to the environmental conditions are described below.

A. Aggregation of Myosin

One of the major causes for the difficulty in determining an accurate molecular weight for myosin has been the strong tendency for myosin to aggregate. This property of myosin is evident even in high concentrations of salt (0.5 M KCl), in which electrostatic interactions would normally be completely eliminated. Phosphate ions help to reduce these interactions (9), but detailed analyses of molecular weight data suggest that a small amount of dimer remains in rapid reversible equilibrium with monomeric myosin (91). As the ionic strength is lowered, high-molecular-weight components can be detected both by analytical ultracentrifugation and by electron microscopy (6,92,93). Below pH 7.3 heterogeneous particles with average sedimentation rates of 1100, 330, and 180 S are encountered, in addition to monomeric myosin (93). These filaments bear some resemblance to native thick filaments, although the lengths and diameters are quite variable and very sensitive to the total amount of protein, salt, and pH.

Above pH 8 the sedimentation profile changes dramatically and only two boundaries are now observed; the slower one corresponds to monomer and the other is a well-defined aggregate of myosin with a sedimentation coefficient of 150 S (92,93). By adjustment of solvent conditions to pH 8.3, 0.137 M KCl, the amount of monomeric myosin is reduced to a minimum and the physical characteristics of the polymer can be determined. (The discussion that follows is taken largely from the work of Josephs and Harrington, 93–95.) Electron micrographs of the polymer show a fairly narrow size distribution with the majority of the particles having a length between 5600 and 7500 Å. The particles have an average diameter of 100–150 Å and look very similar to the synthetic filaments of Fig. 3 which, except for the shorter length, display all the topological features of the native thick filaments (Fig. 1). Estimates of the molecular weight from viscosity and sedimentation measurements suggest a value of about 50×10^6 g/mole or about 100 myosin molecules per filament (93).

The self-association reactions described above are most simply interpreted in terms of a rapidly reversible equilibrium between monomer (M) and polymer (P),

$$n\text{M} \overset{K}{\rightleftharpoons} \text{P}$$

where $n = 100$ and the equilibrium constant (K) can be calculated from measurements of the concentration of slow component in the sedimentation velocity studies, in this case equal to the monomer concentration (94). The value of K calculated for pH 8.3, 0.137 M KCl is about 10^{50} (in concentration units of grams per deciliter). As might be anticipated from more qualitative observations on filament formation, K is very dependent on pH and KCl; it can be shown that about 11 moles of KCl are liberated and about 0.7 moles of hydrogen ions are absorbed per myosin molecule in forming a polymer (94). The standard free-energy change for the reaction is simply $\Delta F = -RT \ln K'$ (where K' is independent of salt and pH). The equilibrium is not affected by temperature; therefore the enthalpy change is zero, and the entropy change reduces to $\Delta S = -\Delta F/T$. A summary of the thermodynamic constants for the association of myosin is given in Table VI.

TABLE VI

Thermodynamic Parameters for the Formation
of Myosin Filaments[a, b]

	$\Delta F,$[c] kcal	$\Delta H,$ kcal	$\Delta S,$ eu	$\Delta V,$ ml	Effect of electrolytes
Per monomer	-2.2	0	$+7.8$	$+384$	Destabilizes
Per polymer	-180	0	650	3.2×10^4	Destabilizes

[a] Reprinted from W. H. Harrington and R. Josephs (95) by courtesy of Academic Press, Inc.
[b] For the reaction n monomer $\rightleftharpoons$ polymer, where $n = 83$.
[c] At 278°K.

The most interesting effect on the equilibrium constant is produced by varying the pressure. The data referred to above were all obtained at relatively low rotor speeds of 9000–11000 rpm where the pressure effect is negligible. However, when the rotor velocity is increased to 30,000 rpm, the pressure across the ultracentrifuge cell varies from 10 to almost 70 atm (at 59,780 rpm the pressure across the cell is 330 atm). At 50 atm, the value of log K is approximately one-half that at 1 atm, indicating a dissociation of the polymer to monomer under the influence of pressure (94). From the slope of log $K(x)$ plotted against $P(x)$ [where the pressure $P(x)$ is a function of speed ω and radial position x in the cell, $P(x) = \omega^2 \rho/2(x^2 - x_0^z)$], the volume change ΔV in forming 1 mole of polymer from n moles of monomer can be determined (Table VI).

What can these thermodynamic parameters tell us about the formation of the thick filaments? One of the principal conclusions from this work is the important role played by water in stabilizing the structure of the myosin filament (95). The myosin molecule has a high percentage of polar residues which presumably lie on the surface of the protein and bind to dipolar water molecules. As myosin aggregates, new protein–protein interactions replace the former protein–solvent interactions and water is released to the extent of about 400 ml/mole myosin. Such volume changes are quite common in associating systems; for example, TMV is dehydrated by about 110 ml per subunit (96) and each G-actin loses 68 ml of water in forming F-actin (97). Thus the reduced rotational and translational freedom of the molecule in the polymer may be more than compensated for by the increased disorder in the released water leading to a net increase in entropy (Table VI). The magnitude of the thermodynamic potentials reported for this equilibrium reaction are characteristic of interactions involving hydrophobic or ionic bonding (98). For both types of bonds, the transfer of an amino acid residue from an aqueous to a nonaqueous environment results in a loss of structured water, with an accompanying gain in entropy. The enthalpy change for such transfers is close to zero, ånd the volume change is about 10–20 ml per bond (98). Since the myosin filament is obviously sensitive to salt, ionic bonds are probably more prevalent in stabilizing the polymer than hydrophobic bonds although the latter undoubtedly contribute to some degree. A rough estimate of the number of bonding sites in myosin can be made by dividing the total volume change ($\Delta V \simeq$ 400 ml) by the volume change per bond ($\Delta V \simeq 15$ ml). This calculation predicts that about 25 ionic (or hydrophobic) residues, that is, less than 1 % of the residues in myosin, are required for filament growth (95). Clearly, with so few sites involved, molecular packing within the filament may be remarkably sensitive to the local environment in the cell.

B. Aggregation of the Subfragments

A thermodynamic description of the monomer $\rightleftharpoons$ polymer equilibrium provides a general picture of the forces influencing the polymerization reaction. These studies do not, however, provide any information on specific bonding regions in the molecule or on the molecular arrangement in the filament. Such insights can only be supplied by application of the more direct methods of electron microscopy and X-ray diffraction. Up to the present, electron microscopy has proved of limited value in regard to the fine structure of the thick filament; the HMM S-1 globules lying at the surface are probably distorted by the negative stain, and the resulting image is rather obscure. The application of new methods in image reconstruction (99) to the thick filament may yield more detailed information in the future.

In this section we approach the problem of molecular interactions by examining the ordered aggregates formed by the helical segments of myosin, namely, the rod, LMM, and HMM S-2. Presumably, many of the bonding sites used in the aggregation of the subfragments occur in the native filament. That this is indeed the case, can be shown by a comparison of the periodicities in the paracrystalline tactoids with those found by X-ray diffraction of whole muscle.

1. LMM

When LMM is dialyzed against a low-ionic-strength buffer, the subfragment, similar to the parent molecule, is precipitated from solution. Unlike myosin, LMM tends to form tactoids and open lattices rather than bipolar filaments (Fig. 24). The most commonly occurring tactoid has an axial period of 430 Å (30), and a band pattern distinguished by dark staining regions alternating with narrow light bands (6). This pattern can be interpreted in terms of molecules staggered by 430 Å, with the light band attributable to a molecular overlap (l) or to an exclusion of stain from one end of the molecule ($l = 0$). The molecular length of LMM from such a model would be $n(430 \text{ Å}) + l$, where $n = 2$ gives the best agreement with length determinations by light scattering and electron microscopy (6,10). Tactoids with a 145-Å repeat are frequently seen in these preparations (100). These could arise from a superposition of the 430-Å arrays shifted by one-third of a period, or they may represent an association of molecules staggered by 145 Å.

Other forms of LMM, present in smaller amounts, include square nets with sides approximately 400 Å in length (41) and hexagonal lattices with sides about 600 Å long (6,41). In most instances these lattices are probably two-dimensional structures, although the existence of limited three-dimensional order or microcrystals seems likely (Fig. 24) (140).

LMM can be precipitated from tris, pH 8 buffers, in which it is soluble, by divalent cations. Despite the unusual nonphysiological solvent conditions, the bonding relations leading to a 145- and 430-Å repeat persist, and tactoids displaying both these periods are observed. A "segment" aggregate, composed of LMM molecules aligned in a lateral direction, is also formed by divalent cations. The length of LMM measured from these segments is about 900 Å (100).

2. HMM S-2

The rodlike tail of HMM is the most soluble member of the highly α-helical fibrous proteins. Only by lowering the pH to the isoelectric region (< pH 5) can precipitation be induced. Unlike the equally soluble HMM S-1 and HMM, which form amorphous precipitates, HMM S-2 forms tactoids with a 145 Å periodicity (Fig. 25) (41). At present, no other ordered aggregate of HMM S-2 has been seen, irrespective of the solvent.

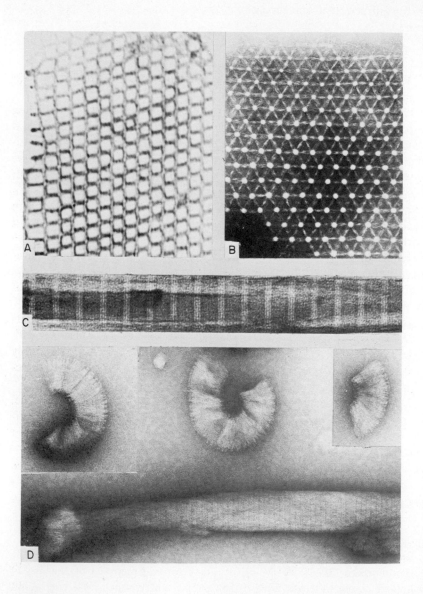

Fig. 24. Polymorphic forms of LMM. (A) 400-Å square net. (B) 600-Å hexagonal array. (C) Periodic tactoid with 430-Å spacing. (D) Periodic tactoid with 145-Å spacing and segments about 900-Å wide. (Some of these forms are reprinted from Reference *41* and Reference *100* by courtesy of Academic Press, Inc.)

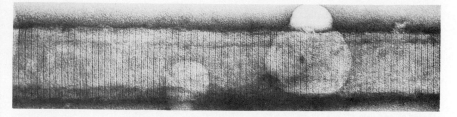

Fig. 25. Paracrystalline tactoid of HMM S-2 at pH 4.5. (Reprinted from Reference *41* by courtesy of Academic Press, Inc.)

3. Rod

As the rod portion of myosin probably contains most of the sites participating in the assembly of myosin, ordered aggregates of this subfragment hold great promise for unraveling the structure of the thick filament. Unfortunately, the tactoids formed at low ionic strength do not have a particularly informative banding pattern and show only the ubiquitous 145-Å repeat. A more promising structure is the segment aggregate formed by divalent cations (*100*). These segments often appear as flat ribbons up to 2 μ long and about 1800 Å wide. The bipolar nature of the aggregate is evident from the symmetrical distribution of the stain (Fig. 26A). A darkly staining "fringe" on either side encloses a lightly stained region about 1300 Å wide. Assuming that the accumulation of stain at the edges arises from a molecular displacement of 145 Å, a length of 1450 Å can be deduced for the rod. This value is in good agreement with that calculated for a two chain α-helical molecule about 220,000 g/mole (*100*). By a slight alteration of solvent conditions, the rod will form a second type of segment aggregate with a molecular overlap of 900 Å and fringes 600 Å in width (Fig. 26C) (*141*). This latter type of segment can also be formed by myosin (Fig. 26B) (*141*). The significance of the "segment" aggregate for the structure of the thick filament is discussed in the following section.

C. Relation of the in Vitro Structures to the Native Thick Filaments

Much of our knowledge concerning the thick filament comes from the low-angle X-ray diffraction studies of relaxed and contracted live muscle (*101*). Together with earlier electron microscopy on sectioned muscle (*102*), these studies show that the myosin molecules in the filament are arranged in the form of a "6/2 helix" with a pitch of 429 Å and a subunit repeat of 143 Å. In other words, the "cross bridges" projecting from the filament are arranged in pairs related by a twofold axis; each pair must be rotated 120° and translated 143 Å to generate the helix (Fig. 27). The cross bridges

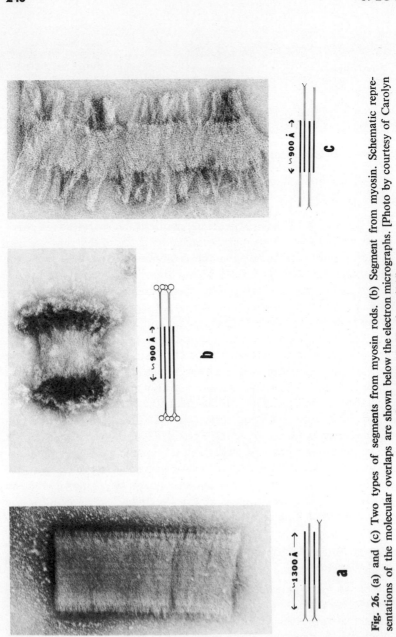

Fig. 26. (a) and (c) Two types of segments from myosin rods. (b) Segment from myosin. Schematic representations of the molecular overlaps are shown below the electron micrographs. [Photo by courtesy of Carolyn Cohen and R. G. Harrison (141).

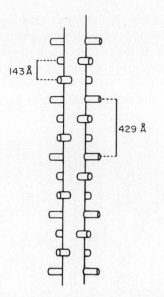

143 Å

429 Å

Fig. 27. Schematic diagram of a myosin filament showing arrangement of cross bridges on a 6/2 helix. (Reprinted from H. E. Huxley and W. Brown (*101*) by courtesy of Academic Press, Inc.)

as we now understand them probably consist of HMM S-1 and a portion of HMM S-2, the extent of the latter's contribution depending on the state of the muscle. In resting muscle the HMM S-2 component of myosin may be visualized as lying, for the most part, parallel to the axis of the filament with the HMM S-1 globules positioned on the surface. Evidence in support of this structure is as follows:

1. The low-angle X-ray diffraction pattern of frog sartorius muscle in the relaxed state is distinguished by a series of meridional and off-meridional reflections arising from the components of the thin and thick filaments. The variation in the X-ray intensity along the layer lines can be accounted for by a model consisting of a helical arrangement of "scattering centers" or subunits placed at a given distance from the axis of the filament. The best fit with the observed intensity was obtained by letting the center of mass of the cross bridges extend to a radius of about 130 Å from the axis of the filament (the radius of the backbone of the myosin filament is about 75 Å from electron microscopy). The computed transforms of models in which the bridges extend all the way to the surface of the actin filaments, that is, to a radius of about 190 Å, do not correspond as well to the observed patterns (*101*). Since the dimensions of HMM S-1 lie in the range

70–90 Å (Table III), these results imply that most of the HMM S-2 is interacting with the backbone of the thick filament.

2. The low-angle, equatorial X-ray reflections from striated muscles arise from the hexagonal arrangements of the thin and thick filaments in the myofibril (Fig. 28). The relative intensities of these reflections are related to the distribution of mass associated with the filaments. A com-

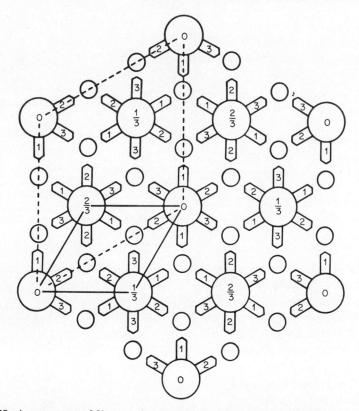

Fig. 28. Arrangement of filaments in a hexagonal lattice. (Reprinted from H. E. Huxley and W. Brown (*101*) by courtesy of Academic Press, Inc.)

parison of the X-ray patterns obtained from relaxed muscles with those from muscles in rigor shows that a considerable amount of protein passes from the myosin filaments to the actin filaments in going from the one state to the other (*103*). This transfer of material is also evident in electron micrographs of cross sections of muscles fixed under similar conditions (*103*). In these experiments, as in those described previously, the conclusion is that the HMM S-1 globules do not touch or interact with the actin filaments until the muscle contracts or goes into a state of rigor.

3. The greater solubility of HMM S-2 as compared to LMM implies that interactions between HMM S-2 and neighboring molecules are of a much weaker nature than those occurring between LMM molecules. That these interactions can and do exist, however, is shown by the ability of HMM S-2 to form tactoids with a 145-Å periodicity. This repeat, a property shared by all the helical subfragments of myosin, and a prominent feature of the X-ray pattern of living muscle, must represent an essential mode of interaction in the myosin filament. Direct evidence for an interaction between HMM S-2 and LMM is provided by the rod segment. The bipolarity of the segment indicates that the LMM portion of the rod packs well with the HMM S-2 portion of the oppositely oriented array. Furthermore, the lack of polar segments suggests that HMM S-2 binds preferentially to LMM rather than to another HMM S-2 fragment. LMM alone can form short segments and a variety of other ordered structures; HMM S-2 is very limited in its interactions. The greater tendency for LMM to form stable aggregates is undoubtedly related to its role in maintaining the organization of the thick filament. In contrast, the function of the less stable HMM S-2 appears to be to position the enzymically active globules at an appropriate distance from the actin filament (*41,101,104*).

It is a striking fact that the two periods displayed by the subfragments, namely, the 145- and 430-Å repeats, are also present in X-ray diagrams of living muscle. As noted above, the 145-Å period probably represents a side-to-side stagger of the molecules which is fundamental to the packing of the thick filament. The relation of the 430-Å period in the LMM tactoid to the thick filament is less clear. If the banding pattern of the LMM tactoid is explained by an overlap of LMM molecules, then "gaps" must simultaneously exist in the structure. Head-to-tail bonding is only possible when the length of the molecule is an integral multiple of 430 Å. These observations suggest that gaps may exist in the arrangement of the myosin molecules in the thick filament (*105*). A very convincing case has been made for the existence of gaps and "overlaps" in the structure of the paramyosin thick filament of molluscan muscles (*106,142*). This mode of packing may be common to the thick filaments of many species of muscle. Since myosin has a globular head which must be positioned on the surface of the thick filament, strict head-to-tail packing of myosin molecules is obviously not possible, and some tilting or bending of the molecule relative to the axis of the filament must occur.

The molecular packing discussed above pertains to the polar portion of the thick filament. Information concerning the packing in the bare zone can also be derived from the studies on the in vitro aggregates. The dihedral symmetry of the bipolar rod segment is characteristic of the native and synthetic myosin filaments. It is possible that the packing relations in the segments are related to the packing in the bare zone of the myosin filaments. If molecules 1450 Å long are positioned on a surface lattice with 6/2 symmetry and an axial repeat of 435 Å, molecular overlaps of 1300 Å and 870 Å are generated in the center of the bipolar array (141). The distance between the globular regions marking the boundaries of the bare zone (the so-called pseudo-H-zone) has been reported to be about 2000 Å. The minimum length of the bipolar segment is about 1800 Å. These observations suggest that filament growth may be initiated by certain dimer relations between the myosin molecules. Further growth requires a different set of interactions of the type described above. The myosin filament is thus a rather unusual example of an aggregate in which identical subunits can assume two entirely different modes of packing within the same structure. This type of polymorphism is quite distinct from the more usual polymorphism displayed by LMM, which can form a variety of structures depending upon the solvent conditions and the length of the subfragment (107).

V. SPECULATIONS ON THE NATURE OF THE INTERACTIONS BETWEEN THE MYOSIN AND ACTIN FILAMENTS

We assume, on the basis of the available data, that during relaxation the helical portion of myosin interacts with the backbone of the thick filament and the HMM S-1 globules lie inactive on the surface. Upon stimulation of the muscle, a "bridge" is formed between the thick and thin filaments. This bridge has the dual task of providing the chemical energy for contraction, in the form of the free energy of ATP hydrolysis, as well as generating a sufficient force to move the actin filament relative to the myosin filament. How this is accomplished is a fundamental problem in muscle research.

A. Interfilament Interactions

Muscle undergoes changes in length while maintaining a "constant-volume" relationship. This means that when the muscle is stretched the filaments move closer together, and when the fiber shortens the filaments move further apart, their separation varying inversely as the square root of the sarcomere length (108). If we accept the concept of a direct physical link between HMM S-1 and actin during contraction, and this assumption seems amply supported by the existing biochemical, electron microscope, and X-ray diffraction data, the question arises how this link can be sustained despite the increasing interfilament distance during contraction. A plausible solution to this dilemma can be found in the unique solubility properties of the myosin molecule. The weak interactions exhibited by the HMM S-2 subfragment suggest that this portion of the myosin molecule may be endowed with the ability to bend reversibly away from the axis of the thick filament—the degree of bending depending on the interfilament distance it must bridge to link with actin (*1,36,104*). Such a scheme would have the cross bridges (consisting of HMM S-1 and anywhere up to 400 Å of HMM S-2) involved in a cyclical process consisting of attachment to the actin filaments, resulting in their movement toward the center of the A-band, and then detachment prior to a repeat of the cycle.

An angled orientation of the cross bridges relative to the filaments can actually be observed in electron micrographs of certain highly ordered muscles. Sections of insect flight muscle fixed in rigor show the cross bridges tilted by an average angle of 45° from the resting position (*109*). This "angling" or movement of the cross bridges is consistent with the X-ray diffraction patterns obtained from muscles in rigor in which a new pattern of off-meridional reflections replaces the spacings based on the 429-Å repeat seen in live muscles (*101,109*). The new helical arrangement of cross bridges is suggestive of an actinlike helix, indicating that in order for the myosin to make the maximum number of contacts with the actin the bridges must bend toward the thin filament.

The diffraction pattern of contracting muscle is distinguished by the disappearance of most of the reflections attributed to relaxed muscle and the absence of the new layer lines associated with rigor. The marked disorder of this pattern has been interpreted to mean that only a small percentage of the cross bridges are attached to actin at any given time during contraction (*101*). Moreover, the bridges that are detached are not in the orientation found in relaxed muscle. It appears that activation affects all the cross bridges and that this movement does not lead to any regularly ordered array. The only feature of the myosin pattern that remains relatively constant throughout changes in state of the muscle is the 145-Å meridional reflection. Since this spacing arises from the longitudinal subunit repeat, it implies that the myosin molecules do not shift their position appreciably relative to the long axis of the filament during contraction or stretch (*110,111*). The invariance of the length of the thick filament, first shown by electron microscopy is of course a basic tenet of the sliding filament hypothesis (*112–114*).

B. Intrafilament Interactions

The behavior of myosin in living muscle, as inferred from X-ray diffraction studies, is in many ways compatible with what we know about myosin from solution studies. We have seen that only slight changes in the ionic environment are needed to influence the associated states of myosin. One may speculate that activation of the muscle might cause a sufficient change in the composition of the myoplasm to shift the equilibrium position of the thick filament from a tightly packed structure to a looser structure in which the HMM S-2 regions no longer bond as strongly to the backbone of the filament but swing out toward the actin filament. It has been shown that pH, ionic strength, and cation valency can significantly alter the lattice dimensions of glycerinated muscle (115). We know that the divalent cation Ca^{2+} exerts a profound influence on both glycogenolysis (116) and the contraction–relaxation system (2). It is not inconceivable that other ions, and this could apply to simple hydrogen ions or monovalent cations and anions, may cause a cooperative transition within the core of the thick filament which expresses itself at the surface of the filament (105,107,117). The interactions leading to the 145-Å stagger must remain relatively invariant, but this does not exclude the possibility of slight rotational movements on the part of the LMM moiety. We have seen that LMM has potential bonding sites for many ordered states. Alternatively, the effects of charge and ionic milieu may be restricted to the surface lattice where the inherently unstable HMM S-2 interactions may be further weakened during stimulation of the muscle. The idea of small variations in protein–protein interactions regulating the movements of the cross bridges has been proposed as a possible explanation for the "catch" mechanism in molluscan muscles. It is postulated that the ability of the myosin molecules to maintain tension for prolonged periods of time is made possible by cooperative transitions of the paramyosin molecules forming the core of the molluscan filament (117).†

C. Does Myosin Have a Hinge?

In discussing the movement of the cross bridges, the question arises whether a specific region in myosin acts as a flexible joint or "hinge", or whether myosin bends uniformly along its length. The concept of a hinge connecting the HMM and LMM regions has arisen quite naturally from the finding that enzymes of widely differing specificities can split myosin to meromyosins. Before seriously postulating a functional hinge, however, the following observations should be considered.

1. The length of the LMM fragment depends on the type of enzyme and the conditions for the digestion.

---

† The molluscan thick filament differs from the vertebrate thick filament in that it contains large amounts of a second protein, paramyosin, in addition to myosin. Paramyosin is a fully α-helical, double-stranded, coiled coil about 1300 Å long (118,142). Its gross structure is in many ways similar to the rod portion of myosin.

2. Prolonged tryptic digestion of LMM has been shown to produce several fragments of a finite size rather than a random distribution of sizes (*119*).

3. Some species of myosin, for instance, cardiac myosin, require much higher concentrations of trypsin to form meromyosins (*120*), and chicken gizzard myosin is reported as not forming meromyosins (*121*).

These experiments suggest that there are regions of varying susceptibility to proteolytic enzymes in myosin; these sites probably arise from perturbations in the helical packing of the coiled coil which in turn is determined by the primary structure of the particular myosin. The present data do not preclude the existence of a discrete functional hinge, but there do not appear to be compelling arguments in its favor. One fact can be unequivocally stated from the amino acid analysis of the rod subfragment: the hinge, if it exists, does not owe its flexibility to helix-disrupting residues. No proline residues are present in the rod portion of the myosin molecule (Table IV).

D. Regulation of Myosin–Actin Interactions

The bulk of the work on myosin during the past decade has been concerned with its structure, enzymic activity, and interaction with actin. The presence of several additional proteins in the thin filament that regulate the activity of the actomyosin was not realized until relatively recently. If the interaction between pure actin and myosin resulting in the hydrolysis of Mg-ATP can be considered as a model system for the contractile event, no other protein appears to be required for full activation. However, if one wishes to consider the mechanism of relaxation, or in biochemical terms, the inhibition of activity, then these "other proteins" become essential. The discovery of regulatory proteins stems from the observation that crude actomyosin, similar to the myofibril but unlike pure actomyosin, is sensitive to the concentration of calcium ions (*2,122,123*). It is now firmly established that calcium ions are released from the sarcoplasmic reticulum upon electrical stimulation and that these ions activate the contractile system; conversely, withdrawal of calcium from the myoplasm by the reticulum induces relaxation. Thus the calcium ion or the "calcium pump" is the physiological regulator of the contraction–relaxation cycle. When a chelating agent such as glycol ether diaminetetraacetic acid (EGTA) is added to crude or "natural" actomyosin, the ATPase activity is depressed; addition of low levels of calcium ($\sim 10^{-6}$) restores the ATPase. Chelating agents have no effect on pure or "synthetic" actomyosin (*124*). The difference in behavior of the two protein preparations apparently resides in the actin fraction, which in the case of the crude preparation was shown to contain a variable amount of tropomyosin and a new globular protein which was subsequently named "troponin" (125). Pure tropomyosin, similar to the rod portion of myosin and the invertebrate protein paramyosin, is a highly α-helical, double-stranded, coiled coil about 400 Å long (126–128). By itself it confers no calcium sensitivity upon actin, but in conjunction with troponin the "tropomyosin–troponin–actin" complex is subject to calcium regulation (129). Detailed physicochemical data on troponin are still lacking, but it has been shown to consist of more than one subunit; troponin A is the calcium receptor, it binds about 1 mole of calcium per 20,000 g molecular weight. Troponin B alone is an inhibitor of the Mg-ATPase of actomyosin (130). Troponin A joined with B removes the inhibition in the presence of calcium and tropomyosin (131). From structural and antibody staining studies, it is thought that the tropomyosin molecules may lie in an end-to-end arrangement in the grooves of the actin helix with the troponin bonded to the tropomyosin every 400 Å along the thin filament (128,132). Since there is an insufficient amount of troponin to regulate every actin molecule individually, one of the functions of tropomyosin may be to serve as an amplification system which can transmit the on-off signals originating from troponin to a whole group of actin molecules.

This picture of the regulatory mechanism tacitly implies that the cross-bridge link between myosin and actin is completely under the control of the thin filament. This may, of course, be a gross oversimplification of the actual events in vivo. One could, for instance, envisage a scheme in which troponin transmits its signals through the thick filament as well as the thin filament. If a direct interaction between a troponin molecule and a myosin molecule occurs, such an interaction could initiate a cooperative transition within the myosin filament which might regulate the behavior of several cross bridges in their interaction with actin (133). Clearly, more extensive experimental data are required before a mechanism for the contraction–relaxation cycle can be postulated.

VI. CONCLUSIONS

Protein chemistry has, until recently, usually been concerned with the more easily extractable, small, soluble, globular proteins. The structural proteins, which by definition are part of the structural framework of the cell, have unusual asymmetric shapes, are fairly insoluble, and tend to associate spontaneously into larger aggregates. Although these characteristics make it difficult to study the monomeric state of the fibrous protein, the insight gained into the principles governing the self-assembly process of biological structures has more than compensated for the intractable nature of the protein.

Myosin has been a difficult but exciting protein to study—it is an enzyme and yet it is also a structural protein. It expresses its biological activity and function not as a monomer but as part of a highly organized multisubunit filament consisting of hundreds of like molecules. We now have a reasonably good picture of the overall substructure of myosin, and yet probably as much remains to be discovered in the future about the molecule as has been found in the past.

Against an extensive background of biochemical and biophysical solution studies of the muscle proteins, and morphological and physiological studies of muscle itself, one can begin to consider how activation and relaxation affect the individual molecules in the filaments, how the link between myosin and actin is achieved, how a sliding force between the filaments is generated, and how the linkages are finally broken in the contraction–relaxation cycle.

This chapter, summarizing as it does some of the more current concepts about the structure and interactions of the myosin molecule, is at best only a prelude to the much more intricate but at present largely unknown story of how this complex molecule carries out its multiple activities in the living muscle.

ACKNOWLEDGMENTS

I wish to express my appreciation and thanks to my colleagues Drs. C. Cohen and D. L. D. Caspar for the many valuable discussions during the years of our association on the concepts and principles underlying structural molecular biology. This work was supported by Public Health Service Grant AM–04762 and Research Career Program Award K3–AM–10630; a grant from the Muscular Dystrophy Associations of America, Inc.; and Grant GB–8616 from the National Science Foundation.

REFERENCES

1. H. E. Huxley, *Science*, **164**, 1356 (1969).
2. S. Ebashi and M. Endo, *Progr. Biophys. Mol. Biol.*, **18**, 123 (1968).
3. J. Gergely, *Ann. Rev. Biochem.*, **35**, 691 (1966).
4. S. V. Perry, *Progr. Biophys. Mol. Biol.*, **17**, 325 (1967).
5. D. M. Young, *Ann. Rev. Biochem.*, **38**, 913 (1969).
6. H. E. Huxley, *J. Mol. Biol.*, **7**, 281 (1963).
7. A. Szent-Györgyi, *The Chemistry of Muscular Contraction*, Academic Press, New York, 1951.
8. A. Holtzer and S. Lowey, *J. Am. Chem. Soc.*, **78**, 5954 (1956).
9. A. Holtzer and S. Lowey, *J. Am. Chem. Soc.*, **81**, 1370 (1959).
10. A. Holtzer, S. Lowey, and T. Schuster, in *The Molecular Basis of Neoplasia*, University of Texas Press, Austin, 1962, p. 259.
11. A. Holtzer and S. Rice, *J. Am. Chem. Soc.*, **79**, 4847 (1957).
12. C. Cohen, *J. Polymer Sci.*, **49**, 144 (1961).
13. R. V. Rice, *Biochim. Biophys. Acta*, **52**, 602 (1961); **53**, 29 (1961).

14. C. R. Zobel and F. D. Carlson, *J. Mol. Biol.*, 7, 78 (1963).
15. P. A. Small, W. F. Harrington, and W. W. Kielley, *Biochim. Biophys. Acta*, 49, 462 (1961).
16. S. Lowey and C. Cohen, *J. Mol. Biol.*, 4, 293 (1962).
17. E. F. Woods, S. Himmelfarb, and W. F. Harrington, *J. Biol. Chem.*, 238, 2374 (1963).
18. D. M. Young, S. Himmelfarb, and W. F. Harrington, *J. Biol. Chem.*, 240, 2428 (1965).
19. H. Mueller, *J. Biol. Chem.*, 240, 3816 (1965).
20. R. V. Rice, A. S. Brady, R. H. De Pue, and R. E. Kelly, *Biochem. Z.*, 345, 370 (1966).
21. H. S. Slayter and S. Lowey, *Proc. Natl. Acad. Sci. U.S.*, 58, 1611 (1967).
22. A. K. Kleinschmidt, D. Lang, D. Jacherts, and R. K. Zahn, *Biochim. Biophys. Acta*, 61, 857 (1962).
23. S. Cohen and R. R. Porter, *Advan. Immunol.*, 4, 287 (1964).
24. J. C. Gerhart and A. B. Pardee, *J. Biol. Chem.*, 237, 891 (1962).
25. F. Lymen, *Federation Proc.*, 20, 941 (1961).
26. J. Gergely, *Federation Proc.*, 9, 176 (1950).
27. J. Gergely, *J. Biol. Chem.*, 200, 543 (1953).
28. E. Mihalyi and A. G. Szent-Györgyi, *J. Biol. Chem.*, 201, 189 (1953).
29. A. G. Szent-Györgyi, *Arch. Biochem. Biophys.*, 42, 305 (1953).
30. A. G. Szent-Györgyi, C. Cohen, and D. E. Philpott, *J. Mol. Biol.*, 2, 133 (1960).
31. C. Cohen and A. G. Szent-Györgyi, *J. Am. Chem. Soc.*, 79, 248 (1957).
32. C. Cohen and A. G. Szent-Györgyi, *Proc. IVth Intern. Cong. Biochem.*, *Vienna*, 8, 108 (1960).
33. C. Cohen and K. C. Holmes, *J. Mol. Biol.*, 6, 423 (1963).
34. F. H. C. Crick, *Acta Cryst.*, 6, 689 (1953).
35. S. Lowey and A. Holtzer, *Biochim. Biophys. Acta*, 34, 470 (1959).
36. S. Lowey, H. S. Slayter, A. G. Weeds, and H. Baker, *J. Mol. Biol.*, 42, 1 (1969).
37. H. Mueller and S. V. Perry, *Biochem. J.*, 85, 431 (1962).
38. A. Holtzer and S. Lowey, unpublished data.
39. D. R. Kominz, E. R. Mitchell, T. Nihei, and C. M. Kay, *Biochemistry*, 4, 2373 (1965).
40. S. Lowey, in *Symposium on Fibrous Proteins, 1967* (W. G. Crewther, ed.), Butter-worths, Australia, 1968, p. 124.
41. S. Lowey, L. Goldstein, C. Cohen, and S. M. Luck, *J. Mol. Biol.*, 23, 287 (1967).
42. S. S. Margossian and S. Lowey, unpublished data, 1970.
43. D. M. Young, W. F. Harrington, and W. W. Kielley, *J. Biol. Chem.*, 237, 3116 (1962).
44. W. W. Kielley and W. F. Harrington, *Biochim. Biophys. Acta*, 41, 401 (1960).
45. T. C. Tsao, *Biochim. Biophys. Acta*, 11, 368 (1953).
46. R. H. Locker, *Biochim. Biophys. Acta*, 20, 514 (1956).
47. R. H. Locker and C. J. Hagyard, *Arch. Biochem. Biophys.*, 120, 454 (1967).
48. H. Oppenheimer, K. Bárány, G. Hamoir, and J. Fenton, *Arch. Biochem. Biophys.*, 120, 108 (1967).
49. W. R. Middlebrook, in *Biochemistry of Muscle Contraction* (J. Gergely, ed.), Little, Brown, Boston, 1964, p. 27.
50. D. R. Kominz, E. R. Carroll, E. N. Smith, and E. R. Mitchell, *Arch. Biochem. Biophys.*, 79, 191 (1959).
51. L. C. Gershman, P. Dreizen, and A. Stracher, *Proc. Natl. Acad. Sci. U.S.*, 56, 966 (1966).

52. E. Gaetjens, K. Bárány, G. Bailin, H. Oppenheimer, and M. Bárány, *Arch. Biochem. Biophys.*, **123**, 82 (1968).
53. M. Harris and H. C. Suelter, *Biochim. Biophys. Acta*, **133**, 393 (1967).
54. E. G. Richards, C. -S. Chung, D. B. Menzel, and H. S. Olcott, *Biochemistry*, **6**, 528 (1967).
55. S. Lowey and L. A. Steiner, unpublished data, 1970.
56. L. C. Gershman, A. Stracher, and P. Dreizen, *J. Biol. Chem.*, **244**, 2726 (1969).
57. J. Gazith, S. Himmelfarb, and W. F. Harrington, *J. Biol. Chem.*, **245**, 15 (1970).
58. A. G. Weeds, *Nature*, **233**, 1362 (1969).
59. A. G. Weeds, *Biochem. J.*, **105**, 25C (1967).
60. D. W. Frederiksen and A. Holtzer, *Biochemistry*, **7**, 3935 (1968).
61. A. G. Weeds, *3rd Intern. Biophys. Congr. Intern. Union Pure Appl. Biophys., Cambridge, Massachusetts, 1969*, p. 184.
62. A. G. Weeds and S. Lowey, unpublished data, 1970.
63. A. G. Weeds and B. S. Hartley, *Biochem. J.*, **107**, 531 (1968).
64. A. G. Weeds, *Biochem. J.*, **104**, 44 P.
65. M. Young, M. H. Blanchard, and D. Brown, *Proc. Natl. Acad. Sci. U.S.*, **61**, 1087 (1968).
66. G. W. Offer, *Biochim. Biophys. Acta*, **111**, 191 (1965).
67. G. W. Offer and R. L. Starr, *Intern. Union Pure Appl. Biophys., Comm. Mol. Biophys., Cambridge, England, 1968*, p. 25.
68. P. Johnson, C. I. Harris, and S. V. Perry, *Biochem. J.*, **105**, 361 (1967).
69. G. Huszar and M. Elzinga, *Nature*, **223**, 834 (1969).
70. W. M. Kuehl and R. S. Adelstein, *Biochem. Biophys. Res. Commun.*, **37**, 59 (1969).
71. R. P. Ambler and M. W. Rees, *Nature*, **184**, 56 (1959).
72. K. Murray, *Biochemistry*, **3**, 10 (1964).
73. R. J. Delange, A. N. Glazer, and E. L. Smith, *J. Biol. Chem.*, **244**, 1385 (1969).
74. M. Bárány, in *Symposium on Sulfur and Proteins, New York*, Academic Press, New York, 1959, p. 317.
75. L. H. Schliselfeld and M. Bárány, *Biochemistry*, **9**, 3206 (1968).
76. K. M. Nauss, S. Kitagawa, and J. Gergely, *J. Biol. Chem.*, **244**, 755 (1969).
77. S. Lowey and S. M. Luck, *Biochemistry*, **8**, 3195 (1969).
78. M. Young, *Proc. Natl. Acad. Sci. U.S.*, **58**, 2393 (1967).
79. H. Mueller, *J. Biol. Chem.*, **240**, 3816 (1965).
80. A. A. Rizzino, E. Eisenberg, K. K. Wong, and C. Moos, *Federation Proc.*, **27**, 519 (1968).
81. E. M. Szentkiralyi and A. Oplatka, *J. Mol. Biol.*, **43**, 551 (1969).
82. P. B. Moore, H. E. Huxley, and D. J. DeRosier, *J. Mol. Biol.*, **50**, 279 (1970).
83. A. Stracher, *Biochem. Biophys. Res. Commun.*, **35**, 519 (1969).
84. S. M. Heywood and A. Rich, *Proc. Natl. Acad. Sci. U.S.*, **59**, 590 (1968).
85. D. F. Steiner, *Proc. Natl. Acad. Sci. U.S.*, **57**, 473 (1967).
86. P. B. Sigler, D. M. Blow, B. W. Matthews, and R. Henderson, *J. Mol. Biol.*, **35**, 143 (1968).
87. M. F. Jacobson and D. Baltimore, *J. Mol. Biol.*, **33**, 369 (1968).
88. M. Bárány, *J. Gen. Physiol.*, **50**, 197 (1967).
89. D. E. Ashhurst, *Tissue Cell*, **1**, 485 (1969).
90. E. G. Richards, C. -S. Chung, P. Appel, and H. S. Olcott, *Federation Proc.*, **26**, 727 (1967).
91. J. E. Godfrey and W. F. Harrington, *Biochemistry*, **9**, 886, 894 (1970).
92. B. Kaminer and A. L. Bell, *J. Mol. Biol.*, **20**, 391 (1966).
93. R. Josephs and W. F. Harrington, *Biochemistry*, **5**, 3474 (1966).

94. R. Josephs and W. F. Harrington, *Biochemistry*, **7**, 2834 (1968).
95. W. F. Harrington and R. Josephs, *Develop. Biol. Suppl.*, **2**, 21 (1968).
96. C. L. Stevens and M. A. Lauffer, *Biochemistry*, **4**, 31 (1965).
97. T. Ikkai and T. Ooi, *Biochemistry*, **5**, 1551 (1966).
98. W. Kauzmann, *Advan. Protein Chem.*, **14**, 1 (1959).
99. D. J. DeRosier and A. Klug, *Nature*, **217**, 130 (1968).
100. C. Cohen, S. Lowey, R. G. Harrison, J. Kendrick-Jones, and A. G. Szent-Györgyi, *J. Mol. Biol.*, **47**, 605 (1970).
101. H. E. Huxley and W. Brown, *J. Mol. Biol.*, **30**, 383 (1967).
102. H. E. Huxley, *J. Biophys. Biochem. Cytol.*, **3**, 631 (1957).
103. H. E. Huxley, *J. Mol. Biol.*, **37**, 507 (1968).
104. F. A. Pepe, *J. Cell. Biol.*, **28**, 505 (1966).
105. C. Cohen, in *Principles of Biomolecular Organization* (G. E. W. Wolstenholme and M. O'Connor, eds.), Churchill, London, 1966, p. 101.
106. C. Cohen, A. G. Szent-Györgyi and J. Kendrick-Jones, *J. Mol. Biol.*, **56**, 223 (1971).
107. D. L. D. Caspar and C. Cohen, in *Symmetry and Function of Biological Systems at the Macromolecular Level* (A. Engström and B. Strandberg, eds.), Wiley, New York, p. 393.
108. G. F. Elliott, J. Lowey, and C. R. Worthington, *J. Mol. Biol.*, **6**, 295 (1963).
109. M. K. Reedy, K. C. Holmes, and R. T. Tregear, *Nature*, **207**, 1276 (1965).
110. G. F. Elliott, J. Lowy, and B. M. Millman, *Nature*, **206**, 1357 (1965).
111. H. E. Huxley, W. Brown, and K. C. Holmes, *Nature*, **206**, 1358 (1965).
112. H. E. Huxley and J. Hanson, *Nature*, **173**, 973 (1954).
113. A. F. Huxley and R. Niedergerke, *Nature*, **173**, 971 (1954).
114. S. G. Page and H. E. Huxley, *J. Cell. Biol.*, **19**, 369 (1963).
115. E. Rome, *J. Mol. Biol.*, **37**, 331 (1968).
116. W. L. Meyer, E. H. Fisher, and E. G. Krebs, *Biochemistry*, **3**, 1033 (1964).
117. A. G. Szent-Györgyi, C. Cohen, and J. Kendrick-Jones, *J. Mol. Biol.*, **56**, 239 (1971).
118. S. Lowey, J. Kucera, and A. Holtzer, *J. Mol. Biol.*, **7**, 234 (1963).
119. M. Bálint, L. Szilágyi, Gy. Fekete, M. Blazsó, and N. A. Biró, *J. Mol. Biol.*, **37**, 317 (1968).
120. H. Mueller, M. Theiner, and R. E. Olson, *J. Biol. Chem.*, **239**, 2153 (1964).
121. J. Kendrick-Jones, A. G. Szent-Györgyi, and C. Cohen, *J. Mol. Biol.*, in press, 1971.
122. A. Weber, in *Current Topics in Bioenergetics* (D. R. Sanadi, ed.), Academic Press, New York, 1966, p. 203.
123. W. Hasselbach, *Prog. Biophys. Biophys. Chem.*, **14**, 167 (1964).
124. S. V. Perry and T. C. Grey, *Biochem. J.*, **64**, 5P (1956).
125. S. Ebashi and A. Kodama, *J. Biochem.* (*Tokyo*), **58**, 107 (1965); **59**, 425 (1966).
126. K. Bailey, *Biochem. J.*, **43**, 271 (1948).
127. A. Holtzer, R. Clark, and S. Lowey, *Biochemistry*, **4**, 2401 (1965).
128. D. L. D. Caspar, C. Cohen, and W. Longley, *J. Mol. Biol.*, **41**, 87 (1969).
129. S. Ebashi, F. Ebashi, and A. Kodama, *J. Biochem.* (*Tokyo*), **62**, 137 (1967).
130. D. J. Hartshorne and K. Mueller, *Biochem. Biophys. Res. Commun.*, **31**, 647 (1968).
131. D. J. Hartshorne, M. Theiner, and H. Mueller, *Biochim. Biophys. Acta*, **175**, 320 (1969).
132. I. Ohtsuki, T. Masaki, Y. Nonomura, and S. Ebashi, *J. Biochem.* (*Tokyo*), **61**, 817 (1967).
133. J. Kendrick-Jones, W. Lehman, and A. G. Szent-Györgyi, *J. Mol. Biol.*, **54**, 313 (1970).
134. H. E. Huxley, *Sci. Am.*, **213**, 18 (1965).
135. A. J. Murphy and M. F. Morales, *Biochemistry*, **9**, 1528 (1970).

136. P. Dreizen and L. C. Gershman, *Biochemistry*, **9**, 1688 (1970).
137. S. Sarkar and P. H. Cooke, *Biochem. Biophys. Res. Commun.*, **41**, 918 (1970).
138. P. Dreizen, *Trans. N. Y. Adac. Sci.*, **32**, 120 (1970).
139. S. Lowey and D. Risby, *submitted for publication*, 1971.
140. M. V. King and M. Young, *J. Mol. Biol.*, **50**, 491 (1970).
141. R. G. Harrison, S. Lowey, and C. Cohen, *J. Mol. Biol.*, in press, 1971.
142. J. Kendrick-Jones, C. Cohen, A. G. Szent-Györgyi, and W. Longley, *Science*, **163**, 1196 (1969).

CHAPTER 6

ACTIN

Fumio Oosawa and Michiki Kasai*

INSTITUTE OF MOLECULAR BIOLOGY AND
DEPARTMENT OF PHYSICS
NAGOYA UNIVERSITY, NAGOYA, JAPAN

* Present address: Department of Biophysical Engineering, Osaka University, Osaka.

I. POLYMERIZATION OF ACTIN

Actin, which was discovered by Straub about 30 years ago (*1*), is one of the muscle proteins, and its interaction with myosin coupled with the splitting of adenosine triphosphate (ATP) is the elementary process of muscular contraction (*2*). It is extracted into water from the dried powder of muscle treated with acetone after removal of myosin. In a salt-free solvent, actin is in the form of dispersed globular molecules, called G-actin, which has a molecular weight of about 45,000. On the addition of neutral salts, it is transformed into long fibrous polymers called F-actin. Under physiological conditions of salt concentration and pH, actin molecules are in the form of F-actin. In fact, F-actin is the main component of the thin filaments running parallel in I- and A-bands in the striated muscle, as shown by electron microscopy by Huxley and Hanson (*3*). In addition to muscle, F-actinlike polymers have been found to be widely distributed in various kinds of cells (*4*).

Thus the G–F transformation of actin is a typical example of the assembly process of protein subunits into a physiologically significant polymer structure.

In this chapter phenomenological analyses of this G–F transformation are described. The most interesting points are the characteristic features of the transformation as a kind of condensation or crystallization phenomena (*5*) and participation of nucleotides, divalent cations, and other muscle proteins as regulators of this transformation.

A. G–F Transformation and the Splitting of ATP

G-Actin molecules in water are polymerized into F-actin by the addition of neutral salts. Polymerization can be followed by measurements of viscosity or flow birefringence (Fig. 1). F-actin is depolymerized into G-actin by dialyzing out salt ions.

Straub and Feuer (*6*) found that G-actin, when extracted by pure water from dried muscle powder, binds ATP and that this ATP is dephosphory-lated into adenosine diphosphate (ADP) and inorganic phosphate during polymerization to F-actin. Their experiment showed that the total number of ATP molecules split is nearly equal to the total number of G-actin molecules polymerized (Fig. 1). F-Actin thus formed tightly binds ADP. However, the reverse transformation, that is, depolymerization, is not associated with phosphorylation of ADP to ATP. Therefore, to obtain G-actin having ATP again, it is necessary that during depolymerization ATP be present in the solvent. Then, ADP bound to actin is replaced by ATP. G-Actin without ATP loses polymerizability rapidly. The whole scheme of the G–F transformation of actin is shown in Fig. 2. Actin

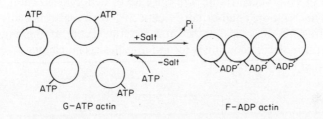

Fig. 1. Polymerization of G-actin followed by measurements of viscosity η_{sp} (—) and inorganic phosphate liberation ΔP (- - -). Solvent conditions: KCl, 45 mM; MgCl$_2$, 3.6 mM; tris–HCl (pH 8.3), 18 mM; urea, 1.36 M; at 18°C. Actin concentration: 1, 1.29 mg/ml; 2, 2.35 mg/ml; 3, 3.86 mg/ml.

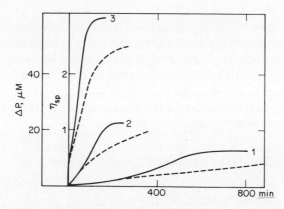

Fig. 2. Reversible polymerization of G-ATP actin to F-ADP actin accompanied by splitting of ATP.

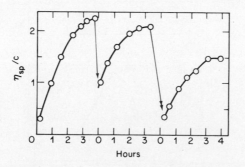

Fig. 3. Reversible G–F transformation of actin by temperature changes. Actin concentration, 2.0 mg/ml; tris–HCl (pH 8.0), 5 mM; MgCl$_2$, 2.0 mM; urea, 1.5 M; ATP, 0.55 mM; Viscosity drops indicated by arrows were induced by incubating the solution at 0°C for 15 and 42 hr. Then the temperature was raised to 25°C again and the viscosity increase was followed.

molecules have the ability to split ATP, but the splitting is accompanied by a great change in the state of actin, namely, polymerization.

Thus polymerization and depolymerization can be repeated many times by changing the salt concentration in the presence of ATP. Similar repetition is caused by changing the temperature at a constant salt concentration as shown in Fig. 3. Raising the temperature favors polymerization; lowering the temperature favors depolymerization.

B. Dynamic Balance between G- and F-Actins

In the above section, two extreme states of actin were considered; dispersed monomers (G-actin) and long fibrous polymers (F-actin), in a salt-free solvent and under physiological salt conditions, respectively.

What happens under intermediate solvent conditions? Is there a state of actin intermediate between G- and F-actin? Or, is a simple mixture of G- and F-actins obtained? If the rate of polymerization increases with increasing salt concentration and the rate of depolymerization decreases, the polymerization equilibrium is expected to be established at certain concentrations of coexisting G- and F-actin. Then, at this equilibrium, each actin molecule continues to undergo the cyclic change from G to F and further to G.

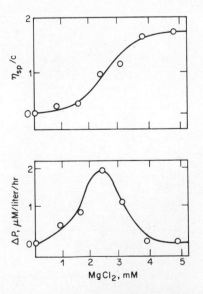

Fig. 4. Rate of ATP splitting ($\Delta P/\Delta t$) and reduced viscosity η_{sp}/c in steady states of actin solutions at various concentrations of $MgCl_2$. Actin concentration, 2.3 mg/ml; tris–HCl (pH 8.2), 5.5 mM; ATP, 0.6 mM; urea, 1.5 M; at 20°C.

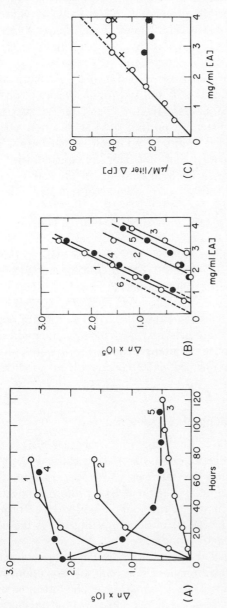

Fig. 5. (A) Approach to equilibrium levels of the G–F transformation at various temperatures starting from different initial states, followed by the change of flow birefringence Δn. Final conditions: actin, 3.36 mg/ml; MgCl₂, 0.4 mM; Veronal–HCl(pH 8.3), 7.5 mM; ATP, 0.75 mM. Actin solutions were incubated at (1) and (4) 20°C; (2) 6°C; and (3) and (5) 0°C; (1), (2), and (3) started from G-actin solutions; (4) and (5) started from F-actin solutions. (B) The temperature effect on the G–F equilibrium of actin solutions of various concentrations, at an intermediate salt concentration. Sample solutions were in the same conditions as in Fig. 5a. The abscissa is the total actin concentration and the ordinate is the degree of flow birefringence. (6) shows the degree of flow birefringence of F-actin solutions completely polymerized. (C) Splitting of ATP during the G–F transformation from various equilibrium states into complete F-actin induced by MgCl₂ 5 mM. Equilibrium solutions obtained in Fig. 5b were used as original solutions. (●) from (2) in Fig. 5b; ○ from (3) in Fig. 5b; × from (5) in Fig. 5b. The ordinate is the amount of inorganic phosphates liberated during polymerization, showing the amount of G-actin that coexisted with F-actin in the original equilibrium states. The results show that the G-actin concentrations above certain critical concentrations were kept constant.

Fortunately, there is a method by which such a cyclic change in the state of each actin molecule can be confirmed. Actin molecules split ATP during the process of polymerization; in the presence of ATP, one G–F–G cycle results in the splitting of one ATP molecule. Therefore continuous splitting of ATP is expected to occur if G- and F-actin coexist in physical equilibrium.

Figure 4 shows the result of experiments designed by Asakura and Oosawa (7) to observe continuous ATP splitting upon the addition of various concentrations of salt ($MgCl_2$) to G-actin. Even after the viscosity reaches final stationary values, steady splitting of ATP is observed at intermediate salt concentrations. The rate of ATP splitting at the steady state assumes a maximum value at the intermediate salt concentration at which the viscosity shows that nearly equal amounts of G- and F-actin coexist. This is reasonable because the cycle is generally expected to have a maximum speed at such a condition.

In this way it has been proved that in actin solutions both polymerization and depolymerization continue to occur and the polymerization balance is established as a result of equilibration of rates of these two reactions. In the presence of ATP, the balance is not a true equilibrium but a steady state associated with the irreversible splitting of ATP.

The G–F transformation of actin is a rare case in which the dynamic balance in the assembly process can be directly proved. The rates of polymerization and depolymerization are determined by the solvent condition. Therefore the final balancing state is independent of the initial state of actin molecules. This has been confirmed under various solvent conditions (8,14) (Fig. 5a).

C. Polymerization as a Condensation Phenomenon

F-Actin is easily oriented by flow. According to the measurement of the extinction angle of flow birefringence at very low shear rates, F-actin is extremely long; the length attains several microns (9). Several hundred actin monomers link together to make F-actin. Even at intermediate salt concentrations and at low actin concentrations, F-actin once formed remains very long. (Therefore the degree of flow birefringence is the best indicator of the amount of F-actin in a solution.)

Oosawa et al. (10) investigated the relation of the final state of actin solutions to the total actin concentration under constant solvent conditions. At an intermediate salt concentration, a remarkable phenomenon was observed. At low actin concentrations no increase of flow birefringence or viscosity was observed. Above a certain value of the actin concentration, flow birefringence or viscosity began to increase, and the increase was proportional to the increase in the actin concentration. The relations between the degree of flow birefringence and the total actin concentration at different salt concentrations or at different temperatures were all expressed by parallel straight lines (Fig. 5b) (*14*). This result was interpreted as follows. Below a critical value of actin concentration no F-actin is formed, and above this critical concentration F-actin is formed and its concentration is equal to the difference between the total actin concentration and the critical concentration, as shown in Figs. 5c and 6 (*10,11,14*).

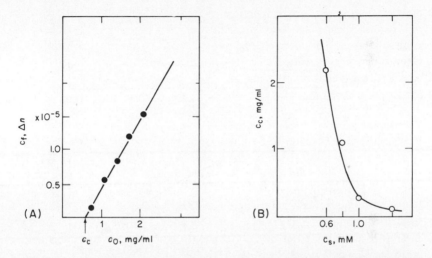

Fig. 6. (A) The relation between the degree of flow birefringence and the total actin concentration. MgCl₂, 0.8 mM; Veronal–HCl (pH 8.2), 8 mM; at 14°C. (B) The relation between the critical actin concentration and the salt (MgCl₂) concentration. Veronal–HCl (pH 8.2), 8 mM; at 14°C.

The concentration of G-actin coexisting with F-actin, which was separately measured by ultracentrifugation and various other methods, was kept constant at the critical actin concentration independently of the amount of F-actin formed (*9,14*).

This phenomenon is similar to the condensation of gas to liquid or crystallization of solute molecules in solution. The critical actin concentration corresponds to the saturated vapor pressure or the saturated solute concentration. The gas molecules in excess of the critical value are all condensed to the liquid state. The excess solute molecules are all crystallized.

As shown in Figs. 5 and 6, the critical actin concentration depends on the solvent conditions: the concentration and the species of salt ions, pH,

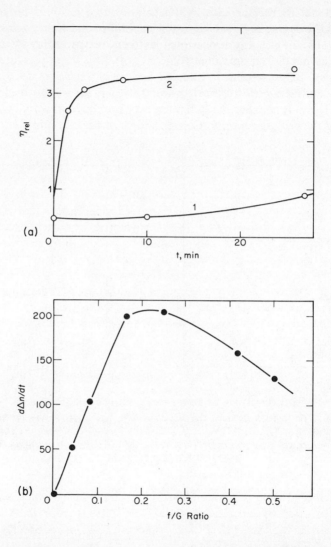

Fig. 7. (a) Explosive polymerization of G-actin induced by the addition of polymer fragments. Actin concentration, 2.8 mg/ml; ADP, 43 μM (no ATP). A solution of polymer fragments was obtained by sonication of the same G-actin solution in the presence of 1 mM MgCl$_2$ for 1 min. The final concentration of salts added for polymerization: KCl, 24 mM; MgCl$_2$, 0.1 mM; tris–HCl (pH 8.0), 5 mM. (1) Salts only; (2) salts and polymer fragments (volume ratio 1/12). (b) The relation between the initial rate of polymerization and the ratio of polymer fragments to G-actin. Actin concentration, 1.5 mg/ml; KCl 10 mM; tris–HCl (pH 8.2), 7.5 mM; ATP, 0.5 mM.

temperature, and so on. Under physiological conditions, the critical actin concentration is so low that almost all actin monomers seem to be polymerized. Even then, however, a very small amount of G-actin must exist in dynamic balance with F-actin (8).

D. Cooperative Nature of Polymerization

The observation noted in the above section suggests that polymerization of actin is a kind of cooperative phenomenon. This was proved by the following kinetic analyses by Kasai and associates (15). First, the initial phase of polymerization was followed as a function of the initial concentration of G-actin by measuring the viscosity increase and the ATP splitting. (Special attention was paid to preparing an initial solution in which all actin molecules were in the form of dispersed monomers.) The initial rate was proportional to the third to fourth powers of the actin concentration. The cooperation of three to four monomers seems to be needed for the initiation of polymerization. Second, the addition of a small amount of F-actin was found to accelerate the polymerization of actin. This is comparable to acceleration of condensation by small drops of liquid added to supersaturated vapor. Later, it was found by Asakura that great acceleration can be induced by adding small fragments of F-actin obtained by sonic vibration (Fig. 7) (16).

II. THEORETICAL ASPECT

A. Theoretical Analysis of Polymerization Equilibrium

Summarizing the above results: (1) In actin solutions, G- and F-actin coexist in a dynamic equilibrium which is accompanied by ATP splitting. (2) Polymerization of actin takes place as a kind of condensation phenomenon having a cooperative nature.

The second point indicates that F-actin is not a simple end-to-end linear polymer of G-actin. Oosawa et al. found that all thermodynamic data described in the previous sections can be understood well by the helical polymer model (5,12,13). Here, the theory of helical polymerization of globular protein molecules presented by Oosawa and Kasai as applied to the G–F transformation of actin is briefly described (5,12).

Let us consider the polymerization equilibrium of protein molecules in which polymers of various degrees of polymerization coexist. Denote the number concentration of i-mers as c_i and the number concentration of monomers as c_1. In the case of simple linear polymers, according to the mass action law, c_i is given by

$$c_i = K^{-1}(Kc_1)^i \tag{1}$$

where $K = \exp(-\Delta F/kT)$; ΔF is the free energy of the bond and assumed to be independent of the degree of polymerization. The binding constant K or ΔF is determined by the solvent condition. The total concentration of protein molecules c_0 is given by the summation

$$c_0 = c_1 + c_l = \sum_{i \geq 1} ic_i = c_1/(1 - Kc_1)^2 \tag{2}$$

where c_l is the total amount of linear polymer ($i \geq 2$). Thus at given values of the total protein concentration c_0 and the binding constant K, c_1 and c_i values are all determined by the use of Eqs. (1) and (2). In this case the relation between c_0 and c_1 or c_l does not show behavior as a condensation phenomenon.

Now let us assume the formation of helical polymers in which each monomer is bound with four or more neighboring monomers; an example is shown in Fig. 8, in which the minimum size (or the nucleus) of the helical polymer is assumed to be four monomers. When a simple linear tetramer

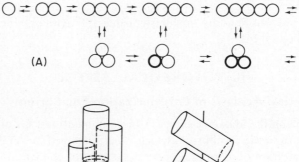

(A)

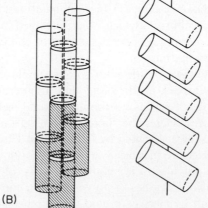

(B)

Fig. 8. (A) Schematic diagram of equilibria between monomer, linear polymers, and helical polymers. (B) Helical polymer (left) and linear polymer (right). The nucleus of the helical polymer is composed of three or four monomers.

is deformed to make an additional bond between the first and the fourth monomers, the basic structure of the helical polymer is obtained. By successive addition of monomers to the end of this polymer making bonds with two neighboring monomers, a long helical polymer is formed.

Denoting the number concentration of helical i-mers as c_{ih}

$$c_{3h} = sc_3 = sK^{-1}(Kc_1)^3$$

and

$$c_{4h} = K_h c_{3h} c_1 = sK^{-1}K^3 K_h c_1{}^4 \tag{3}$$

$$c_{ih} = K_h\, c_{(i-1)h} c_1 = s(K/K_h)^2 K_h^{-1}(K_h c_1)^i$$

$$= AK_h^{-1}(K_h c_1)^i$$

$$A = s(K/K_h)^2$$

where $s = \exp(-\Delta F^*/kT)$, which is smaller than unity because ΔF^* is the excess free energy necessary for deformation of a linear trimer to make a nucleus of the helical polymer. $K_h(= \exp(-\Delta F_h/kT))$ is the binding constant for the addition of a monomer to the end of the helical polymer. Since in the helical polymer the number of bonds per monomer is larger than that in the simple linear polymer, K_h is larger than K. Therefore the constant $A = s(K/K_h)^2$ must be very much smaller than unity. For example, if $\Delta F^* = +5$ kcal/mole and $\Delta F_h - \Delta F = -5$ kcal/mole, then A is of the order of 10^{-8}. From the above equation, the total concentration of helical polymers c_h is given by the summation

$$c_h = \sum_{i \geq 3} i c_{ih} = Ac_1/(1 - K_h c_1)^2 - (Ac_1 + 2AKc_1{}^2) \tag{4}$$

Thus the total concentration c_0 is given by

$$c_0 = c_1 + c_l + c_h \tag{5}$$

If the concentration of simple linear polymers c_l is negligible, we have the approximate formula:

$$c_0 = c_1 + c_h = c_1 + Ac_1/(1 - K_h c_1)^2 \tag{6}$$

Let us increase the total protein concentration c_0 at a constant solvent condition, namely, at constant values of A and K_h. Since A is very much smaller than unity, the second term of Eq. (6) is negligible, so that c_1 is nearly equal to c_0. Almost all molecules are in the form of dispersed monomers. However, when c_0 and therefore c_1 approach K_h^{-1} very closely, the second term can not be neglected. The helical polymer begins to be formed. When c_0 exceeds K_h^{-1}, c_1 does not exceed K_h^{-1} and is kept nearly constant at K_h^{-1}.

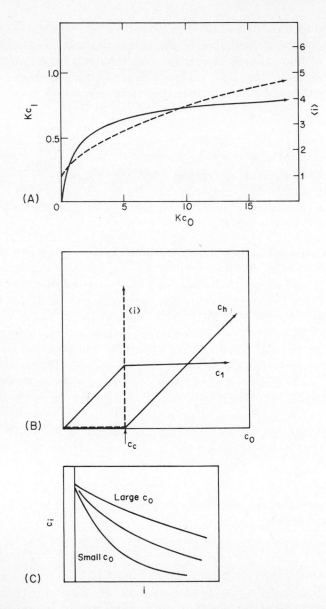

Fig. 9. (A) Physical features of simplel inear polymerization. Ordinate: monomer concentration in equilibrium with linear polymers (solid line) and the number-average degree of polymerization (dotted line). Abscissa: the total concentration of protein. (B) Physical features of helical polymerization. Ordinate: monomer concentration c_1 (and linear polymer concentration) and helical polymer concentration c_h, and the number-average degree of polymerization. (C) The size distribution of helical polymers in the equilibrium at different total amounts of polymers and a constant solvent condition.

Above the critical concentration given by $c_c = K_h^{-1}$, the excess amount of molecules $c_0 - c_c$ are all transformed to helical polymers. Monomers of the critical concentration c_c (and small linear polymers) coexist with these helical polymers.

The average degree of polymerization of helical polymers is given by

$$\langle i \rangle = \sum ic_{ih} / \sum c_{ih} = 1/(1 - K_h c_1) = (c_h/c_c \, A)^{1/2} \qquad (7)$$

For very small values of A, $\langle i \rangle$ becomes very much larger than unity even near the critical concentration where the concentration of helical polymers is small. For example, when $A = 10^{-8}$ and $c_h/c_c = 10^{-2}$, $\langle i \rangle = 10^3$. As summarized in Fig. 9, experimental data on the G–F transformation of actin can be understood very well on the basis of the helical polymerization scheme (5,12,13).

The relation between the number concentration of polymers and the degree of polymerization is given by Eq. (3), which is rewritten as

$$c_i \propto \lambda^i = e^{-\alpha i} \qquad (8)$$

where λ is a little smaller than unity or α is a small positive number. Therefore c_i decreases with increasing i, as shown in Fig. 9 (12,17). In the case of actin, this relation was confirmed in the electron microscope analysis of the length distribution of F-actin by Kawamura and Maruyama (18,19).

In spite of such a broad exponential distribution, the G–F transformation can be regarded as a kind of condensation or crystallization because of the small value of A which is connected with difficulty of the nucleation process.

B. Kinetics of Polymerization

Polymerization of actin consists of two processes: nucleation and growth (Fig. 10). The following kinetic equations can be approximately applied to the change of the concentration of i-mers with time t (12,17):

$$\frac{d}{dt} c_{i_0} = -k_+ c_1 c_{i_0} + k_- c_{i_0+1} + k^* c_1^{i_0}$$

$$\frac{d}{dt} c_i = k_+ c_1 c_{i-1} - k_- c_i - k_+ c_1 c_i + k_- c_{i+1}$$

$$i > i_0 \qquad (9)$$

The last term of the first equation gives the rate of nucleation; k^* is the rate constant of nucleation; k_+ and k_- are rate constants of polymerization and depolymerization in the growth process. Nuclei are assumed to be formed as a result of interaction of i_0 monomers, or deformation of linear

i_0-mers. These equations can be integrated if the rate of depolymerization k_- can be neglected. Then, the change of the concentration of dispersed monomers is given by (12)

$$\ln \frac{1+(1-(c_1/c_0)^{i_0})^{1/2}}{1-(1-(c_1/c_0)^{i_0})^{1/2}} = (2i_0)^{1/2}(k_+k^*)^{1/2}c_0^{i_0/2}\, t \qquad (10)$$

The observed relations between c_1 and t during polymerization of actin agree well with this equation (Fig. 10). Moreover, Kasai (20) found that

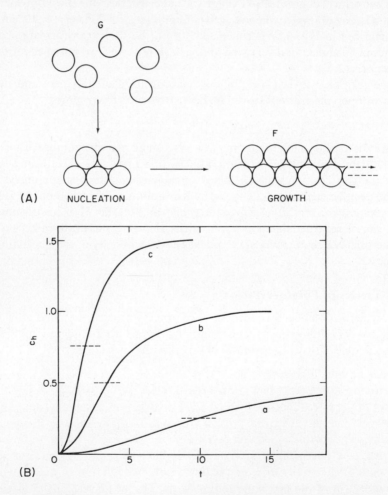

Fig. 10. (A) The process of polymerization composed of nucleation and growth. (B) The kinetics of polymerization at different total concentrations of protein and a constant solvent condition. (In calculation, the rate of depolymerization was neglected.)

the relations at different solvent conditions can be superimposed on each other by displacement along the time axis if the axis is expressed as the logarithm of time, as expected from Eq. (10). This means that under different solvent conditions the polymerization of actin takes place through the same nucleation and growth processes.

Recently, the kinetics of the length distribution of F-actin was also treated theoretically (*17*). It is shown that after the monomer–polymer equilibrium was established, slow rearrangement of polymer lengths takes place toward the exponential length distribution. The time necessary for this rearrangement is proportional to the square of the average length. F-Actin in vivo has a definite length. The mechanism which makes the length of F-actin constant is not known.

III. POLYMER STRUCTURE

A. Ultrastructure of F-Actin

By electron microscopy using a negative-staining technique, Hanson showed that F-actin is a double-stranded helical polymer of actin molecules (Fig. 11) (*21*). In electron micrographs of F-actin, monomers appear as globular particles of the diameter of about 55 Å, and about 13 monomers are contained in two strands per half-pitch of about 355 Å. The model in Fig. 11 proposed by Hanson is consistent with data from small-angle X-ray diffraction of F-actin. Thus the helical polymer structure of F-actin, first suggested by the X-ray analysis by Selby and Bear (*22*) and then predicted on the basis of thermodynamic analyses of G–F transformation, was directly observed by electron microscopy.

The same polymer structure as F-actin constructed in vitro was found in thin filaments isolated from the striated muscle (*21,23*). The X-ray analysis also showed the presence of a periodic structure similar to F-actin in the living muscle (*24,25*). Therefore it is very likely that the main component of thin filaments is actin and their basic structure is similar to F-actin in vitro. On the basis of his recent works, Ebashi (*26*) presented a model of the thin filament in which tropomyosin polymer binds along the actin polymer and the third protein troponin locates periodically on the tropomyosin polymer. (See Section IX,C.)

F-Actin is not completely rigid but is flexible. Its flexibility must be between the flexibility of a simple linear polymer and that of a helical or tubular polymer composed of many strands. The length of the kinetic unit of F-actin estimated from angular distribution of the intensity of scattered light suggested such flexibility (*27*). Recently, the Brownian motion of F-actin in solution was analyzed by Fujime by the measurement of quasi-

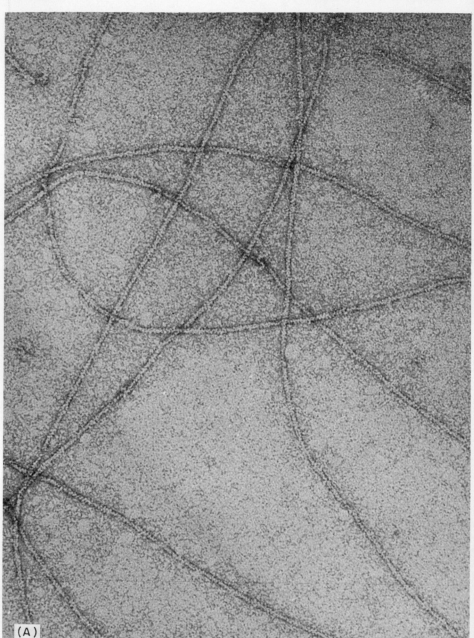

Fig. 11. (A) An electron micrograph of F-actin (by S. H. Fujime). (B) The model of F-actin as a two-stranded helical polymer proposed by J. Hanson on the basis of the electron micrograph.

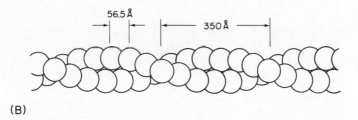

Fig. 11 (*Cont.*).

elastic scattering of laser light (*28*). The Doppler broadening of the scattered light attributable to the slowest mode of intrapolymer bending motion of F-actin was successfully detected. In combination with electron microscope observations, F-actin was found to have the flexibility of a worm-like filament, of which the average end-to-end distance is about 2.2 μ for the contour length 2.5 μ (*28,29*).

Each actin monomer in the two-stranded helical polymer seems to be in an equivalent position with respect to neighboring monomers. The actin monomer is not a symmetrical sphere but has head-and-tail directionality in itself. The actin polymer is expected to have polarity. This polarity was confirmed by electron microscopy by Huxley (*23*) on the addition to F-actin of myosin or heavy meromyosin (HMM) molecules (Fig. 12). These molecules are bound to F-actin parallel to each other in a definite direction relative to the polymer axis. The two strands of actin seem to run parallel, having polarity in the same direction.

Here, the unusual results of the electric birefringence measurements on F-actin by Kobayashi must be noted briefly (*30,31*). In the electric field the polymer axis of long F-actin is oriented perpendicular to the direction of the field; F-actin behaves as if it has a large dipole moment perpendicular to the polymer axis. This phenomenon is not understandable on the basis of a helical polymer structure. A reasonable interpretation has not yet been proposed. The bending motion of F-actin may produce the perpendicular component of the dipole. The binding of heavy meromyosin to F-actin eliminates this dipole moment and the separation of heavy meromyosin from F-actin by ATP restores it.

B. Construction of the Helical Polymer

Each monomer in the two-stranded helical polymer binds with four neighboring monomers. The bonds in the direction of the polymer axis between neighboring monomers in the same strand are probably all equivalent. However, there may be one or two kinds of bonds between monomers in different strands (Fig. 13). The principle of construction of the helical polymer can be understood in the following way.

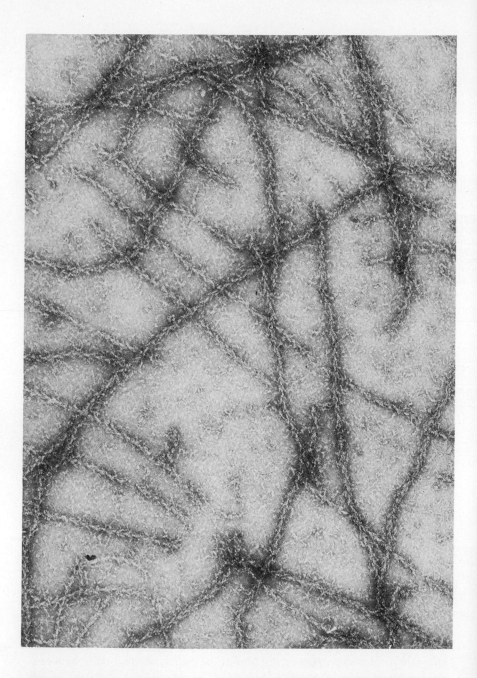

Fig. 12. The arrowhead structure of the complex of F-actin and HMM showing the polarity of F-actin (by S. H. Fujime).

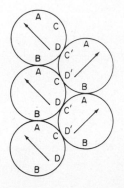

Fig. 13. The bond pattern in F-actin.

First, single strands are assumed to be formed by successive head-to-tail binding of monomers at some surface areas of interaction, not in parallel but at a certain angle. Then, two strands are assumed to be bound side by side at some definite interaction sites in the same areas of monomers. For such binding the strands must be twisted to a certain extent. As a result, we have in general a helical polymer (Fig. 14).

The fact that the two-stranded polymer is stable and that polymers having more than two strands are not formed suggests that binding between strands takes place on the same side of strands. If two strands were bound by interaction between opposite sides of strands, successive attachment of strands would be possible to form polymers composed of many strands. Tubular polymers of globular protein molecules could be constructed in such a way. In the case of actin, however, it is likely that monomers of neighboring strands interact with each other on the same side of strands or monomers (Fig. 14). This type of interaction is usually expected in van der Waals or hydrophobic interaction. However, the interaction in the same strand may be between the opposite sides of monomers. This type is expected in ionic interaction. If the above assumption concerning the polymer structure is true, monomers in the two strands are oriented in the same direction along the polymer axis but in the opposite direction in the perpendicular plane (Figs. 13 and 14). Such an arrangement of monomers may be important for interaction of F-actin with other proteins, for example, myosin and tropomyosin, as described later.

IV. REGULATION OF POLYMERIZATION

A. Polymerization without Participation of Nucleotides

As mentioned in Section I, the splitting of ATP seemed to be obligatorily coupled with the polymerization of G-actin having bound ATP. Depolymerization of F-actin having bound ADP in a salt-free solvent produces G-actin having ADP. G-Actin having ATP is stable, but G-actin having

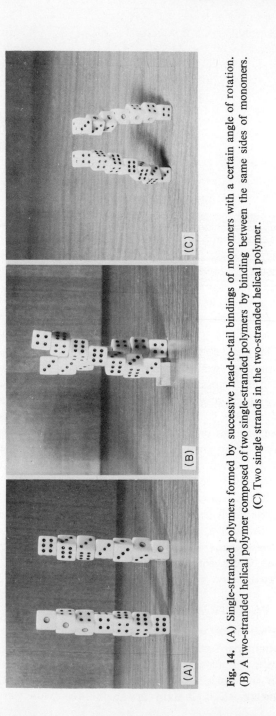

Fig. 14. (A) Single-stranded polymers formed by successive head-to-tail bindings of monomers with a certain angle of rotation. (B) A two-stranded helical polymer composed of two single-stranded polymers by binding between the same sides of monomers. (C) Two single strands in the two-stranded helical polymer.

ADP loses polymerizability gradually. However, it was found by Hayashi (32) and later by Weber (33) that if neutral salts are added immediately after G-actin having ADP was obtained by rapidly dissolving F-actin into a solvent of very low salt concentration, polymerization occurs without splitting of ATP. That is, splitting of ATP is not absolutely necessary for polymerization.

Moreover, ADP tightly bound to F-actin was found to be removable by prolonged dialysis (34). One of the methods of checking whether or not the polymer structure is changed by removal of ADP is the measurement

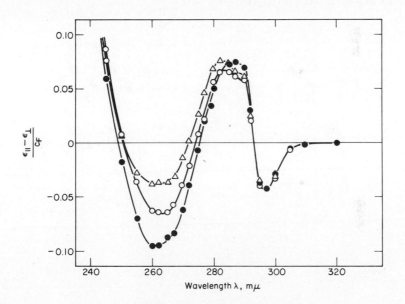

Fig. 15. UV dichroism of F-actin oriented by flow. The negative dichroism at 260 mμ decreased with removal of ADP from F-actin by long dialysis, while the other peak and trough were not changed.

of UV dichroism of F-actin oriented by flow. Higashi et al. (35) found that F-actin having ADP shows a large negative dichroism near 260 mμ, a large positive one near 280 mμ and a small negative one near 300 mμ (Fig. 15). In the process of dialysis, the first one attributable to ADP decreased but the latter two attributable to amino acid residues did not change. Therefore removal of ADP has no influence on the arrangement of monomers in F-actin.

If this F-actin without nucleotides is depolymerized, we obtain G-actin having no nucleotides, which loses polymerizability very rapidly. However,

Kasai et al. (*36*) succeeded in stabilizing this G-actin by adding a high concentration of sucrose and polymerizing it by salts without the participation of nucleotides.

Recently, Mihashi found that G-actin unfolded at a high concentration of urea can be renatured after a suitable procedure including chemical modification. Renatured G-actin can polymerize to F-actin without nucleotides.

Thus the complete scheme of G–F transformation of actin can be rewritten as shown in Fig. 16 (*37*). Nucleotides do not directly participate in the bonds between monomers in F-actin.

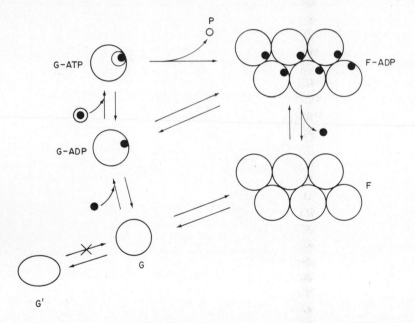

Fig. 16. The complete scheme for the G–F transformation of actin.◉, ATP; ●, ADP.

B. Nucleotides as Rate Regulators

What is the role of nucleotides in polymerization of actin? Hayashi (*32*) pointed out that replacement of ADP bound to G-actin by ATP greatly accelerates polymerization. As shown in Fig. 17, the polymerization rate of actin increases in the order: G-actin without nucleotides, G-actin having ADP, and G-actin having ATP (or ITP) (*38*). Only in the case of G-actin having ATP or ITP is dephosphorylation associated with polymerization.

It is likely that nucleotides accelerate polymerization by making the structure of G-actin favorable for polymerization. On actin monomers the binding site of nucleotides is distant from the site of interaction with other actin monomers; but the binding of nucleotides has some influence on the interaction site. Nucleotides are "allosteric regulators".

Higashi and Oosawa (38) measured UV difference spectra between G-actin without nucleotides and G-actin having ADP or ATP, and also between G- and F-actin having nucleotides. The result showed that the UV absorption spectrum of G-actin having ATP or ADP is between the spectra of G-actin without nucleotides and F-actin having ADP (Fig. 18). Therefore it seems that binding of nucleotides induces the change of the intramolecular structure of G-actin toward the state of monomers in F-actin.

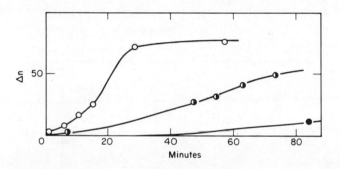

Fig. 17. Polymerization rates of various kinds of G-actins. ○, G-ATP; ◑, G-ADP; ●, original G-actin at a low concentration of ADP. G-Actin concentration 2.8 mg/ml. The solvent condition; KCl, 20 mM; tris–HCl (pH 8.0), 5 mM; at room temperature.

There is a remarkable difference in the intramolecular structure of actin monomers having no nucleotides between the dispersed state and the state incorporated into the polymer. The G–F transformation of actin presents an example of coupling between intramolecular and intermolecular conformations.

The measurement of the UV absorption spectrum, however, showed no appreciable difference between G-actin having bound ADP and that having bound ATP (38). Circular dichroism or optical rotatory dispersion showed no difference either (39). Nevertheless, these two kinds of G-actin show a great difference in polymerization rate. No data on the structural origin of this phenomenon have been obtained. There may be a direct coupling of the splitting of ATP with the reaction between actin monomers in the transient state of bond formation. (The difference between G- and F-actin having nucleotides was also found in the absorption of dye molecules attached to actin (38a) and in the resonance spectrum of electron spin labelled

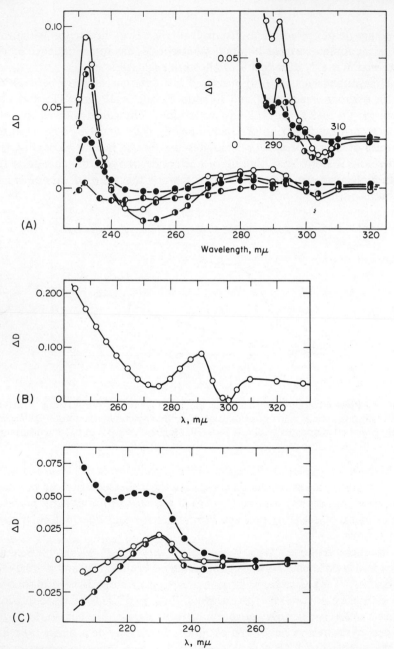

on actin (*38b*). This is due to the environmental change and/or the intra-monomer change associated with the actin–actin binding. The small spectral difference between G-actins having ATP and ADP was also reported very recently (*38c*).)

Various attempts have been made to find a condition necessary for decoupling of ATP splitting and actin–actin interaction. As noted previously, if we start with G-actin having ADP instead of ATP, polymerization occurs without splitting of ATP. However, on starting with G-actin having ATP, splitting of ATP cannot be inhibited during polymerization. Thus far, we have not succeeded in separating the ATP-splitting step from the step during which bond formation takes place between actin molecules. In relation to this point, it is worthwhile citing the following preliminary result (*40*). G-Actin treated by Salyrgan can be polymerized to a certain extent by the addition of salts, although the final viscosity is not as high. When magnesium ions are added, polymerization is accompanied by ATP splitting as usual, however, when calcium ions are added, polymers are formed without ATP splitting. When magnesium ions are added to these polymers, ATP is split.

C. Exchangeability of Nucleotides

The other important point to be noted concerning the behavior of nucleotides is the exchangeability of bound nucleotides with free ones. By the use of radioactive nucleotides, Gergely (*41*) found that nucleotides (ATP) bound to G-actin are rapidly exchanged with those in the solvent, while nucleotides (ADP) bound to F-actin are not easily exchanged. This difference in exchangeability is not attributable to a difference in salt concentration or other solvent conditions. Nucleotides in F-actin seem to be

---

Fig. 18. (A) Spectral change associated with the nucleotide binding to G-actin. G-Actin having various nucleotides were compared spectroscopically with G-actin to which no excess ADP were added. The absorption of added nucleotides was corrected by the double-cell method. Actin concentration, 3.1 mg/ml; tris–HCl (pH 8.0), 5 mM. ●, 100 μM ADP; ◑, 100 μM ATP; ○, 100 μM ITP; ◐, 100 μM deoxy ATP. (In the insert, absorption of added nucleotides was not subtracted.) (B) The difference spectrum of UV absorption between F-ADP actin and G-ADP actin in the same salt condition. Actin concentration, 3.5 mg/ml; KCl, 20 mM; tris–HCl (pH 8.2), 5 mM. (C) The difference spectrum of UV absorption at short wavelengths measured by the double-cell method with very thin cells. ●, the difference between F-ADP actin and G-ADP actin; ○, the difference between G-ATP actin and G-ADP actin; ◑, the difference between G-ADP actin and G-actin to which no excess ADP added. Actin concentration, 3.3 mg/ml; tris–HCl (pH 8.0), 5 mM; for F-ADP actin, MgCl$_2$, 1 mM.

buried inside the polymer structure, although they are not included in the bonding areas between actin monomers.

Therefore the exchangeability of bound nucleotides can be used as an indicator of the state of actin monomers.

D. Role of Divalent Cations

A situation similar to that involving nucleotides was found concerning the role of divalent cations in the G–F transformation of actin. G-Actin extracted from dried muscle powder by the usual procedure binds calcium ions. These calcium ions can be easily exchanged with magnesium or other divalent cations in the solvent (42,43). After polymerization these divalent cations tightly bound to F-actin are buried inside the polymer structure so that they are not easily exchanged (42). However, by prolonged dialysis, they can be removed without destruction of the polymer structure. They do not directly participate in the bonds between actin monomers in F-actin. Their role is to regulate the polymerization rate. G-Actin having bound magnesium ions can be polymerized very much faster than that having calcium ions (44,45). Without bound divalent cations, G-actin loses polymerizability gradually. However, since the rate of loss of polymerizability is not high, it is rather easy to obtain polymerization without bound divalent cations.

V. MONOMER STRUCTURE

A. Structure of G-Actin

Since the discovery of actin by Straub, its molecular weight was long believed to be between 60,000 and 55,000. However, recent chemical and physical analyses have led to the conclusion that it must be between 40,000 and 50,000 (46). Actin has 1 mole of 3-methylhistidine (47) and 1 mole of bound nucleotide (48) per about 43,000 g. Light scattering also gave the same molecular weight (49). The monomer consists of a single polypeptide chain which has no disulfide linkages.

The amino acid composition of actin is shown in Table I (50). It has no special composition as compared with other globular proteins. The amino acid sequence is presently being determined. Sequences in several peptide fragments have been determined by Elzinga (47). We can expect to have complete information on the whole sequence within a few years.

The helix content of G-actin is small (about $15 \sim 20\%$) (39) and, according to the value of the intrinsic viscosity and that of the diffusion constant, it is not a compact sphere (51,52), although actin monomers in electron micrographs of F-actin seem to be spherical spots. Dielectric dispersion measurement showed that G-actin has a large (permanent) dipole moment of about 700 Debye units and a rotary diffusion constant

TABLE I

Amino Acid Composition of Rabbit Striated Muscle Actin and Plasmodium Actin

Amino acid	Moles per 10^5 g protein	
	Rabbit striated muscle actin	Plasmodium actin
Lysine	52	41
Histidine	19	17
Ammonia	66	47
Arginine	38	40
Aspartic acid	82	84
Threonine	59	61
Serine	56	56
Glutamic acid	101	114
Proline	44	61
Glycine	67	74
Alanine	71	68
Cysteine/2	11	11
Valine	42	32
Methionine	30	28
Isoleucine	57	34
Leucine	63	61
Tyrosine	32	32
Phenylalanine	29	29

corresponding to a sphere of the radius of about 62 Å (53). However, G-actin gives only small electric birefringence (54). Therefore it has no large optical anisotropy. A possible interpretation is that G-actin is a swollen particle (52), not largely deviating from the spherical shape, in which distribution of ionic groups and/or α-helices is anisotropic. Spectrophotometric titration of tyrosine residues in the process of unfolding by urea or guanidine hydrochloride showed that G-actin has a central core resistant to denaturing reagent (55).

Thus far, information on the intramolecular conformation of G-actin is so incomplete that the conformation can not be discussed in relation to the location of amino acid residues in the molecule. It seems very difficult to make a three-dimensional crystal of actin for an X-ray diffraction study; consequently, the three-dimensional structure of actin molecules will not easily be made clear by the usual method. Whereas the specific one-dimensional crystal, F-actin, is easily formed under a wide range of solvent conditions, F-actin paracrystals can be used for the structural analysis.

Actin molecules must have several physiologically important sites or regions of interaction:

1. Four (different) sites of interaction with four neighboring actin molecules in the polymer structure of F-actin

2. Sites of specific binding of nucleotide and divalent cation

3. Sites of interaction with other muscle proteins, myosin and tropomyosin.

These sites have no overlapping because there is no direct interference among the binding of these different kinds of molecule. However, the binding of nucleotides influences the sites of interaction with other actin monomers; the binding between actin monomers influences the site of interaction with myosin.

Various attempts have been made to elucidate the structure of these interaction sites or areas by the use of a number of chemical reagents. For example, in relation to the role of sulfhydryl groups it was found (56,57) that p-chloromercuric benzoate (PCMB), when bound to more than two sulfhydryl groups per actin monomer, inhibits polymerization, while N-methylmaleimide (NEM), which can not be bound to more than two, has no effect on polymerizability. These results, however, do not necessarily mean that some sulfhydryl groups are involved directly in the bond to form F-actin. Actually, as described later, actin molecules in which almost all sulfhydryl groups are carboxymethylated can make normal F-actin after suitable procedures.

In spite of accumulated data concerning the action of various chemical modifiers on actin, there is not sufficient information to discuss from the molecular structural point of view. Therefore only phenomenological aspects of interactions between actin molecules and interactions of actin molecules with nucleotides, divalent cations, and other muscle protein molecules are described below.

B. Interaction with Nucleotides

One mole of G-actin binds 1 mole of nucleotide. The binding is essentially reversible and the exchange of bound nucleotide with nucleotides in the solvent occurs very quickly (*41*). For the analysis of stoichiometry of

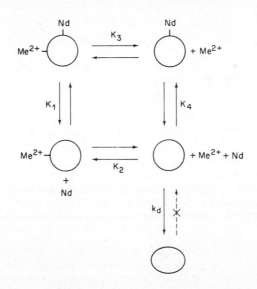

Fig. 19. The reaction scheme among actin monomer, nucleotide and divalent cation. The values of binding constants K_1, K_2, K_3, and K_4 at room temperature and pH = 8.0 were: $K_1 = 2.7 \times 10^6 \, M^{-1}$; $K_2 = 3.6 \times 10^6 \, M^{-1}$; $K_3 = 7.7 \times 10^7 \, M^{-1}$ and $K_4 = 1.3 \times 10^5 \, M^{-1}$; $k_d = 0.6 \, \text{min}^{-1}$.

the binding between ATP and G-actin, it was useful to remove free nucleotides in the G-actin solution by the resin treatment (*58*). Since the binding constant of ATP to G-actin is very large, almost all remaining ATP is bound to G-actin. However, the fact that actin molecules without bound nucleotides lose polymerizability spontaneously complicates the situation. Therefore the reaction scheme of Fig. 19 is assumed, in which denaturation is irreversible. The binding constant between G-actin and nucleotide and the kinetic constant of the denaturation process can both be determined by following denaturation in the presence of various concentrations of nucleotides. By this method, on the basis of a simplified scheme, Asakura found these constants to be (58)

$$K_{\text{ATP}} = 2.3 \times 10^6 \ M^{-1} \quad \text{and} \quad k'_d = 0.96 \ \text{hr}^{-1}$$

The binding of ATP is exothermic and the binding enthalpy $\Delta H_{\text{ATP}} = -24$ kcal/mole. The activation enthalpy for denaturation $\Delta H_d^* = 6$ kcal/mole.

Since G-actin molecules show different UV absorption spectra depending on whether ATP is bound or not, the degree of binding can be estimated from difference spectrum measurements (38). As shown in Fig. 20, the

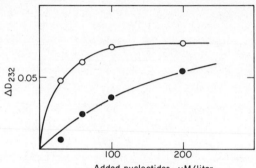

Fig. 20. The relation between the magnitude of difference spectrum at 232 mμ and the concentration of added ATP ($\bigcirc$) or ADP ($\bullet$). The difference ΔD is considered to be proportional to the increase of the amount of G-actin having bound nucleotide.

binding constant of the same order as that from the denaturation experiment was obtained. A similar experiment indicated that the binding constant of ATP is about 30 times greater than that of ADP. According to an examination of the exchangeability of different kinds of bound nucleotides, the binding constant decreases in the order shown in Table II (59). The binding depends greatly on the structure of the nucleotide bases.

TABLE II

Interaction between Actin and Nucleotides

	Nucleotides and their binding constants (M^{-1})					
	ATP	ITP	ADP	CTP	IDP	GTP
G-Actin	4×10^7	4×10^6	4×10^5	4×10^5	10^5	5×10^4
F-Actin	1.3×10^7	3×10^6	3×10^5	4×10^4	10^5	1.5×10^5

C. Interaction with Divalent Cations

One mole of G-actin binds 1 mole of divalent cation. The binding of divalent cations with G-actin stabilizes the binding of nucleotides (58). By taking such a correlation into consideration and assuming that denaturation occurs only in the state of G-actin having neither nucleotides nor divalent cations, all binding constants are determined in a certain solvent condition as summarized in the legend of Fig. 19 (60).

TABLE III

Binding Constants between Actin and Divalent Cations[a]

Salt	G-Actin	F-Actin	Salt	G-Actin	F-Actin
Be	0.03	0	Ni	0.1 ~ 0.2	0.4
Mg	0.2 ~ 0.3	0.4	Fe	0.1 ~ 0.2	0.1
Ca	1	1	Mn	0.6 ~ 1.0	0.7
Sr	0.03	0	Zn	0.1 ~ 0.2	0.5
Ba	0.01	0	Cd	0.4 ~ 0.8	1.0
Co	0.15 ~ 0.3	0.8			

[a] The binding constant of Ca^{2+} was normalized as unity.

The binding constants of different kinds of divalent cations were compared, and the strongest binding was found in cations of an intermediate size, namely, calcium and manganese ions (Table III) (43).

VI. EFFECT OF ENVIRONMENTS

A. Effects of Salts and pH on Interaction between Actin Monomers

Under the conditions in which the denaturation rate is negligible, the final state of equilibrium or dynamic balance between G- and F-actins is determined by the solvent condition independently of the initial state of actin.

With increasing concentration of monovalent salts, the polymerization rate increases and the depolymerization rate decreases. The balance shifts to the polymer state, and at the optimal concentration, that is, at about 0.1 N in monovalent salts, almost all actin is polymerized. When the salt concentration exceeds the optimal value, the depolymerization rate increases and the balance returns to the monomer state $(8,61)$. I^- and SCN^- strongly depolymerize actin. In other words, the critical actin concentration for the helical polymerization of actin decreases with increasing salt concentration, reaching a minimum at the optimal salt concentration and then increasing with further increase in the salt concentration.

The polymerizing action of salt ions is probably attributable to screening of the electric charge of actin monomers. Their apparent charge estimated from the electrophoretic velocity decreases with the addition of salts (10). However, the salt ion itself is not required in the bond between actin monomers.

A decrease in pH from alkaline to acidic increases the polymerization rate of actin, and when the pH approaches the isoelectric point (4.7), the random globular aggregation of actin is superimposed on the polymerization to F-actin. The depolymerization rate seems to be minimum at neutral pH (62).

As previously mentioned, divalent cation tightly bound to G-actin has a specific effect depending on the species of the cation. However, divalent cations additionally bound have a nonspecific effect favoring polymerization $(44,45)$.

B. Effects of Temperature

The G–F equilibrium depends on the temperature. On raising the temperature the polymerization rate increases, reaching a maximum at a temperature higher than room temperature. A further increase in temperature brings about acceleration of the denaturation process, which makes it difficult to follow the G–F equilibrium. In a wide temperature range, however, polymerization is found to be endothermic (14).

From the dependence of the critical actin concentration on temperature, the enthalpy change associated with polymerization is determined by the use of the relation

$$c_c K_h = 1$$

or (11)

$$\ln c_c = -\ln K_h = \Delta H/RT - \Delta S/R$$

A linear relation was found between $\log c_c$ and $1/T$, and the value of ΔH was estimated to be about $10 \sim 15$ kcal/mole, under a wide range of the solvent conditions (14). It must be noted here that the balance between G- and F-actin is usually established in the presence of ATP, where the polymerization is accompanied by irreversible splitting of ATP. Therefore under these conditions, ΔH determined by the above method is not necessarily the amount of heat actually absorbed in the process of polymerization. However, in the latest investigation the enthalpy change of the same order as above was obtained for polymerization of G-actin having ADP instead of ATP (*20*). This ΔH may be considered to be the true enthalpy of the interaction between actin monomers in F-actin.

Thus the decrease in free energy associated with bond formation between actin monomers in F-actin comes not from an energy decrease but from

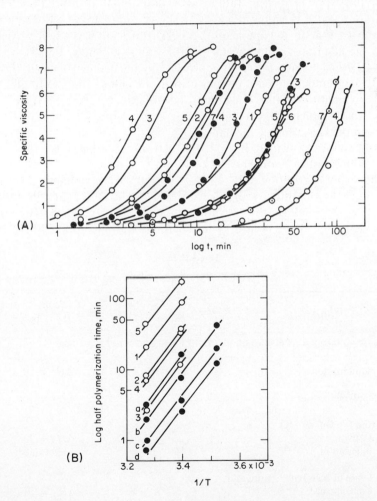

Fig. 21. (A) The time course of polymerization of actin at various ionic conditions and various temperatures, where the abscissa is expressed in the logarithmic scale of time. The lines with the same number correspond to the same ionic condition at different temperatures and the same symbol to the same temperature. (B) The relation between the half polymerization time obtained in Fig. 21A and the reciprocal of temperature. ◯, Various concentrations of KCl; ●, various concentrations of MgCl$_2$. The gradients of the straight lines at different ionic conditions were all equal.

an entropy increase. Such a property of the binding has been found in the polymerization of various globular protein molecules (*63,64*). The importance of hydrophobic interaction and the change of the structure of surrounding water molecules associated with polymerization must be considered.

The binding constant K_h is equal to the ratio of the polymerization rate k_+ to the depolymerization rate k_-. It is desirable to measure these rates separately, but it is rather difficult. As shown in previous sections, in the G-actin to F-actin polymerization process, nucleation is involved aside from the addition of monomers to preexisting polymers. Equation (10) indicates that polymerization is controlled by the geometric mean of two kinetic constants of nucleation and growth, k^* and k_+. For example, the half-polymerization time must be inversely proportional to $(k^*k_+)^{1/2}$.

Polymerization is faster at higher temperatures. From the linear relations between the logarithm of the half-polymerization time and the reciprocal of temperature shown in Fig. 21, the activation enthalpy for $(k^*k_+)^{1/2}$ was estimated to be about 20 $\sim$ 25 kcal/mole.

The depolymerization rate k_- was measured by suddenly bringing F-actin to the condition favorable for depolymerization. From the dependence on temperature, the activation enthalpy for depolymerization was estimated to be about 10 kcal/mole. Therefore the activation enthalpy for

TABLE IV

Thermodynamic Data^a

Reaction	Rate constant	Enthalpy change (kcal/mole)
(a) G–F equilibrium (condensation)	$G = k_-/k_+$	$\Delta H = 10$
(b) Polymerization	$(k_+ k^*)^{1/2}$	$\Delta H\ddagger = 20\text{–}25$
(c) Depolymerization	k_-	$\Delta H\ddagger = 10$
(d) From (a), (b), and (c)		
Growth	k_+	$\Delta H\ddagger = 20$
Nucleation	k^*	$\Delta H\ddagger = 20$
(e) ATPase at high temperature and exchange of Ca^{2+} or ADP	$k\dagger$ or $(k_+ GF_N, k_- F_N)$	$\Delta H\ddagger = 20\text{–}25$
(f) Sonic ATPase	$(k_+ G, k_-)$ (and $k\dagger$)	$\Delta H\ddagger = 10$

^{a} G, concentration of G-actin; F_N, number concentration of F-actin or the number concentration of the ends of polymers; $k\dagger$, rate constant of intrapolymer structure change; ΔH, equilibrium enthalpy difference; $\Delta H\ddagger$, activation enthalpy.

polymerization or, more exactly, growth reaction k_+, must be about $+20$ kcal/mole and that for nucleation k^* must also be of the same order. These results are summarized in Table IV (*20*).

A most interesting finding was that the activation enthalpy for polymerization is independent of the ionic condition of the solvent, that is, of the concentration and the species of mono- and divalent salts and pH. The dependence of the polymerization rate on ionic condition comes not from the change in activation enthalpy, but from the change in activation entropy. The activation entropy assumes a minimum at the optimal salt concentration. The equilibrium enthalpy difference ΔH is also insensitive to the ionic condition (*8*). The reason for these phenomena is not known, but it must be noted that this does not necessarily exclude the role of coulomb interaction and its screening in the binding between actin monomers. Coulomb interaction involves both energetic and entropic characters if the change in the dielectric constant of water with temperature is taken into consideration. (The product of the dielectric constant and the absolute temperature is insensitive to the change in temperature.)

C. Effects of Pressure

Ikkai and Ooi found that hydrostatic pressure shifts the polymerization equilibrium of actin to the monomer state and finally induces irreversible denaturation (*65*). Thermodynamically, this means that polymerization is associated with an increase in the volume of the actin solution. Actually, Noguchi et al. confirmed by dilatometry a large volume increase of about 400 ml/mole of actin in the polymerization process (*66*). This may be attributed to a change in the water structure and/or the void volume of the protein molecule associated with polymerization.

D. Effects of Organic Solvents

Alcohols and acetone favor polymerization at low temperatures and at room temperature (*45*). Even at very low salt concentrations these solvents induce polymerization. It is remarkable that they change the temperature dependence of polymerization (*20*). That is, activation enthalpy decreases in proportion to the increase in the concentration of alcohols and becomes negative, which means that the polymerization rate decreases with increasing temperature in the presence of high concentrations of alcohols. There is a specific temperature at which the polymerization rate does not depend on the alcohol concentration. At higher temperatures alcohols decelerate polymerization. Such specific effects of alcohols also suggest the importance of the hydrophobic interaction between actin monomers.

Urea favors depolymerization. It does not change the activation enthalpy for polymerization.

Thus the polymerization of actin is associated with increases in enthalpy, entropy, and volume. The effects of various reagents show that ionic and hydrophobic interactions both take part in the polymerization. Although F-actin is a polymer having a special regular structure constructed by bonds between specific sites of actin monomers, the following two phenomena suggest the importance of the general property of the polypeptide chain. Nagy and Jencks (67) found by comparing many different kinds of reagents that there is very good parallelism between the depolymerizing action on F-actin and solubilizing action on a small model peptide. There is also good parallelism between the polymerization–depolymerization of native G-actin and the random aggregation and disaggregation of denatured actin under various solvent conditions (11,68).

E. Renaturation of Carboxymethylated Actin

In order to elucidate the structural origin of the interaction between protein molecules, it is useful to start with completely unfolded molecules and examine their interaction at different states of refolding. Actin, however, spontaneously denatures when bound nucleotides are lost. This suggests that nucleotides are indispensable for maintenance of the native structure of actin molecules. The polypeptide chain can not maintain its functional site by itself. In the absence of bound nucleotides, the unfolded structure seems to be more stable than the native structure. Therefore renaturation of unfolded actin without nucleotides is impossible. Even in the presence of nucleotides, renaturation has never been performed successfully. However, nucleotides are not absolutely necessary to form F-actin and the polypeptide chain of actin molecules is responsive to the binding between them. Very recently, Mihashi (69) found that unfolded actin molecules can recover their polymerizability after chemical modification. His experiment was carried out in the following way.

G-Actin was dissolved in a high concentration of urea (5 M), where its α-helical structure was almost lost, according to the optical rotation measurement. Then, in the presence of 2.5 M urea, almost all sulfhydryl groups of actin were carboxymethylated by the reaction with iodoacetate. The reagent and urea were removed by dialysis against water at neutral pH.

Carboxymethylated actin was collected by isoelectric precipitation and dissolved again at neutral pH. Then the actin gave a viscous solution, indicating the formation of polymers. These polymers were not F-actin, but after a long incubation at a little lower pH, they were gradually transformed into the two-stranded helical polymer. At various points this polymer has the same structure and function as normal F-actin.

The modified actin is not exactly the same protein molecule as native actin. Actually, it has properties different from the original actin; for example, it is not in a monomeric state even in a salt-free solvent. However, it maintains the most important function of actin, that is, the ability to polymerize to F-actin. This function was recovered after the intramolecular three-dimensional structure was destroyed. The nucleotide-binding ability was also recovered after renaturation.

From this result it is definitely concluded that polymerizability is wholly attributable to the polypeptide chain of actin. No additional substance is necessary.

Thus in near future it will be possible to start with the unfolded polypeptide chain of actin and follow the process of obtaining the function in connection with the formation of the three-dimensional structure.

VII. POLYMORPHISM OF POLYMER

A. Possible Polymorphism of Actin Polymer

Now let us return to the problem of the dynamic equilibrium between G- and F-actin. Each actin molecule undergoes the cyclic change from G to F, and then to G again. The main process in the solution of coexisting G- and F-actin is association and dissociation of monomers at the ends of polymers. The other process may be association between polymers and their fragmentation. When, however, equilibrium and kinetic analyses suggested that F-actin is a helical polymer of globular actin monomers, the following idea was presented concerning the possible polymorphism of actin polymers (5,12,70).

In the structure of F-actin, each monomer binds with probably four neighboring monomers through two or three different kinds of bonds. Therefore we can assume that the process of formation and detachment of each different kind of bond occurs separately. If one kind of bond, for example, the bonds between two monomers in the same strand along the polymer axis, were all broken without breaking the bonds between monomers in different strands, the two-stranded helical polymer would be transformed into a single-stranded (zig-zag) polymer, as shown in Fig. 22. Or, if bonds between two strands were all broken, the two-stranded helical polymer would be separated into two single-stranded polymers. Thus it can be presumed the dynamic change in the polymer structure occurs without its fragmentation.

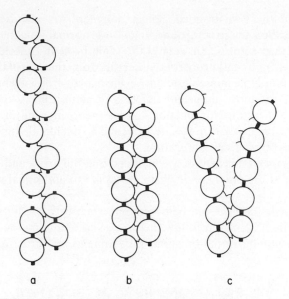

Fig. 22. Polymorphism of the helical polymer (5).

If such dynamic phenomena, including the G ⇌ F cycle and interpolymer and intrapolymer cyclic processes, occurred in the presence of ATP, the splitting of ATP would take place and bound nucleotides and/or divalent cations that were not easily exchangeable in the state of F-actin would be exchanged with those in the solvent. Some experiments undertaken to examine these possibilities are described below.

B. Effects of Sonic Vibration

Asakura (71) found that sonic vibration induces rapid splitting of ATP in an F-actin solution. At the salt concentration optimal for the stability of F-actin, ATP splitting continues during sonic vibration and ceases when the vibration stops. The splitting rate is of the order of 1 mole of ATP per 1 mole of actin every several minutes. This is very much faster than the spontaneous splitting in the steady state of G- and F-actin coexistence at the intermediate salt concentration. A fast exchange of ADP and divalent cations bound to F-actin was also observed with sonic vibration (72). For example, if F-actin having Ca^{2+} is placed under sonic vibration in the presence of Mg^{2+}, bound Ca^{2+} are all exchanged by Mg^{2+}, and if in the presence of EDTA, bound Ca^{2+} are all removed (62,73).

The splitting rate under sonic vibration was found to be proportional to the concentration of F-actin under constant solvent conditions. Since at the optimal salt condition no appreciable amount of G-actin was present in the solution, ATP splitting during the vibration was considered to be exhibited by F-actin itself. The intrapolymer structure change of F-actin, that is, the partial interruption of the helical polymer structure, was assumed to be caused by sonic vibration, as shown in Fig. 23. At the interrupted point actin monomers have a local structure equivalent to G-actin,

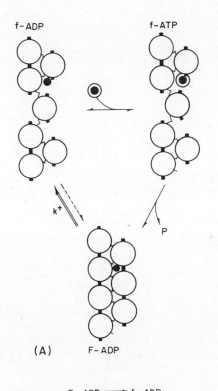

F – ADP ⇌ f – ADP
f – ADP ⇌ f + ADP
f + ATP ⇌ f – ATP
(B) f – ATP ⟶ F – ADP + P

Fig. 23. (A) Schematic representation of a cycle of conformational change in the actin polymer. ●, ADP; ⊙, ATP. (B) The reaction scheme corresponding to Fig. 23A.

where bound ADP is exchanged with ATP in the solvent and this ATP is dephosphorylated in the process of re-formation of the bond between actin monomers to recover the helical polymer structure. In fact, exchange and splitting were found to take place in parallel with each other. Thus the reaction scheme in Fig. 23 seems to be reasonable.

This experiment was extended by Asakura et al. (*16*) and later by Nakaoka and Kasai (*74*) to an intermediate salt concentration. Under these conditions, in which G- and F-actin coexist, the rapid splitting of ATP continues not only during sonic vibration but also after the vibration is stopped, as shown in Fig. 24. The relation between the splitting rate during vibration and the actin concentration is shown in Fig. 25. Splitting is found only above the critical actin concentration and the rate is always

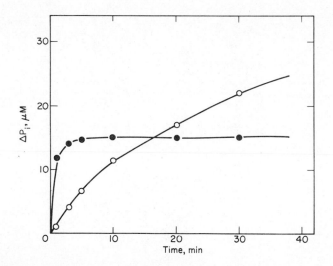

Fig. 24. The splitting of ATP after stopping the sonic vibration. Actin concentration 1.5 mg/ml; KCl, 5 mM; tris–HCl (pH 8.0), 5 mM; ATP, 0.5 mM; at 20°C. Sonic vibration was given for 5 min. ○, in the solution left standing after the vibration; ●, in the solution to which salts (KCl, 60 mM and MgCl$_2$, 5 mM) were added just after stopping the vibration.

proportional to the F-actin concentration. The rate per unit weight of F-actin depends on the salt concentration. It is remarkable that the critical concentration for ATP splitting during sonic vibration is nearly equal to the critical concentration for the final steady state of the G–F balance without the vibration.

The concentration of G-actin coexisting with F-actin is not changed by the vibration. This phenomenon would be expected from the theory of helical polymerization described previously. It is similar to a situation in which the concentration of saturated solute molecules coexisting with crystals is not changed by a vibration that breaks down large crystals to

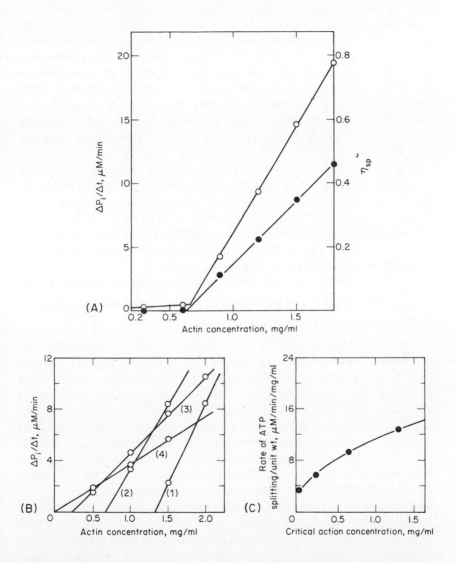

Fig. 25. (A) Rate of ATP splitting during sonic vibration and the final viscosity after stopping vibration, at various concentrations of actin. The solvent condition is the same as in Fig. 24. ●, ATP splitting; ○, viscosity. (B) The relations between the rate of ATP splitting during sonic vibration and the total actin concentration, at various salt concentrations. The concentration of KCl was varied from 3 to 100 mM. (C) The relation between the rate of ATP splitting per unit weight of F-actin under sonic vibration and the critical actin concentration obtained from the data of Fig. 25B.

small powders. In the case of actin also, sonic vibration breaks F-actin into small polymers, as confirmed by electron microscopy (74).

The G $\rightleftharpoons$ F cycle is increased by sonic vibration because of the increase in the number of (the ends of) actin polymers. Therefore rapid ATP splitting can be expected, even if the intrapolymer structure change does not occur. Nakaoka and Kasai tried a few experiments to determine which mechanism operates during vibration as the cause of the rapid ATP splitting, the G $\rightleftharpoons$ F cycle at the ends of polymers or the cycle of some intrapolymer structure change. Their results showed that at intermediate salt concentrations rapid ATP splitting is mainly attributable to increased G $\rightleftharpoons$ F cycles.

However, it is not certain whether or not the splitting rate can be quantitatively explained by the increased number of the ends of polymers. It should also be mentioned that splitting during vibration takes place even if the critical actin concentration approaches zero (see Fig. 25). Moreover, electron micrographs suggest that actin polymers with interrupted structures are produced during the vibration. Therefore the intrapolymer cycle proposed earlier by Asakura et al. is also probable.

C. Splitting of ATP and Exchange of Nucleotides and Divalent Cations under Various Conditions

The rapid splitting of ATP and exchange of bound nucleotides and divalent cations were also observed by Asai and Tawada at an elevated temperature (Fig. 26) (75). At about 50°C the rate attained the same order as that during sonic vibration. (Above about 65°C F-actin begins to be irreversibly degraded to monomers.) For this ATP splitting also, two mechanisms can be supposed. Taking into consideration the temperature dependence of the rate constants of nucleation, polymerization, and depolymerization, it is difficult to explain the rapid ATP splitting at high temperatures by the G $\rightleftharpoons$ F cycle at the ends of F-actin (20). It is likely that a cycle involving the intrapolymer structure change takes place at high temperatures. If so, detachment of the bond between actin monomers would be the rate-limiting process. In the presence of EDTA and a high concentration of ATP, the rate of ATP splitting becomes very large. This condition is favorable for the detachment of the bond without denaturation of actin molecules.

The other condition for the rapid exchange and splitting of nucleotides was the hydrostatic pressure of about several hundred atmospheres (65).

Thus we found that sonic vibration, high temperature, and high pressure caused the rapid splitting of ATP and the rapid exchange of bound nucleotides and divalent cations in F-actin solutions. They were considered to be associated with the G ⇌ F cycle and/or the intrapolymer cycle. In principle, these cycles might be expected under ordinary solvent conditions even though they were very slow. Therefore the rate of exchange of bound nucleotides and divalent cations was extensively investigated by Kasai over a wide range of solvent conditions: pH, the concentrations of monovalent and divalent salts, and the concentration of nucleotides (62).

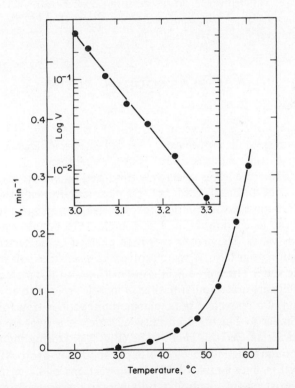

Fig. 26. The rate of ATP splitting in F-actin solutions at high temperatures. Actin concentration 2.4 mg/ml; KCl, 60 mM; tris–HCl (pH 8.0), 10 mM; MgCl₂, 1 mM; ATP, 0.8 mM. The insert is the logarithmic plot of the splitting rate against the reciprocal of temperature.

The mean life-time of the exchange is longest at neutral pH and 0.1 M monovalent salts; it is of the order of 1 week at low temperature and 1 or 2 days at room temperature. For example, if F-actin having Ca^{2+} is long incubated in the presence of Mg^{2+}, Ca^{2+} are slowly replaced by Mg^{2+} even without sonic vibration. In relation to this, it should be noted that divalent cations bound to F-actin in muscle are mainly Mg^{2+}, not Ca^{2+} (76).

Summarizing the above results, we have no direct evidence for the occurrence of a cycle of intrapolymer structure change such as a partial helical to linear transformation, although we have some suggestive data. The interrupted structure would be formed only as a transient state having a life-time too short to be detected by usual physicochemical methods. Positive results on the stable polymorphism of actin polymers have not been obtained yet. However, an interesting phenomenon was found in the case of an actinlike protein obtained from plasmodium. This protein was extracted and purified by Hatano (77).

VIII. PLASMODIUM ACTIN

A. Plasmodium G-Actin

Plasmodium of a myxomycete consists of an inner sol-like protoplasm and an outer gel-like protoplasm. The sol shows a vigorous streaming which reverses its direction at intervals of a few minutes.

An actinlike protein was selectively absorbed by muscle myosin from a water extract of the dried powder of this plasmodium treated with acetone (77,78). The actin-myosin complex formed was treated again with acetone and from the dried powder of the complex the actinlike protein was extracted by water. The protein was then purified by isoelectric precipitation and salting out (78). When dissolved in a salt-free solvent, it is in a monomeric state. The amino acid composition and the molecular weight of this protein are quite similar to that of muscle actin (Table I) (78). Upon the addition of monovalent salts, plasmodium actin is transformed into a polymer similar to F-actin from muscle, which is a two-stranded helical polymer under an electron microscope (Fig. 27). During the transformation ATP bound to actin molecules is dephosphorylated into ADP (79).

The electron micrograph of fixed thin sections of living plasmodium shows the existence of F-actinlike polymers in the protoplasm, especially in the region exhibiting streaming (80). A myosinlike protein was also extracted from the plasmodium (81–83) and the interaction between these two proteins coupled with the ATP splitting is considered to be the elementary process of the force generation for protoplasmic streaming, as in the case of muscular contraction.

B. Two Types of Polymers of Plasmodium Actin

The whole situation concerning plasmodium actin seemed to be very similar to the case of muscle actin. However, investigation of the similarity between two actins led Hatano to the discovery that plasmodium actin can make the other type of polymer (79). On the addition of $MgCl_2$ to plas-

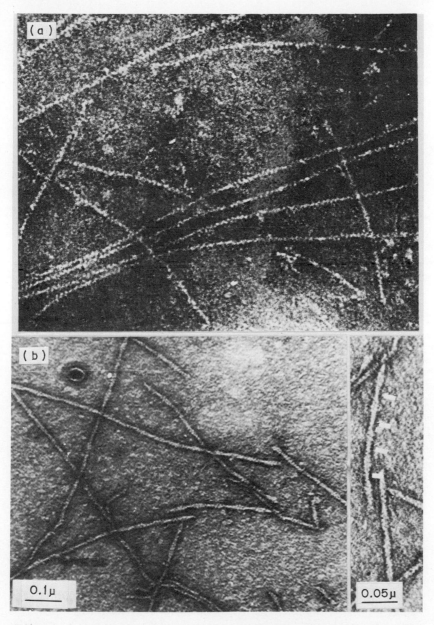

Fig. 27. (a) An electron micrograph of plasmodium F-actin reconstructed in vitro. (b) An electron micrograph of plasmodium F-actin separated from actomyosin complex extracted from plasmodium.

modium G-actin, polymers are formed; they have a high sedimentation coefficient of the same order as F-actin but show lower viscosity (Fig. 28). In electron micrographs they appear to be flexible polymers, or sometimes globular polymers, which are clearly distinguishable from normal F-actin. This new type of polymer was tentatively named Mg-polymer by Hatano. Magnesium ions are the most effective in making this polymer.

The relation between the state of actin polymers and the concentration of magnesium ion was investigated in the following way. Plasmodium G-actin was first incubated in various concentrations of magnesium ions in the absence of monovalent salts at a low temperature; no polymerization took place and calcium ions bound to G-actin were (partially) exchanged with magnesium ions. Then monovalent salts were added in the presence

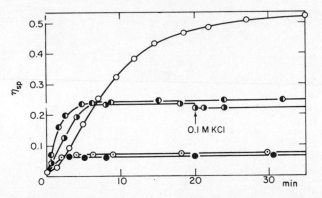

Fig. 28. The polymerization of plasmodium actin at various salt conditions. Actin concentration, 0.92 mg/ml; ATP, 0.05 mM; cysteine, 3 mM (pH 8.0); at 21°C. $\bigcirc$, KCl 0.1 M; $\odot$, MgCl$_2$, 2 mM; $\bullet$, MgCl$_2$, 2 mM and KCl 0.1 M; $\leftmoon$, CaCl$_2$, 2 mM; $\rightmoon$, CaCl$_2$, 2 mM and KCl, 0.1 M.

of magnesium ions and the temperature was elevated to room temperature. After polymerization the final value of viscosity was measured. As shown in Fig. 29, the viscosity decreased linearly with the increasing amount of magnesium ions, reaching a constant value at approximately equimolar amounts of magnesium ion and actin monomer. The transformation between normal F-actin and Mg-polymer takes place quantitatively, depending on the amount of actin monomers having bound Mg^{2+}.

Transformations among three states of plasmodium actin, G-actin, F-actin, and Mg-polymer are reversible. Mg-polymer can be slowly transformed into F-actin by dialyzing out magnesium ions in the presence of monovalent salts. F-actin can be slowly transformed into Mg-polymer by adding magnesium ions. A most interesting observation is that Mg-polymer shows the steady splitting of ATP while F-actin does not. The ATP-splitting rate of Mg-polymer is of nearly the same order as that of muscle F-actin during sonic vibration or at high temperatures. The fact that this ATPase activity is characteristic of the structure of Mg-polymer was proved by comparing the ATP splitting between Mg-polymer and F-actin in the same solvent condition. The relation between ATP splitting and the exchange of radioactive ADP bound to Mg-polymer suggests that splitting takes place on actin monomers with equal probability everywhere along the polymer, not only at the end of the polymer. Therefore, splitting is not associated with the polymerization–depolymerization cycle at the end but probably with the intrapolymer cycle.

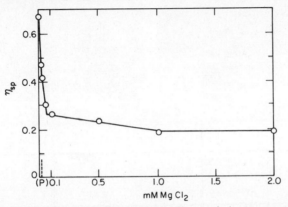

Fig. 29. The formation of Mg-polymer of plasmodium actin in the presence of various concentrations of $MgCl_2$. Actin was previously incubated in the solution of $MgCl_2$ at the concentration indicated in abscissa. Actin concentration, 1.1 mg/ml; the final salt concentration: KCl, 0.1 M; ATP, 0.05 mM; cysteine, 3 mM (pH 8.2).

Rapid transformation of Mg-polymer to normal F-actin was found under two conditions. Mg-polymer is transformed into F-actin by raising the temperature to about 50°C (*84a*). The rate of the rapid viscosity increase after the rise in temperature is proportional to the concentration of Mg-polymer, suggesting that the rate-limiting process is an intrapolymer structure change.

The other remarkable finding is that a high concentration of ATP induces the transformation of Mg-polymer to normal F-actin (*85*). Electron micrographs show that flexible Mg-polymers become stiff (in Fig. 30), and the increment in the viscosity after the addition of ATP is consistent with the change from a flexible chain to a rigid rod. [The laser light scattering data also support this interpretation (*85a*).] The steady ATP splitting is lost with this transformation. In other words, the activity of ATP splitting of actin polymers is inhibited by the substrate ATP.

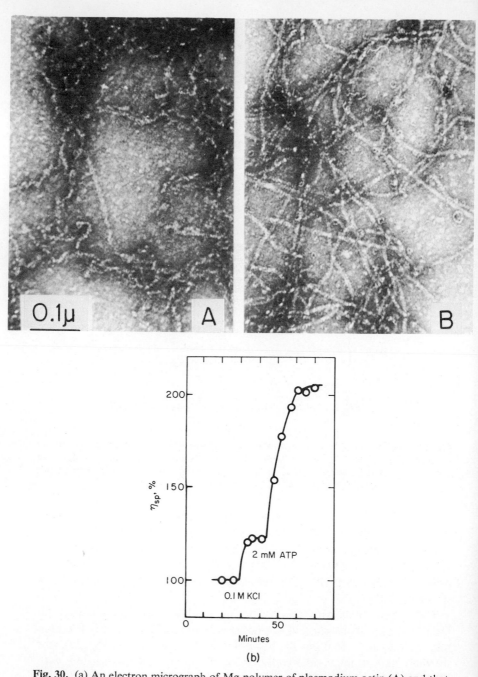

Fig. 30. (a) An electron micrograph of Mg-polymer of plasmodium actin (A) and that of F-actinlike polymer obtained by the addition of ATP to the Mg-polymer (B). (b) The increase of viscosity indicating the transformation of Mg-polymer to F-actin induced by ATP. Actin concentration: 1 mg/ml; $MgCl_2$, 2 mM; tris–maleate (pH 7.0), 5 mM; cysteine, 3 mM; ATP, 0.05 mM; at room temperature.

Thus on the basis of electron microscope observation and steady ATP splitting, it may be assumed that Mg-polymer is an interrupted (partially linear) polymer having a flexibility greater than normal F-actin, and that the intrapolymer transformation is induced by ATP (Fig. 31). Rapid reverse transformation from normal F-actin to Mg-polymer is also possible by reduction of ATP concentration (*85a*). It is interesting that the concentration of ATP required for the transformation of Mg-polymer to F-actin is of the same order as its physiological concentration in plasmodium.

The stable polymorphism and the transformation of polymers described above are specific to actin from plasmodium. Actin from muscle does not exhibit such phenomena. Recently, an attempt was made to copolymerize

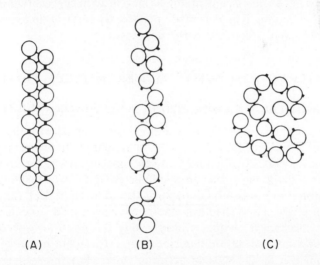

(A)   (B)   (C)

Fig. 31. The polymorphic transition assumed for plasmodium actin polymers.

these two kinds of actin at various ratios (*84*). The copolymer formed in the presence of magnesium ions had the activity of steady ATP splitting, which decreases rapidly with an increasing fraction of muscle actin in the copolymer (*84*). The relation between the ATP-splitting activity and the ratio of the two actins suggests that the interaction between neighboring (plasmodium) actin monomers is necessary for the splitting.

C. Polymers of Chemically Modified Muscle Actin

In the case of muscle actin, the stable polymorphism of its polymer was not found. However, in chemically modified muscle actin, polymers having a structure different from normal F-actin can be formed. An example is

muscle actin treated with Salyrgan. When polymerized in the presence of
Mg ions, the polymer showed the activity of steady ATP splitting (40).

Another example is actin renatured after carboxymethylation in the
unfolded state (69). As described previously, this actin can form two types
of polymer. When dissolved at neutral pH after isoelectric precipitation,
it forms globular or flexible polymers, which are converted into two-
stranded helical polymers by long incubation at lower pH. These two
polymers are very different in the property of interaction with myosin.
They can both bind with myosin. However, the former does not show
activation of myosin ATPase, while the latter shows high activation
depending on the species of divalent cations present in the solution.

These results on chemically modified muscle actin suggest that the polymorphism of actin polymers is made possible by very subtle regulation of
the intramolecular structure of actin monomers.

IX. INTERACTION WITH OTHER MUSCLE PROTEINS

A. Interaction of Actin with Myosin in the Absence of ATP

The interaction between F-actin and myosin coupled with the splitting
of ATP is the elementary process of muscular contraction and other
cellular movements. At physiological salt concentration, myosin strongly
binds with F-actin and forms an insoluble complex, which shows superprecipitation (in vitro contraction) upon the addition of ATP (2). A great
number of physical and chemical studies have been made to elucidate the
molecular mechanism of this process. Only the physical aspects of the
actin–myosin interaction are described here. The main interest is in the
effect of myosin on the state of actin polymers. For the analysis of this
interaction, it is often more convenient to use HMM—the product of
limited tryptic digestion of myosin—which is soluble at the physiological
salt concentration (86).

Myosin or HMM does not bind or weakly binds with G-actin, while it
strongly binds with F-actin in the absence of ATP. In electron micrographs
Huxley found that the complex shows an "arrowhead" structure, giving
the evidence of polarity in the structure of F-actin (Fig. 12) (23). Why
HMM or myosin binds with F-actin but not with G-actin is an important
but unsolved problem. At saturation, one HMM molecule binds with one
actin monomer in F-actin. Accordingly, it is likely that polymerization
makes the structure of the interaction site in each actin monomer favorable
for binding with myosin. The stronger binding of HMM with F-actin than
with G-actin indicates that HMM can shift the G–F equilibrium toward
F-actin. Actually, the critical actin concentration for polymerization is

decreased (*88*) and the rate of polymerization increased by myosin (*2*). HMM promotes the nucleation of actin polymers (*87*).

The properties of the HMM–F-actin complex were examined by Tawada at various ratios of the two proteins (*88*). With increasing binding of HMM to F-actin, the degree of flow birefringence decreases and the UV absorption of the complex gives a remarkable difference from the sum of the absorptions of the two components. The decrease of birefringence and the difference of absorption have maximum values at the HMM/actin molar ratio of about 1 : 6 (Fig. 32). With further increase in HMM, the birefringence recovers to the value of F-actin and the spectrum becomes equal to the sum of the components.

A similar behavior of F-actin was recently found in the experiment on the quasielastic scattering of laser light by Fujime and Ishiwata (*88a*). As described before, F-actin is not a rigid rod but its intrapolymer Brownian movement brings the Doppler broadening of frequency of scattered light. The flexibility of F-actin estimated by this method remarkably increases with the addition of HMM, reaching a maximum at a molar ratio of about 1 : 6, and then returns to the original value. (See footnote *a*, p. 322.)

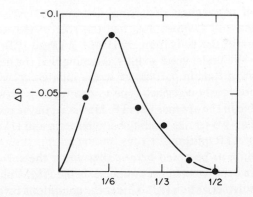

Fig. 32. The difference spectrum of the UV absorption of the HMM–F-actin complex against the simple sum of HMM and F-actin at various ratios in the complex. F-actin concentration, 0.5 mg/ml; KCl, 60 mM; tris–HCl (pH 8.0), 10 mM.

According to electron microscope observation, when the number of bound HMM molecules is small, the angle between the long axis of HMM molecule and the polymer axis of F-actin is larger than in the case of the arrowhead structure (*89*). The type of binding between HMM and F-actin depends on the ratio of the two proteins. At small HMM/F-actin ratios, the binding of HMM may induce some distortion in the structure of F-actin and increase its flexibility which is the origin of the difference spectrum.

B. Interaction of Actin with Myosin in the Presence of ATP

At high salt concentrations the F-actin–myosin complex is dissociated by ATP. At low salt concentrations superprecipitation of the complex is brought about by ATP. Under these conditions the ATPase activity of myosin or HMM is activated by combination with F-actin. If an F-actin pellet is gently dissolved in a solvent of very low salt concentration, F-actin is not readily dispersed, but by sonic vibration it is completely dispersed to G-actin. Therefore solutions of F-actin and G-actin can be obtained under the same solvent conditions. When HMM is added to these solutions and the ATPase activity is measured in the presence of Mg ions, high ATPase activity is observed in the case of F-actin but no activity is observed in the case of G-actin (*90*). The polymer structure of F-actin and/or the structure of actin monomer in F-actin is required for the activation of ATPase of HMM.

The binding constant of HMM and F-actin in the presence of ATP can be determined from the relation between the ATPase activity of HMM and the concentration of F-actin added (*91*). It is much smaller than that in the absence of ATP.

By the use of copolymers of normal G-actin and carboxymethylated G-actin, Tawada and Oosawa analyzed the role of the arrangement of actin monomers in the activation of HMM ATPase (*92*). Their experimental results were understood well by assuming that the neighboring two actin monomers in F-actin participate in the interaction with one HMM molecule. As previously described, one-to-one binding between actin and HMM is possible in the absence of ATP. However, in the presence of ATP it is probable that two-to-one binding between actin and HMM takes place in the process of ATP splitting.

Another finding to be noted here is that under the solvent conditions in which F-actin depolymerizes spontaneously the HMM plus ATP system accelerates depolymerization (*93*). Under the conditions favorable to polymerization, the same system accelerates polymerization, without a shift in the G–F equilibrium.

From these facts it is concluded that the interaction of HMM molecules having ATP with actin monomers in F-actin is influenced by the state of neighboring actin monomers, and that the interaction between neighboring actin monomers in F-actin is influenced by their interaction with HMM molecules.

The interaction between HMM and F-actin in the presence of ATP probably is not permanent but cyclic. During this cycle some cyclic change is expected to occur in the state of F-actin. For the purpose of detection of such a change, the exchange of nucleotides and divalent cations bound to F-actin was followed; but in the case of HMM no appreciable exchange was observed (62).

In the case of myosin, however, a large amount of rapid exchange was found. In relation to the molecular mechanism of muscular contraction, it has been an important problem whether or not the polymer structure of F-actin or thin filaments changes during superprecipitation or contraction by ATP. In the in vitro experiment using the myosin–F-actin complex, Kasai et al. observed the release of radioactive divalent cations bound to F-actin under conditions of superprecipitation (Table V) (37,62,94). Szent-Györgyi carried out the same kind of experiment extensively by following

TABLE V

Exchange or Release of Calcium (^{45}Ca) Bound to F-Actin during Superprecipitation[a]

Sample No.	Divalent cations and/or EDTA	ATP	Super-precipitation	Exchange or release of bound ^{45}Ca, %
1	None	No	−	0
2	Mg^{2+}, 1 mM and Ca^{2+}, 1 mM	No	−	5
3	EDTA, 1 mM	No	−	9
4	None	1.3 mM	+	9
5	Ca^{2+}, 1 mM	1.3 mM	+	14
6	Mg^{2+}, 1 mM	1.3 mM	+ +	23
7	Mg^{2+}, 1 mM and Ca^{2+}, 1 mM	1.3 mM	+ +	34
8	Mg^{2+}, 1 mM and EDTA, 1 mM	1.3 mM	−	14

[a] The weight ratio of F-actin to myosin was 1 : 4 and the total protein concentration 1.16 mg/ml. The salt concentration was: KCl, 50 mM; tris–HCl (pH 7.2), 10 mM; and divalent cations and/or EDTA as shown in the table (62).

both release and incorporation of bound nucleotides in F-actin (95). More than 50% of bound nucleotides were found to be exchanged instantaneously upon the addition of ATP to the myosin–F-actin complex. The rapid exchange was observed only under the conditions in which superprecipitation occurred. Figure 33 is an example of the results of this kind of experiment (62). This phenomenon was confirmed in other laboratories also (96).

Therefore loosening of the polymer structure of F-actin takes place in the process of superprecipitation. This loosening is transient. According to electron microscope observations, no permanent change remains in the polymer structure of F-actin after superprecipitation. Concerning the molecular mechanism of this phenomenon, no explanation has been sug-

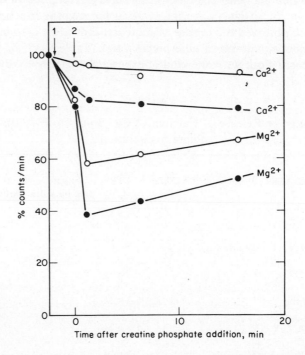

Fig. 33. Incorporation of ^{14}C ADP into F-actin during superprecipitation. F-actin and myosin were mixed at a ratio of 1 : 4 and 50 μg/ml of creatine kinase was added. Arrow 1, addition of 5.6 μM ^{14}C ATP; arrow 2, addition of 85 μM creatine phosphate. At indicated times after the addition of creatine phosphate, the sample solution was centrifuged at 10,000 rev/min for 1 min. ^{14}C in the supernatant was counted. ○, F-actin, 0.179 mg/ml, with bound ADP, 4.6 μM; ●, partially ADP-free F-actin 0.173 mg/ml with bound ADP, 1.2 μM. Tris–maleate (pH 7.1), 10 mM; KCl, 50 mM; and MgCl$_2$, 4.2 mM or CaCl$_2$, 4.2 mM at 20°C (62).

gested. We do not know why the HMM plus ATP system can not induce the exchange of bound nucleotides while the myosin plus ATP system can.

The experiment by Gergely et al. (97), which was designed to detect the loss of radioactive nucleotides bound to thin filaments during in vivo contraction, gave negative results. The X-ray analysis of contracting striated muscle by Huxley (25) gave no indication of the change in the periodic structure attributable to F-actin during contraction. Consequently, the physiological significance of the intrapolymer structure change in F-actin during superprecipitation is not clear.

In summary, it must be emphasized that myosin has a dual action on the state of actin, depending on the absence or the presence of ATP (see Table VI) (70).

TABLE VI

Rate Regulators of Polymerization of Actin

Polymerization process (bond formation)	Depolymerization process (bond breaking)
Nucleotides	—
Divalent cations	—
—	Mechanical agitation
—	Thermal agitation
Myosin	Myosin + ATP
Tropomyosin	—
Polymer nuclei	—

C. Interaction of Actin with Tropomyosin

According to Ebashi's experiment, it is likely that thin filaments in muscle are formed by the binding of tropomyosin (and troponin) with F-actin filaments (26). The close relation of tropomyosin with actin has been recognized in various phenomena. In the water extract from dried muscle powder, actin is often contaminated by tropomyosin, and tropomyosin once coprecipitated with F-actin is not easily separated from F-actin at the physiological salt concentration (98). Although the parallel association between F-actin and tropomyosin expected in vivo has not yet been confirmed in vitro, the binding of tropomyosin with F-actin has been examined by measuring the increase in the degree of flow birefringence and the amount of coprecipitation in ultracentrifugation (99,100,101).

These measurements showed that the binding is reversible and becomes weaker with a decrease in salt concentration and an increase in temperature. At the physiological salt concentrations tropomyosin is almost completely dissociated from F-actin at about 45°C (102) (Fig. 34). A low concentration of urea can also dissociate tropomyosin from F-actin (102).

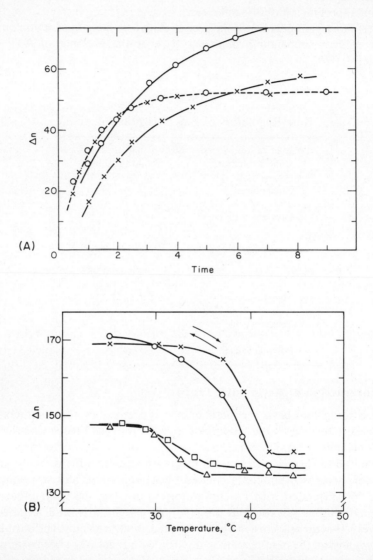

Fig. 34. (A) Time course of polymerization of actin in the presence (○) and in the absence (×) of tropomyosin. Actin concentration, 0.8 mg/ml; tropomyosin concentration, 0.6 mg/ml; KCl, 100 mM; tris–HCl (pH 8.0), 5 mM; ATP, 0.5 mM. —, at 25°C; - - -, at 45°C. (B) The temperature dependence of binding between F-actin and tropomyosin. ×, KCl, 100 mM; ○, KCl, 60 mM; □, 400 mM; △, 40 mM.

The binding of tropomyosin suppresses the steady ATP-splitting activity of F-actin at room temperature and the exchange of nucleotides and divalent cations bound to F-actin (62,96,102). This suggests that association of tropomyosin makes the polymer structure of F-actin rigid. Actually, the mixture of tropomyosin and F-actin shows higher viscoelasticity than pure F-actin (99).

The latest experiment of the laser light scattering also showed that the binding of tropomyosin decreases the flexibility of F-actin until the molar ratio of tropomyosin to actin reaches 1 : 6 (102a). The further addition of tropomyosin does not decrease the flexibility but the flow birefringence and the ultracentrifugation suggest that tropomyosin can bind with F-actin up to a molar ratio of about 1 : 3. Taking into consideration the result of the exchange experiment of bound tropomyosin (102b), it may be probable that one mole of tropomyosin tightly binds with about six moles of actin along F-actin and the next one mole of tropomyosin loosely binds onto the tightly bound tropomyosin (102a).

The analysis of the Doppler broadening of scattered light further revealed that the addition of troponin to the complex of F-actin and tropomyosin makes the F-actin filament most rigid in the absence of calcium ions, and that this rigidity is decreased by calcium ions to the level of the original complex. Summarizing the experimental results, the flexibility increases or the rigidity decreases in the following order (102a): the complex of F-actin, tropomyosin and troponin in the absence of calcium ions; the complex of F-actin, tropomyosin and troponin in the presence of calcium ions and the complex of F-actin and tropomyosin; pure F-actin; the complex of F-actin and heavy meromyosin at a molar ratio of about 1 : 6 (102a). These results strongly suggest the importance of the flexibility or the intrapolymer Brownian movement of F-actin in the force generation process in muscle.

The other experimental result indicating the regulation of the F-actin polymer structure by tropomyosin was obtained by an ESR study (103). The resonance signal of spin-labeled F-actin was not changed by simple binding of tropomyosin, but when troponin interacted with tropomyosin the signal was found to be influenced by the addition and the removal of calcium ions.

CONCLUDING REMARKS

In this article we have described mainly the physical aspects of equilibrium and kinetics of polymerization of actin and the polymorphism of actin polymers. F-Actin is a helical polymer in which each monomer is bound to many neighboring monomers through different kinds of interactions. There are many regulators of formation and detachment or strengthening and loosening of bonds between actin monomers. Among them the actions of nucleotides, divalent cations, myosin, and tropomyosin are most remarkable, as summarized in Table VI.

What is the physiological function of actin polymers? First, F-actin is one of the principal participants in muscular contraction. According to the sliding theory, contractile force is generated in the elementary cycle of interaction between actin and myosin molecules in thin and thick filaments. In this cycle F-actin may behave as if it is simply a rigid rod having sites of interaction with myosin or it may undergo an intermonomer or intramonomer conformational change. Such a conformational change, if it takes place, must be local and transient. The change in F-actin does not remain after the cycle and therefore it is difficult to prove experimentally. No indication of the change in the structure of thin filaments has been found during contraction in vivo. Only in the case of in vitro contraction are there some data suggesting a transient change in the polymer structure of F-actin. However, the significance of this transient change in generating force is not known.

In cellular movements other than muscle also, there may be an elementary cycle between actin and myosin similar to that occurring in muscle. In most of the cells showing protoplasmic movements, F-actinlike thin filaments of the diameter of about 60 Å ~ 80 Å have been found with parallel arrangements. However, thick filaments composed of myosin are not always found. Myosin extracted from plasmodium, for example, does not form thick filaments by itself at the physiological salt concentration (*81*). Therefore it is not certain that the same sliding mechanism as that in muscle can operate in such lower organisms. The other possibility may be that the polymorphism of actin polymer is directly related to the mechanism of force generation and movement.

Recent progress in the study of regulation of muscular contraction by Ebashi et al. has revealed that the tropomyosin–troponin system is the acceptor of calcium ion, which is the key substance for contraction (*104*). Since this system is probably located on thin filaments and only F-actin in thin filaments interacts with myosin in thick filaments it is most probable that the intermonomer or intramonomer conformation of F-actin is regulated by the tropomyosin–troponin system, depending on the presence and absence of calcium ions (Fig. 35). Some specific structure of actin polymer necessary for interaction with myosin is inhibited by a tropomyosin–troponin system having no calcium ions and/or the specific structure is prepared by the same system having bound calcium ions. Some results supporting this idea have been obtained. Difficulty arises from the fact, however, that the same thin filament occurring in vivo has not yet been reconstructed in vitro from actin, tropomyosin, and troponin.

In the case of muscle the amplitude of the conformational change of thin filament in the regulatory cycle may be very small. However, in the case of ameboid movement its regulation seems to be performed through the sol–gel transformation of protoplasm. A large conformational change of actin polymer such as that found for plasmodium actin in vitro may be involved in the regulatory cycle (Fig. 35). The polymorphism of actin polymers may be the elementary process in regulation.

Finally, it must be emphasized that actinlike filaments are found in various kinds of cells, not only in cells having motility but also in other

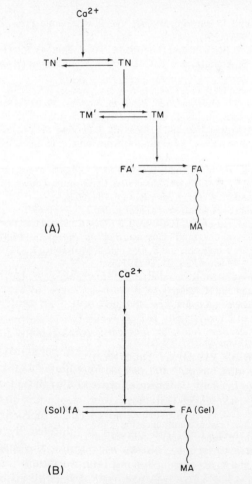

Fig. 35. (A) The schematic representation of the regulatory mechanism of muscular contraction. MA, Myosin; FA, F-actin; TM, tropomyosin; TN, troponin. (B) The regulatory mechanism of protoplasmic flow.

kinds of cells, for example, in the nerve cell (*105*), the sea urchin egg cell (*106*), and so on. The filaments are identified as actin polymer by the interaction with HMM given from outside and, actinlike proteins can actually be extracted and purified from these cells (*107,108*). We do not know the physiological function of these actins.

REFERENCES

1. F. B. Straub, *Studies Inst. Med. Chem. Univ. Szeged.*, **3**, 23 (1943).
2. A. Szent-Györgyi, *Chemistry of Muscular Contraction*, Academic Press, New York, 1951.
3. H. Huxley and J. Hanson, *Muscle*, Vol. 1, Academic Press, New York, 1960; *Nature*, **173**, 973 (1954).
4. H. Ishikawa, R. Bischoff, and H. Holtzer, *J. Cell. Biol.*, **43**, 312 (1969).
5. F. Oosawa, S. Asakura, and T. Ooi, *Prog. Theoret. Phys.* (*Kyoto*) *Suppl.* No. 17, 14 (1961).
6. F. B. Straub and G. Feuer, *Biochim. Biophys. Acta*, **4**, 455 (1950).
7. S. Asakura and F. Oosawa, *Arch. Biochem. Biophys.*, **87**, 273 (1960).
8. M. Kasai, S. Asakura, and F. Oosawa, *Biochim. Biophys. Acta*, **57**, 13 (1962).
9. M. Kasai, H. Kawashima, and F. Oosawa, *J. Polymer Sci.*, **44**, 51 (1960).
10. F. Oosawa, S. Asakura, K. Hotta, N. Imai, and T. Ooi, *J. Polymer Sci.*, **37**, 323 (1959).
11. S. Asakura, K. Hotta, N. Imai, T. Ooi, and F. Oosawa, *Proceedings of a Conference on the Chemistry of Muscular Contraction*, Igakushoin, Tokyo, 1957, p. 57.
12. F. Oosawa and M. Kasai, *J. Mol. Biol.*, **4**, 10 (1962).
13. F. Oosawa and S. Higashi, *Prog. Theoret. Biol.*, **1**, 79 (1967).
14. S. Asakura, M. Kasai, and F. Oosawa, *J. Polymer Sci.*, **44**, 35 (1960).
15. M. Kasai, S. Asakura, and F. Oosawa, *Biochim. Biophys. Acta*, **57**, 22 (1960).
16. S. Asakura, M. Taniguchi, and F. Oosawa, J. Mol. Biol., **7**, 55 (1963).
17. F. Oosawa, *J. Theoret. Biol.*, **27**, 69 (1970).
18. K. Maruyama and M. Kawamura, *Zool. Mag.* (*Tokyo*), **76**, 420 (1967).
19. M. Kawamura and K. Maruyama, *J. Biochem.* (*Tokyo*), **66**, 619 (1970).
20. M. Kasai, *Biochim. Biophys. Acta*, **180**, 399 (1969).
21. J. Hanson and J. Lowy, *J. Mol. Biol.*, **6**, 46 (1963).
22. C. Selby and R. Bear, *J. Biophys. Biochem. Cytol.*, **2**, 71 (1956).
23. H. Huxley, *J. Mol. Biol.*, **7**, 281 (1963).
24. J. Hanson, *Quart. Rev. Biophys.*, No. 2 (1968).
25. H. Huxley and W. Brown, *J. Mol. Biol.*, **30**, 383 (1967).
26. S. Ebashi and F. Ebashi, *J. Biochem.* (*Tokyo*), **58**, 7, 2 (1965).
27. T. Ooi, *J. Phys. Chem.*, **84**, 984 (1960).
28. S. Fujime, *J. Phys. Soc. Japan*, **28**, 267 (1970); *J. Phys. Soc. Japan*, **29**, 751 (1970).
29. M. Kawamura, private communication, 1969.
30. S. Kobayashi, *Biochim. Biophys. Acta*, **88**, 541 (1964).
31. S. Kobayashi, H. Asai, and F. Oosawa, *Biochim. Biophys. Acta*, **88**, 528 (1964).
32. T. Hayashi and R. Rosenbluth, *Biol. Bull.*, **119**, 290 (1960).
33. N. Grubhoffer and H. Weber, *Z. Naturforsch.*, **16b**, 435 (1961).
34. M. Kasai and F. Oosawa, *Biochim. Biophys. Acta*, **75**, 323 (1963).
35. S. Higashi, M. Kasai, A. Wada, and F. Oosawa, *J. Mol. Biol.*, **7**, 421 (1963).
36. M. Kasai, E. Nakano, and F. Oosawa, *Biochim. Biophys. Acta*, **94**, 494 (1965).

37. F. Oosawa, S. Asakura, S. Higashi, M. Kasai, S. Kobayashi, E. Nakano, T. Ohnishi, and M. Taniguchi, *Molecular Biology of Muscular Contraction* (S. Ebashi, F. Oosawa, T. Sekine, and Y. Tonomura, eds.), Igakushoin, Tokyo, and Elsevier, Amsterdam, 1965.

38. S. Higashi and F. Oosawa, *J. Mol. Biol.*, **12**, 843 (1965).

38a. K. Tawada, H. Asai, and B. Gerber, *Biochim. Biophys. Acta*, **194**, 486 (1969).

38b. D. Stone, S. Prevost, and J. Botts, *Biochemistry*, **9**, 3937 (1970).

38c. J. West, *Biochemistry*, **9**, 3847 (1970).

39. B. Nagy and W. Jencks, *Biochemistry*, **1**, 987 (1963).

40. Y. Nakaoka, private communication.

41. A. Martonosi, M. Gouvea, and J. Gergely, *J. Biol. Chem.*, **235**, 1700 (1960).

42. M. Bárány, F. Finkelman, and T. Therattil-Antony, *Arch. Biochem. Biophys.*, **98**, 28 (1962).

43. M. Kasai and F. Oosawa, *Biochim. Biophys. Acta*, **154**, 520 (1968).

44. S. Asakura and M. Taniguchi, *Biochemistry of Muscular Contraction* (J. Gergely, ed,), Little Brown, Boston, 1964, p. 158.

45. K. Mihashi and T. Ooi, *Molecular Biology of Muscular Contraction* (S. Ebashi, F. Oosawa, T. Sekine, and Y. Tonomura, eds.), Igakushoin, Tokyo, and Elsevier, Amsterdam, 1965, p. 77.

46. M. Rees and M. Young, *J. Biol. Chem.*, **242**, 4449 (1967).

47. M. Elzinga, *Abstr. Intern. Congr. Biophys.* (*MIT*), 186 (1969); *Biochemistry*, **10**, 224 (1970).

48. K. Tsuboi, *Biochim. Biophys. Acta*, **160**, 420 (1968).

49. I. Sakakibara and K. Yagi, *Abstr. Ann. Meeting Biophys. Japan*, 1969.

50. D. Kominz, A. Hough, P. Symonds, and K. Laki, *Arch. Biochem. Biophys.*, **50**, 148 (1954).

51. M. Lewis, K. Maruyama, W. Carroll, D. Kominz, and K. Laki, *Biochemistry*, **2**, 34 (1963).

52. K. Mihashi, *Arch. Biochem. Biophys.*, **107**, 441 (1964).

53. A. Minakata, *Biochem. Biophys. Acta*, **126**, 570 (1966).

54. S. Kobayashi, private communication, 1970.

55. K. Mihashi and T. Ooi, *Biochemistry*, **4**, 805 (1965).

56. A. Katz and W. Mommaerts, *Biochim. Biophys. Acta*, **65**, 82 (1962).

57. C. Lusty and H. Fasold, *Biochemistry*, **8**, 2933 (1969).

58. S. Asakura, *Arch. Biochem. Biophys.*, **92**, 140 (1961).

59. D. Seidel, D. Chak, and H. Weber, *Biochim. Biophys. Acta*, **140**, 93 (1967).

60. M. Kasai, unpublished data, 1968.

61. M. Bárány, N. Biro, and J. Molnar, *Acta Physiol. Acad. Sci. Hung.*, **5**, 63 (1954)

62. M. Kasai and F. Oosawa, *Biochim. Biophys. Acta*, **172**, 300 (1969).

63. D. Waugh, *J. Cell. Comp. Physiol.*, **49**, 145 (1957).

64. B. Gerber, S. Asakura, and F. Oosawa, *J. Mol. Biol., Adv. Biophysics*, **1**, 99 (1970).

65. T. Ikkai and T. Ooi, *Biochemistry*, **5**, 1551 (1966).

66. T. Ikkai, T. Ooi, and H. Noguchi, *Science*, **152**, 1756 (1966).

67. B. Nagy and W. Jencks, *J. Am. Chem. Soc.*, **87**, 2480 (1965).

68. F. Oosawa, *J. Polymer Sci.*, **26**, 29 (1957).

69. K. Mihashi, submitted to *Biochim. Biophys. Acta* and *J. Biochem.* (*Tokyo*).

70. F. Oosawa, M. Kasai, S. Hatano, and S. Asakura, in *Principles of Biomolecular Organization*, Ciba Foundation Symposium (G. Wolstenholme, ed.), Churchill, London, 1966, p. 273.

71. S. Asakura, *Biochim. Biophys. Acta*, **52**, 436 (1961).

72. S. Asakura, M. Taniguchi, and F. Oosawa, *Biochim. Biophys. Acta*, **74**, 140 (1963).

73. M. Bárány and F. Finkelman, *Biochim. Biophys. Acta*, **63**, 98 (1962).
74. Y. Nakaoka and M. Kasai, *J. Mol. Biol.*, **44**, 319 (1969).
75. H. Asai and K. Tawada, *J. Mol. Biol.*, **20**, 403 (1966).
76. M. Kasai, *Biochim. Biophys. Acta*, **172**, 171 (1969).
77. S. Hatano and F. Oosawa, *J. Cell. Physiol.*, **68**, 197 (1966).
78. S. Hatano and F. Oosawa, *Biochim. Biophys. Acta*, **127**, 488 (1966).
79. S. Hatano, T. Totsuka, and F. Oosawa, *Biochim. Biophys. Acta*, **140**, 109 (1967).
80. K. Wohlfarth-Bottermann, in *Primitive Motile Systems in Cell Biology* (R. Allen and N. Kamiya, eds.), Academic Press, New York, p. 79.
81. S. Hatano and M. Tasawa, *Biochim. Biophys. Acta*, **154**, 507 (1968).
82. S. Hatano and J. Ohnuma, *Biochim. Biophys. Acta*, **205**, 110 (1970).
83. M. Adelman and E. Taylor, *Biochemistry*, **8**, 4964, 4976 (1969).
84. T. Totsuka and S. Hatano, *Biochim. Biophys. Acta*, **223**, 189 (1970).
84a. T. Totsuka, *Biochim. Biophys. Acta.* **234**, 162 (1971).
85. S. Hatano, submitted to *J.* Mechana-chemistry and Cell. Motility.
85a. S. Fujime and S. Hatano, private communication.
86. A. G. Szent-Györgyi, *Arch. Biochem. Biophys.*, **42**, 305 (1953).
87. K. Yagi, T. Mase, I. Sakakibara, and H. Asai, *J. Biol. Chem.*, **240**, 7448 (1965).
88. K. Tawada, *Biochim. Biophys. Acta*, **172**, 311 (1969).
88a. S. Fujime and S. Ishiwata, *J. Phys. Soc. Japan*, **29**, 1651 (1970).
88b. T. Tsao. *Biochim. Biophys. Acta*, **11**, 227 (1953).
89. K. Takahashi and K. Tawada, in preparation.
90. F. Oosawa, unpublished data, 1969.
91. E. Eisenberg and C. Moos, *Biochemistry*, **7**, 1486 (1968).
92. K. Tawada and F. Oosawa, *J. Mol. Biol.*, **44**, 311 (1969).
93. K. Tawada and F. Oosawa, *Biochim. Biophys. Acta*, **180**, 199 (1969).
94. M. Kasai, E. Nakano, and F. Oosawa, in *Biochemistry of Muscular Contraction* (J. Gergely, ed.), Little, Brown, Boston, 1964, p. 158.
95. A. G. Szent-Györgyi and G. Prior, *J. Mol. Biol.*, **15**, 515 (1966).
96. S. Kitagawa, W. Drabikowsky, and J. Gergely, *Arch. Biochem. Biophys.*, **125**, 706 (1968).
97. A. Martonosi, M. Gouvea, and J. Gergely, *J. Biol. Chem.*, **235**, 1707 (1960).
98. W. Drabikowsky and J. Gergely, *J. Biol. Chem.*, **237**, 3412 (1962).
99. A. Martonosi, *J. Biol. Chem.*, **237**, 2795 (1962).
100. K. Maruyama, *Arch. Biochem. Biophys.*, **105**, 142 (1964).
101. W. Drabikowsky and E. Nowak, *European J. Biochem.*, **5**, 37 (1968).
102. H. Tanaka and F. Oosawa, *Biochim. Biophys. Acta*, in press.
102a. S. Ishiwata and S. Fujime, *J. Phys. Soc. Japan*, **30**, 302, 303 (1971).
102b. W. Drabikowski, D. Kominz, and K. Maruyama, *J. Biochem.* (*Tokyo*), **63**, 802 (1968).
103. Y. Tonomura, S. Watanabe, and M. Morales, *Biochemistry*, **8**, 2171 (1969).
104. S. Ebashi and M. Endo, *Prog. Biophys. Mole. Biol.*, **18**, 123 (1968).
105. F. O. Schmitt, *Proc. Natl. Acad. Sci. U.S.*, **60**, 1092 (1968).
106. R. Harris, *Exptl. Cell Res.*, **52**, 677 (1968).
107. S. Hatano, H. Kondo, and T. Miki-Noumura, *Exptl. Cell Res.*, **55**, 275 (1969).
108. T. Miki-Noumura and F. Oosawa, *Exptl. Cell Res.*, **55**, 275 (1969).

---

[a] Tsao's classical experiment on fluorescence depolarization of dye molecules attached to actin also suggested the increase of freedom of actin monomers in F-actin by binding with myosin (*88b*).

CHAPTER 7

STRUCTURAL COMPONENTS OF THE
STRIATED MUSCLE FIBRIL†

Frank A. Pepe

DEPARTMENT OF ANATOMY
UNIVERSITY OF PENNSYLVANIA MEDICAL SCHOOL
PHILADELPHIA, PENNSYLVANIA

I. INTRODUCTION

The muscle fibril is a highly organized structure made up of macro-molecular aggregates arranged in such a way that interactions between them lead to shortening of the fibril. The complexity of the problem of contractility becomes obvious when it is realized that in order to rigorously describe contractile behavior we must know the contributions to con-tractility made by each of the protein constituents of the myofibril in conjunction with every other protein constituent, where each constituent modifies in some way the contribution of every other. Our concern in this chapter is to discuss what is known about the macromolecular organization of the myofibril and to consider the possible relationship of this organiza-tion to the expression of contractility in striated muscle. An exhaustive review of the literature on muscle is not attempted since several excellent reviews have appeared recently on various aspects of muscle and the muscle proteins (*1–9*).

† This work was supported by U.S. Public Health Service grant R01-AM-04806.

The main structural components of the striated fibril are the thin actin-containing filaments and the thick myosin-containing filaments. The actin-containing filaments are structurally similar in all muscles studied except that they have different lengths in different muscles. The myosin-containing

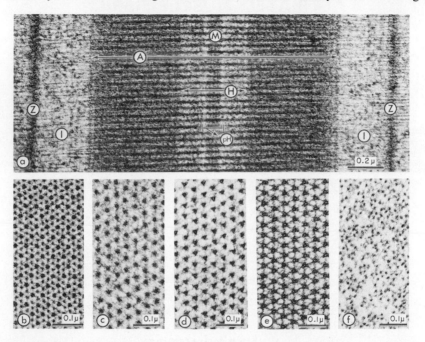

Fig. 1. The sarcomere (repeat unit of the striated myofibril); lateral muscles of the freshwater killifish (*Fundulus diaphanus*). (a) Longitudinal section through the sarcomere which extends from Z-band to Z-band; Z, Z-band; I, I-band; A, A-band; H, H-zone; pH, pseudo-H-zone; M, M-band. (b) Cross section through the A-band in the region of overlap of thin and thick filaments. Note that the thick filaments are solid and have circular profiles. Some appear to have triangular profiles. (c) Cross section through the A-band in the region of the H-zone. Only thick filaments are present. Note that the thick filaments are solid and have circular profiles. (d) Cross section through the A-band in the region of the pseudo-H-zone. Note that the thick filaments are solid and have triangular profiles. (e) Cross section through the A-band in the region of the M-band. Note that each thick filament is connected to each of its six neighboring thick filaments by M-bridges. Also, some of the filaments are clearly hollow and have circular profiles. (f) Cross section through the I-band. Only thin filaments are present.

filaments may vary in length as well as thickness in different muscles and may also vary in macromolecular organization. The repeat unit of the striated fibril is the sarcomere (Fig. 1). The A-band is characterized by the presence of thick filaments and the I-band is characterized by the presence of thin filaments. The thin filaments are linked together in the middle of the I-band by attachment to the proteins that make up the structure of the Z-band. The M-band, which is present in some striated fibrils, is in the middle of the A-band. In the M-band the thick filaments are joined together by protein which attaches to the thick filaments and bridges between them. The thin filaments extend from the Z-band and continue part way into the A-band where they interdigitate with the thick filaments. The part of the A-band into which the thin filaments do not reach is the H-zone. Interaction occurs between the thick and thin filaments by means of cross bridges, which extend from the thick filaments and interact with the thin filaments. As a result of this interaction, sliding of the thin filaments with respect to the thick filaments occurs without change in length of the filaments, and the sarcomere shortens. Shortening thus results in a decrease in the width of both the I-band and the H-zone. These are the essential features of the sliding filament hypothesis of muscle contraction (10,11). The proteins making up the structural elements of the fibril may have other proteins associated with them. It has been shown that the thin filaments have tropomyosin (12,13) and troponin (13,14) associated with them. These additional proteins are responsible for the 400-Å repeat periodicity seen in the I-band in electron microscopy (5,15).

The fibril may then be considered as consisting of highly organized protein aggregates which make up the basic structural components of the fibril and of possibly less highly organized but specifically related proteins associated with these basic structures. The interactions between these various components lead to the functional characteristics observed for muscle fibrils. The differences observed in the properties of different muscle fibrils depend not only on the differences observed in the basic structural components but also on the differences in the properties of both the protein molecules making up these structures and the protein molecules associated with them.

II. THICK FILAMENT

A. Structural Components

The thick filament is the structural component characteristic for the A-band and it is continuous from one end of the A-band to the other (16). The A-band protein has been identified as myosin using differential extraction of the A-band from fibrils (17) and by specific antibody staining of the A-band with fluorescent antimyosin (9,12,18). Also, artificial filaments can be grown from myosin solutions by decreasing the ionic strength of the solution (19). These artificial filaments are similar in structure to the naturally occurring filaments of the A-band. Therefore the evidence is strongly in favor of myosin as the structural protein of the thick filament. Although there is no evidence as yet for the presence of another protein in the structure of the thick filament, it has been suggested that such a component might account for the presence of an unexplained 442-Å meridional reflection in the X-ray diffraction pattern associated with the myosin filament and may be involved in the precise length-determining mechanism of the myosin filament (19,20). In the detailed model of the myosin filament discussed below (21,22), no component other than myosin is considered to be present. This model is made up of structural units each of which may consist of one or more than one myosin molecule, and each structural unit gives rise to one cross bridge. In this model a scheme has been proposed for the precise determination of the length of the filament based on the relationships between the myosin molecules themselves (21).

The myosin molecule is essentially a rod with a globular region at one end. By controlled trypsin digestion of myosin, two fragments, the L-meromyosin (LMM) fragment and the H-meromyosin (HMM) fragment are obtained (23–26). The LMM fragment consists of a portion of the rod part of the molecule and this fragment retains the solubility characteristics of the myosin (insolubility at low ionic strength). The HMM fragment consists of the remaining portion of the rod and the globular portion of the molecule, and this fragment, soluble at low ionic strength, retains the ATPase activity and actin-combining ability of the myosin. The HMM fragment can be further split into two fragments using more extensive trypsin digestion (27–28) or papain digestion (29–31). These two fragments consist of the rod portion of the HMM (S_2-fragment) and the globular region (S_1-fragment), both of which are soluble at low ionic strength and one of which (the S_1-fragment) retains the ATPase activity and actin-combining ability of the myosin. Therefore the myosin molecule consists of three different regions, the LMM (carrying the solubility characteristics), the S_1 (carrying the ATPase activity and actin-binding ability), and the S_2 which is between the LMM and the S_1.

Huxley's (19) observations on the growth of myosin filaments from myosin solutions lead to some general information concerning the aggregation of myosin molecules into myosin filaments. He found that as the ionic strength of a myosin solution was decreased the first aggregates to form were short filaments with a smooth surface and projections protruding from each end. The smooth surface was approximately 0.2 μ in length and remained constant with further increase in length of the filament from both ends. That is, the longer filaments had projections all along their surface except for the smooth central portion. He concluded from this that the first aggregation of myosin molecules occurred by aggregation of the rod portions in a tail-to-tail fashion with the globular or head regions sticking out at each end. Further aggregation, he concluded, took place in a head-to-tail fashion with the tail or LMM part of the molecule (insoluble at low ionic strength) forming the core of the filament and the head region of the molecule (soluble at low ionic strength) forming the projections or cross bridges on the surface available for interaction with the surrounding thin filaments in the myofibril. Therefore the molecules in one half of the filament are pointed in a direction opposite to those in the other half of the filament and, based on the orientation of the myosin molecules, the filament is symmetrical with respect to the center. This symmetry is consistent with the observation that the sarcomere shortens by a sliding of the thin filaments from each side of the A-band toward the center during shortening of the fibril. The smooth central portion of the filaments devoid of cross bridges corresponds to the relatively less dense short zone of constant length in the middle of the A-band. This is the pseudo-H-zone (32), and the M-band is in the middle of the pseudo-H-zone (Fig. 1a).

From sections of vertebrate muscle it was originally observed that one cross bridge occurred between a thick and thin filament approximately every 400 Å along the thick filament (16). This repeat corresponds to the 430-Å repeat period of bridges later obtained from X-ray diffraction studies (20,33). The detailed arrangement of the cross bridges on the surface of the myosin filament can be obtained from X-ray diffraction studies (20). The arrangement consists of pairs of bridges every third of a period (430 Å/3) along the filament. The two bridges are on diametrically opposite sides of the filament with a rotation of 120° between successive pairs of bridges. This same arrangement of bridges was derived independently, using both the antimyosin staining pattern obtained in electron microscopy and the structural characteristics of the A-band and of individual myosin filaments (21,22). This latter approach, in addition, led to a model for the detailed packing arrangement of the myosin molecules in the core of the filament. In this model each structural unit gives rise to one cross bridge, in which each structural unit may consist of one or more myosin molecules. For simplicity, one myosin molecule per bridge was considered. In this model the myosin molecules are all aggregated in parallel with those in one half of the filament oriented opposite to those in the other half of the filament. In order to satisfy the requirements imposed by the facts and assumptions used to derive the model, it was necessary to have 860 Å of the rod portion of the molecule, which includes the entire LMM portion, aggregated in the central core of the filament, and 430 Å of the molecule, which corresponds to the S_2-portion, overlapping successive molecules along the filament in a head-to-tail fashion (Fig. 2a). The LMM portions are closely packed in the filament so that 3 in the center are surrounded by

9, making a total of 12 in a cross section through the filament (Fig. 2b and c). Therefore the model consists of 12 parallel rows of molecules and each row is a linear aggregate of myosin molecules with the molecules in a row overlapping successive molecules in a head-to-tail fashion. Taking 1 myosin molecule per cross bridge there are 192 molecules in a single filament. A cross section through the filament includes the LMM portion of 12 molecules and the overlap portion of 6 molecules (Fig. 2b). The overlapping portions from the three centrally located molecules project between

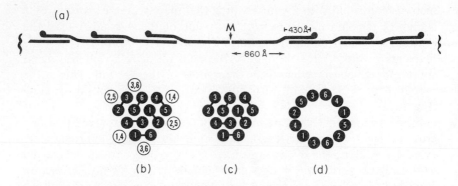

Fig. 2. Diagrammatic representations of the arrangement of myosin molecules in the filament model (*21,22*). (a) Relationship of myosin molecules in one of the 12 rows (linear aggregates) of moysin molecules making up the filament. Note tail-to-tail abutment of the two myosin molecules in the middle (M) of a row. (b) to (d) Relationship of myosin molecules as seen in cross sections through different regions of the filament. In the diagrams solid circles represent cross sections of the LMM portion of molecules and empty circles represent cross sections of the overlap portion. Note that the overlap portions from all rows of molecules are on the surface of the filament. A difference of 1 in the numbers represents a stagger of 430 Å/3 between the rows (connected by solid lines). (b) Cross section corresponding to the solid circular profiles seen in Fig. 1b and c. (c) Cross section corresponding to the solid triangular profiles seen in Fig. 1d. (d) Cross section corresponding to the hollow circular profiles seen in Fig. 1e.

the LMM portions surrounding them so that all the overlaps are on the surface of the filament. The arrangement of the overlaps in rows on the surface of the filament is restricted by both the position of the LMM portion from which they originate and by the stagger between the molecules. The overlaps contributing to one row of overlaps on the surface of the filament come from neighboring molecules staggered by 430 Å. The stagger relationship between the rows of molecules is such that there are three sets of four rows. In each set there are two pairs. The stagger between the rows of a pair is 430 Å/3 and the stagger between the two pairs is 430 Å (Fig. 2b and c). In the middle of the filament, where the orientation of the molecules is reversed on going from one side to the other, there is a tail-to-tail abutment of myosin molecules in a single row (Fig. 2a). Also, because of the stagger between rows, tail-to-tail overlap of myosin molecules from different rows occurs in this region of the filament and there is a helical twist in the rows (22). The helical twist results in the hollow circular cross-sectional profiles seen in this region of the filament (Figs. 1e and 2d). Everywhere else, where head-to-tail overlap occurs, the rows are in parallel. All of these relationships result from the facts and assumptions used to derive the model (21,22). The observation that the S_2-fragment (which corresponds to the overlapping portion of the molecule) is soluble under conditions in which LMM is insoluble suggests that any interactions that occur between the overlap portions and LMM portions must be weak (31), and this correlates nicely with exclusion of the overlap portion from the core of the filament model (18,22). The first indication that the myosin molecules could bend out of the filament as part of the cross bridge came from fluorescent antibody staining studies (9,12,18). The fluorescent antibody staining studies indicated that in some parts of the filament some of the LMM portion is included in the part that can bend out of the filament. Later studies using X-ray diffraction (20,34) also led to the conclusion that a portion of the myosin molecules can bend out of the filament. In these studies movement was considered to be limited to the HMM portion of the myosin molecules. A more detailed treatment of the movement of the cross bridges and the length of the myosin molecule involved is presented later (Section II.C).

In the presentation of the filament model (21,22), it was emphasized that for simplicity the model was derived considering one myosin molecule per cross bridge but that in fact the model could apply just as well for any number of molecules per cross bridge. Therefore the model sets up relationships that exist between structural units that may be variable in their composition. Probably, at least some of the basic relationships involved are shared by all myosin filaments. Small variations in these basic relationships from muscle to muscle might lead to very different looking filaments. An example may be the differences between the myosin filaments of vertebrate and insect flight muscle. In vertebrate muscle the cross bridges are arranged in diametrically opposite pairs every 143 Å along the filament (20), and in insect flight muscle cross-bridge origins occur in diametrically opposed pairs every 146 Å along the filament (35). However, in vertebrate muscle the helical arrangement of the bridges is such that the helix repeats after one turn, or 3 × 143 Å, along the filament, while in insect flight muscle the helix repeats every one and one-half turn, or every 8 × 146 Å, along the filament. Therefore the bridges occur in pairs along the filament at similar intervals in both filaments, but the helical arrangement of the bridges on the surface of the filaments is quite different. In insect flight muscle, in addition, each cross-bridge origin gives rise to two bridges so that there are really four bridges every 146 Å along the filament, each of which is attached to one of the surrounding actin filaments (*35*). So far there is no evidence for such a splitting of the bridge in vertebrate muscle. Other structural differences and similarities between myosin filaments from different muscles have been found. From observations of cross sections of myosin filaments from a variety of species, the internal structure of the myosin filament has been described by Baccetti (*36,37*) to consist of nine peripheral subunits and two central ones. All the filaments examined were in the range 140–180 Å in diameter. The central subunits were not clearly seen, sometimes only one or none being visible. The peripheral subunits were reported as being 25 × 33 Å, whereas the central subunits, when seen, were reported to be about 25 Å in diameter (*36*). The larger dimension of peripheral subunits could be accounted for by fusion of an LMM and an overlap portion of the myosin molecules in the model (*21,22*). This would account for a larger dimension in only six of the nine peripheral subunits since there are only six overlaps in a cross section (Fig. 2b). However, since the central subunits were not clearly seen and sometimes were missing, when missing they might contribute to the mass of the other peripheral subunits. Therefore, considering these difficulties, the observations are entirely consistent with the filament model (*21,22*). Gilev (*38,39*), also from cross sections of myosin filaments, reported a maximum of 18 subunits in the internal structure of the myosin filaments of crab muscle. In this case the diameter of the filaments was over 200 Å. The finding of 18 subunits suggests that in this case both the LMM portions and the overlap portions of the structural units in the model (*21,22*) were visualized. In thick filaments of larger diameter, each structural unit may consist of a greater number of myosin molecules (*21*), which might make it more easily visualized. It is significant that any correlation at all can be made with the model considering the variety of thick filaments involved.

The cross-sectional profiles in different regions along the length of the filament ought to be related to the internal structure of the filament. In vertebrate muscles such as chicken breast muscle and the lateral muscles of fish, the cross-sectional profiles in different regions of the A-band vary from hollow circles (in the middle of the M-band region), to solid triangles (in the pseudo-H-zone), to solid circles (in the remainder of the A-band) (Figs. 1 and 2) (*22*). The triangular profiles in the pseudo-H-zone in chicken breast muscle are not oriented relative to each other in any discernible pattern. However, in the fish muscle the triangular profiles in the pseudo-H-zone on one side of the M-band are all oriented in the same direction (*22*). Therefore, in slightly oblique sections through the pseudo-H-zone in fish muscle, it can be seen that the triangular profiles on one side of the M-band are oriented opposite to those on the other side of the M-band, indicating that the hollow circles in the middle of the M-band represent a short region of twist in the filament. The solid triangular profiles, even in slightly oblique sections, are not distorted, thus being consistent with a parallel arrangement of molecules in this region of the filament. Throughout the rest of the filament where successive myosin molecules in a row overlap in a head-to-tail fashion, the cross-sectional profiles are solid and circular. The presence of overlaps around a triangular central core would be expected to round out the cross-sectional profile. Interaction of the cross bridges with the surrounding thin filaments results in bending of the myosin molecules out of the filament (*12,18,20,22,34*). As least in some parts of the filament, this should result in removal of the overlap portions, revealing the triangular central core. In the region of the A-band where both thin and thick filaments are present, it is possible in some cases to make out some triangular cross-sectional profiles (Fig. 1b).

The differences in the cross-sectional profiles in different parts of the thick filaments of one muscle are not necessarily the same as those of a different muscle. For instance, Halvarson and Afzelius (*40*) showed that the middle of the thick filaments in the body muscles of the arrowworm also have triangular cross-sectional profiles, but in this case the triangles are hollow. Throughout the rest of the A-band, the cross-sectional profiles are hollow circles. Other variations in cross-sectional profiles in different parts of the thick filament have been observed in the flight muscles of insects. Reger and Cooper (*41*) and Auber (*42,43*) observed elongated cross-sectional profiles of the middle of the thick filaments of the flight muscles of lepidopterans. These elongated profiles were made up of two units each about 60–80 Å in diameter. Along the rest of the filament, hollow, circular cross-sectional profiles were observed. In the flight muscle of hymenopterans, Auber (*43*) found that the cross-sectional profile of the middle of the thick filament was a solid square and that the square profiles of neighboring filaments were all similarly oriented. The rest of the filament gave a hollow, circular cross-sectional profile. It is most likely that these differences in different parts of the same filament represent alternate arrangements of the same basic structural relationships as seems to be the case for myosin filaments from chicken and fish muscles (*21,22*). It is also reasonable to suppose that it will be possible to establish some common basic structural relationship between the differences observed between myosin filaments from different muscles.

In chicken and fish muscle, the position of the M-band in the fibril reflects the part of the filament where tail-to-tail abutment of myosin molecules occurs (*21,22*). In a single row of molecules, one half of the row of molecules points in one direction and the other half points in the other direction with a single tail-to-tail abutment in the middle of the filament (Fig. 2a). Therefore in a single row of molecules there is no tail-to-tail overlap in the center of the filament. Tail-to-tail overlap occurs only in the center of the filament between molecules in different rows where the rows are staggered with respect to each other (Fig. 3a). Suppose that in some muscles the properties of the myosin molecules were such that instead of abutting tail-to-tail in the center of the filament one half of a single row of molecules (pointing in one direction) overlaps the other half of that same row of molecules (pointing in the opposite direction) in the center of the

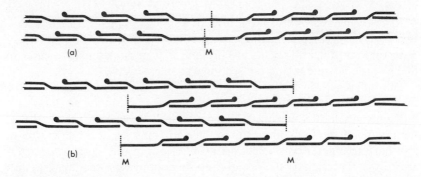

Fig. 3. Diagrammatic representation of two rows (linear aggregates) of myosin molecules staggered by 430 Å/3 with respect to each other. M, Position of the M-material attached to the row of molecules. (a) Case in which tail-to-tail abutment of myosin molecules, in the same row, occurs in the middle of the filament and M-material attaches to the point of tail-to-tail abutment. (b) Possible case in which overlap occurs between one half of a row of molecules (pointing in one direction) and the other half of the same row of molecules (pointing in the opposite direction); and in which M-material attaches to the free end of each half of the row in the middle of the filament (see text).

filament (Fig. 3b). In this case we would expect to see a filament that is thicker in the middle portion than at the ends. The length of this thicker portion in the middle of the filament would depend on the extent of overlap of the two halves of the row of molecules. Also, if M-material can still attach to the middle of the filament, in this case to the free end of each half of the row, one would expect to obtain a split M-band with the space between it depending on the extent of overlap of the two halves of the row of molecules (Fig. 3b). Myosin filaments have been described in crab muscle fibers (*7,44–46*) in which the central portion of the filament is thick and the rest of the filament is much thinner. In the crab *Pinnixia* sp. Reger (*44*) found that the central portion of the filament had a diameter of 300–400 Å and was solid, whereas everywhere else the filament had a diameter of 150–200 Å and was hollow. In the crab *Podophthalmus vigil*, Hoyle and McNeill (46) and Hoyle (45) found that the central portion of the thick filament was solid and had a diameter of 250–260 Å that tapered to a diameter of 160 Å, at which point it abruptly became hollow. The thicker region was about 0.4 μ long in the middle of the filament, tapering over approximately 0.4 μ on each side of the middle region, and the hollow region continued for up to 3 μ from this point. In this muscle two M-bands were found and each M-band was described as being "part-way along each half of the central shaft" (46). The diagrammatic representation in Fig. 3b relates the structure of these filaments to the model previously discussed (21,22). Note that between the two M-bands in Fig. 3b the myosin molecules point in opposite directions in the same portion of the filament. Paramyosin, which is a constituent of the very large thick filaments found in molluscan smooth muscles, has been shown to form paracrystals in which there is an antiparallel arrangement of molecules (47) as suggested for the myosin molecules between the two M-bands in Fig. 3b. This may indicate that similar structural relationships exist even in the most widely different thick filaments. In order to properly interpret the observed structural variations of the thick filament in terms of the relationships proposed in the model (21,22), considerably more detailed observations of the M-band structure and its relation to the thick, middle portion of the filament would have to be made.

B. Antimyosin Staining

As already discussed, the myosin molecules in the middle of the filament (pseudo-H-zone region) are aggregated tail-to-tail, while those along the rest of the filament are aggregated in a head-to-tail fashion. One of the most significant conclusions to come from antimyosin staining of the myofibril is that in addition to these differences there are differences in the lateral interactions between myosin molecules along the region of the filament where head-to-tail aggregation occurs (9,12,18,21). This conclusion is based on (1) the nonuniformity of antimyosin staining in the A-band, and (2) analysis of the staining patterns with respect to the antigenic specificities involved. As already discussed, the evidence for continuity of the filament throughout the A-band and for the presence of myosin throughout the filament is compelling. Lack of antimyosin staining in parts of the A-band may therefore be attributable to: (1) blockage of antigenic sites as a result of their involvement in the aggregation of myosin molecules that leads to the structure of the filament, (2) blockage of antigenic sites as a result of their involvement in interactions with the thin filaments, or (3) blockage of antigenic sites as a result of steric hindrance to the antibody molecules. The antimyosin staining patterns were analyzed in terms of the antigenic specificities involved in different parts of the staining pattern. This was done by using the fragments of the myosin molecule obtained by trypsin digestion to absorb the antibody prior to staining the fibrils. In this way it was shown that the antimyosin contained antibody specific to: (1) the LMM portion of the myosin molecule (LMM-specific antigenic sites), (2) the HMM portion of the myosin molecule (HMM-specific antigenic sites), and (3) a portion of the myosin molecule destroyed by the action of trypsin (β-specific antigenic sites). It was also shown that HMM-specific antibody staining could be correlated with the region of the A-band in which there was no overlap of thin and thick filaments. In stretched fibrils where this region (the H-zone) is wider, the HMM staining correspondingly increased in width. In contrast to this, the LMM- and β-specific antibody always stained the same region of the A-band regardless of the sarcomere length. That is, the presence of thin filaments did not interfere with either the LMM- or β-staining. This eliminates the possibility of blockage of antigenic sites as a result of steric hindrance attributable to the presence of thin filaments in the A-band. Therefore the blockage of the HMM sites in the region of overlap must result from the interaction or association of those sites with the thin filaments. The LMM-specific antibody always stained the region of the A-band near the A–I junction and the β-specific antibody always stained the middle third of each half of the A-band. These differences in staining can be explained in terms of differences in the lateral aggregation of the molecules. In proposing a length-determining mechanism from the detailed molecular model of the myosin filament, Pepe (*21*) considered termination of the length to be attributable to cumulative mismatching of laterally interacting sites on the LMM portions of the molecules, until the mismatching was severe enough to prevent further growth. This means that there would be weaker lateral aggregation between the molecules at each end of the filament, and this is consistent with exposure of LMM antigenic sites at the two ends of the filament. That is, where LMM antigenic sites are available, the LMM portions of the myosin molecule are not held as tightly in the core of the filament and a part of the LMM portion of the molecule can project from the core of the filament and thus contribute to the cross bridge.

It has been shown that LMM fragments from early trypsin digests of myosin retain some antigenic sites that are not present on LMM fragments obtained by longer digestion (*18*). These are the trypsin-sensitive or β-antigenic sites referred to earlier. That this β-antigenicity is not entirely confined to the trypsin-sensitive part of the molecule can be seen from the fact that under conditions in which it is completely removed from the LMM fragment some is still present on the HMM fragment. Since the trypsin-sensitive region is in the middle of the molecule, between the LMM and HMM portions, exposure of LMM antigenic sites should also result in exposure of β-antigenic sites. However, the β-antigenic sites are available for staining only in the middle third of each half of the A-band, where no LMM staining occurs, and they are not available where LMM sites are available. The original hypothesis concerning the availability of the β-antigenic sites was that the inability of the LMM portion of the myosin molecules to bend out of the filament imposes a stress on the β-region of the molecule when the bridge attaches to the actin filament. This stress results in exposure of β-antigenic sites which in the absence of stress (when the LMM portion can bend out of the filament) are not available for staining (18). However, β-staining has been shown to occur even when the thin filaments are absent from this region of the A-band (22). This means

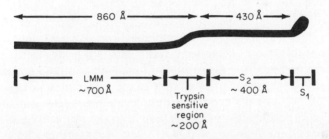

Fig. 4. Diagrammatic representation of the relationship between the myosin fragments (LMM, S_2, and S_1) and the linear aggregation of myosin molecules in the filament. The 860-Å portion of the molecule is in the core of the filament and the 430-Å portion represents the overlap of successive molecules in a row (see Fig. 2).

that β-staining does not occur solely as a result of the interaction of the thin filaments with the thick but must result, at least in part, from the inherent structural organization of the myosin filament. Consider that the beginning of the overlap region of myosin molecules in the model corresponds to the trypsin-sensitive (or β-) region (18,22). For fully digested myosin the LMM portion of the myosin molecule is approximately 700 Å in length (28,31,48). The trypsin-sensitive region of the molecule extends over approximately the next 200 Å of the rod portion of the molecule (28,31,48) and the S_2-portion of the rod extends over approximately the next 400 Å (28). According to the model the overlap of successive molecules in a row begins 860 Å from the LMM end. Therefore the overlap would begin in the trypsin-sensitive region approximately 160 Å from the end of the LMM region (Fig. 4). The tight packing of the LMM portions of the myosin molecules in this part of the filament, resulting in no exposure of LMM sites, might be responsible for stress being imposed on the beginning of the overlap and thus result in exposure of β-antigenic sites. Further stress can then be imposed by interaction of the bridge with the thin filaments and by the increase in distance between the filaments that occurs with a decrease in sarcomere length (49). A decrease in sarcomere length is accompanied by an increase in brightness of the β-staining consistent with increased exposure of β-sites. The availability of LMM sites also increases with a decrease in sarcomere length.

C. Cross-Bridge Action

Most speculations concerning the action of the cross bridge during contraction consider movement of the bridge to provide the driving force for the sliding of the filaments (2,35,50,51). These speculations have been concerned with the changes in orientation of the bridges, the fraction of all the bridges that are attached at any one time or the distance along the actin filament between successive attachments of a bridge as the thin filament slides past it. From the results of fluorescent antimyosin staining, it was suggested that the major function of the bridge may be to serve as a connection between the actin and myosin filaments and that the actual translation producing interaction is limited to the contact region between the bridge and the actin filament (12,18,22). From further studies of antibody staining in the presence of relaxing medium, it was concluded that the end of the bridge may be restricted in some relation to the actin filament even when the actin–myosin link is dissociated (22). This conclusion was based on the fact that the presence of relaxing medium during staining of the myofibril did not result in exposure of HMM antigenic sites in the region of overlap of the filaments. In the presence of relaxing medium, the actin–myosin link is dissociated, therefore unless something else blocks the HMM sites they should become available for staining. The presence of the thin filaments between the thick ones does not block LMM- and β-staining and therefore should not block available HMM sites. It was suggested that the proteins tropomyosin and troponin, which are bound to the actin filaments (13) and serve a regulatory function in the interaction between the actin and myosin filaments (4), may be involved in some way with the end of the bridge leading to the blockage of HMM antigenic sites. Native tropomyosin, which consists of a mixture of tropomyosin and troponin (52), does not bind to myosin (53) although it binds strongly to actin and actomyosin (4). Therefore the interaction of the tropomyosin and troponin with the end of the bridge, which leads to blockage of HMM antigenic sites, must be a weak one. The hypothesis suggested was that the bridge, in conjunction with the proteins associated with the actin filaments, forms a micellar aggregate which changes from one stable configuration to another through the successive stages in a cycle of activity of the bridge (22). Therefore even under conditions in which the actin–myosin link is broken, the bridge may still be related in some way to the regulatory proteins on the actin filament. It was argued that such a relation would facilitate the successive interactions of the bridge along the thin filament and would be especially advantageous in directing the action of a very long bridge. The 400-Å repeat periodicity along the actin filament representing the distribution of the proteins that have a regulatory function on the actin–myosin interaction (4) suggests that this repeat may represent the turning-on and turning-off points for the translation-producing interaction. The sliding might be turned on at the beginning of the 400-Å interval and turned off at the end and then turned on again beginning the next cycle, and so on. One cycle of the bridge in this case would represent sliding over the 400-Å interval, the successive interactions of the bridge with the monomeric units of the actin in this interval being modified by any changes in the micellar aggregate occurring during the cycle. The changes occurring during the cycle would include: (1) any changes in conformation imposed on the bridge and the micellar proteins by the translation of the end of the bridge along the actin filament, (2) any changes in conformation imposed on the bridge and the micellar proteins by ion binding or by the steric, electrostatic, and hydrophobic interactions occurring between them, and (3) any changes in the flexibility of the bridge caused by these interactions and by the ionic environment in the immediate vicinity of the bridge. The ionic environment in the immediate vicinity of the bridge may be affected by the products of enzymic activity associated with the translation process, or by the ion-binding characteristics of the micellar proteins and the sarcoplasmic reticulum. These changes would be synchronized with the translation occurring during one cycle of activity in such a way as to give the desired contractile characteristics.

The length of the myosin molecule involved in the cross bridge is of primary concern in any hypothesis concerning the action of the bridge during contraction. From the results of fluorescent antibody staining, the length of the cross bridges, based on the exposure of antigenic sites, varies along the length of the filament. In the region of the A-band where LMM antigenic sites are exposed, the entire trypsin-sensitive region, about 200 Å in length (48), and the HMM, approximately 600 Å in length (26,48), project out of the filament in addition to a small portion of the LMM, giving a total length of approximately 800 Å or more as part of the bridge. In the region of the A-band, where no LMM antigenic sites are available, less than 800 Å of the molecule would contribute to the bridge. What could be the possible function of such long bridges and of the variation in length of the bridge in different parts of the filament? The actin filament has a 360- to 370-Å crossover repeat attributable to the actin helix itself (5) and a 400-Å repeat attributable to the proteins associated with the actin filament (5,15). The proteins associated with the actin filament have a regulatory function in the interaction between the actin and the myosin (4). The myosin filaments have a 430-Å repeat of bridges associated with them, that is, between an actin and a myosin filament there is one myosin cross bridge every 430 Å along the myosin filament. Suppose that during the cycle of action of the bridge during contraction the 430-Å repeat of myosin cross bridges has to accommodate to the 400-Å repeat associated with the actin filament. This would mean that when there was complete overlap of the thin filaments with the thick, 16 myosin bridges spaced 430 Å apart (along one half of the filament) would have to accommodate to a spacing of 400 Å. In order to do this, the length of the myosin bridges would have to increase on going from the center of the A-band to the edge of the A-band, the last bridge at the edge of the A-band being the longest. The difference (16×430 Å $- 16 \times 400$ Å $= 480$ Å) would have to be made up by the length of the last bridge at the edge of the A-band. Accommodation to the 360- to 370-Å crossover repeat of the actin helix would require 1120–960 Å to be made up by the length of the last bridge at the edge of the A-band. A bridge 800 Å or more in length at the edges of the A-band could easily accommodate to the 400-Å repeat associated with the actin filament but would have difficulty accommodating to the 360- to 370-Å crossover repeat of the actin helix, especially considering the extra length needed to bridge the gap between the two filaments. X-Ray diffraction data (20,34) has indicated that in rigor there is attachment of material onto the outside of the units along the actin helices (20), which have a 360- to 370-Å crossover repeat (5). This may indicate that in rigor a portion of the end of the bridge binds to or associates with the units along the actin helices. The ends of successive bridges between a myosin and actin filament could still occur every 400 Å along the actin filament. In the scheme presented: (1) the cross bridge bridges the gap between the composite actin filament (actin + tropomyosin + troponin) and the myosin filament at all times, even under relaxing conditions, (2) between a myosin and actin filament the ends of the myosin cross bridges that occur every 430 Å along the myosin filament accommodate to the 400-Å repeat along the composite actin filament, and (3) the 400-Å repeat along the composite actin filament represents the interval of a cycle of bridge activity. In this scheme the changes observed in X-ray diffraction data represent changes in the conformation of the bridge as reflected by changes in the ordered regions of the bridge. If a portion of the bridge is disordered so that it no longer contributes to periodic repeats of protein density along the filament, it would not be picked up by the X-ray diffraction analysis.

A simple two-dimensional diagrammatic representation of the possible conformation of the bridge in the relaxed and rigor states is shown in Fig. 5a and b. The actual conformation of the bridge at any point in the bridge cycle would depend on the previously discussed micellar organization at any point in the cycle. In Fig. 5 the angling of the bridge with respect to the actin filament is consistent with the arrowhead structures seen on thin filaments decorated with HMM (19) and the arrowheads seen in sections of muscle in rigor (54,55). The conformation of the relaxed bridge is consistent with the orientation of the bridge perpendicular to the actin filament seen in sections of relaxed muscle (54,55). Therefore the bridge as visualized in electron microscopy represents a part of the effective total length of the bridge. As can be deduced from the simple diagram in Fig. 5, the total distance that the thin filament is moved with respect to the thick filament at any point in the cycle depends on both the distance traversed by the end of the bridge along the thin filament (A to B) and on where along the bridge the hairpin loop (see Section III) is located. The position of the loop is determined by both the organization of the micellar aggregate, which includes the bridge, and by the stiffness of all or part of the bridge at any point in the cycle. The bridge has been assumed to be inextensible throughout the cycle of activity. This does not exclude the possibility that the translation producing interaction at the point of contact between the bridge and the actin filament may involve an extensible polypeptide portion of the myosin molecule as proposed by Davies (56). In Fig. 5 the distance from A to B represents the distance traversed in one cycle of activity, the rigor configuration representing the stopping point of a cycle. In the relaxed state the position of the end of the bridge along the actin filament probably is not limited to the starting point of the cycle. Only the relationship that the end of the bridge has with the actin filament is determined since this is imposed by the organization of the micelle. Activation may involve positioning of the end of the bridge on the actin filament by slight movement to the nearest starting point for a cycle of activity. In the cycle of activity between the starting and stopping points, the conformation of the bridge, the flexibility or stiffness of the bridge, and the organization of the micellar aggregate including the bridge, are changed by interdependent influences, as previously discussed. The test and more rigorous definition of this scheme will come from being able to interpret available data and being able to design experiments in terms of this simplified basic scheme.

As analyzed by X-ray diffraction (20,34) the helical arrangement of the cross bridges is lost on going into rigor with movement of mass from the

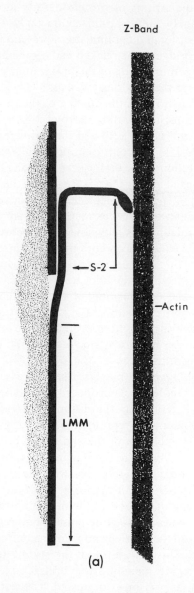

Fig. 5. Diagrammatic representation of the possible conformation of the bridge in the relaxed (a) and rigor (b) states. A cycle of activity of the bridge would result in translation of the end of the bridge over 400 Å along the thin filament from A to B (see text for details).

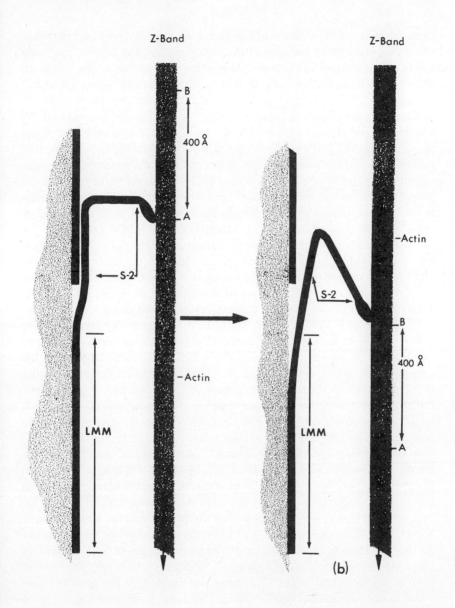

(b)

thick filament to the vicinity of the thin filaments. It was suggested that this could result from bending of the HMM part of the myosin molecule out from the core of the myosin filament on attachment of the bridge to the actin filament. This is similar to the conclusions previously arrived at from fluorescent antibody staining (12,18,22), except that the fluorescent antibody staining results suggest that in some parts of the filament (at the ends) a portion of the LMM part of the molecule can also bend out of the filament (Fig. 5b). If the entire rod portion of the molecule were aggregated into the filament in the same helical arrangement as the overlaps, one might expect that even though the myosin molecules bend out of the filament the source of the bridges on the backbone of the filament might still give the helical repeat. However, according to the model for the myosin filament previously described (21,22), the overlap regions or HMM portions of the molecule are arranged in the same helical arrangement as the cross bridges but the LMM portions are not. Since fluorescent antibody staining (9,12,18) indicates that the overlap region and more can be included in the part of the bridge that bends out of the filament in rigor, the model is consistent with loss of the helical cross-bridge repeat in rigor. It was also concluded from X-ray diffraction studies (20,34) that at any given moment during contraction only a small percentage of the cross bridges appear to be attached to the actin filament. The criterion for attachment was the change in the X-ray diffraction pattern occurring when the muscle goes into rigor, when all the bridges are presumed to be attached. In the scheme for the action of the bridge presented (Fig. 5), the rigor-type interaction would correspond to the turning-off point of the cycle, and so would represent a small portion of the entire cycle.

In the working hypothesis presented for the action of the cross bridge, the translation producing interaction that results in a sliding of the end of the cross bridge along the 400-Å interval of the actin filament for each cycle has not been considered. Different models have been presented for the translation producing interaction of the bridge with the actin filament. Most of these involve changes in the conformation of the bridge as the driving mechanism for translation. In the model proposed by A. F. Huxley (50), the bridge with its reactive site oscillates about an equilibrium position. Attachment of the bridge to an active site on the actin filament which is to one side relative to the equilibrium position of the bridge is maintained with a high probability and results in movement toward that side, whereas attachment of the bridge to a site on the other side has a low probability of being maintained. This results in movement in one direction as links are made and broken. In Pringle's model (2), which is particularly related to insect flight muscle, the translation is produced by attachment of the bridge to the actin filament followed by a change in the angle that the bridge makes with the actin filament. Release and reattachment of the bridge at a small angle with respect to the actin filament begins the next cycle. The active stroke for translation therefore consists of angling of the bridge. A similar model has recently been proposed by H. E. Huxley (34) in which the active stroke for translation is confined to angling of the two globular heads of the myosin with respect to the actin filament. The S_2-portion of the myosin molecule provides an inextensible link between the globular head and the LMM region of the molecule, the LMM always being confined to the core of the filament. In a model proposed by Loewy (51), the interaction between actin and myosin is considered to involve formation of a covalent bond during a portion of the cycle of action of the bridge. The formation of the covalent bond leads to a conformational change in the bridge, which results in translation between the filaments. Two models have been presented in which the translation producing inter-action involves a small region of the bridge and therefore can be limited to the contact region between the end of the bridge and the actin filament, therefore not involving the angled part of the bridge in translation. The model described by Davies (56) for the translation producing interaction involves an extended asymmetrically arranged polypeptide chain which is part of the cross bridge. Extension of the chain results from electrostatic repulsion between bound ATP at the free end of the chain and a fixed charge inside the HMM portion of the cross bridge. Attachment of the end of the extended chain to the actin filament through calcium results in neutralization of charge on the bound ATP and contraction of the extended polypeptide chain to form an α-helix. This results in translation between the filaments and brings the end of the chain into a region of the cross bridge having ATPase activity. Hydrolysis of the bound ATP results in breaking the actin–myosin link. Rephosphorylation then restores the poly-peptide chain to the extended form and the cycle is repeated. Perry (3) proposes that the translation producing interaction involves a series of make-and-break processes, each of which results in an increment of move-ment, with the movement being directed by the positions of the active centers on the myosin molecules. The active centers are the actin-binding site and the ATPase site and are in the region of contact of the bridge with the actin filament. The position of the actin-binding site being closer to the middle of the A-band effects translation by binding to the actin when the local concentration of ATP is decreased by activation of the nearby ATPase site. Release is effected by restoration of the ATP concentration by rephos-phorylation and the cycle then repeats. Neither of these two models involves a large part of the bridge as the active part of the translation-producing mechanism and could therefore be incorporated into the scheme for the action of the bridge already described.

III. THIN FILAMENT

The detailed molecular organization of the thin actin-containing filament is the same in all muscles so far investigated (*57,58*). The monomeric form of actin is G-actin and the polymerized form is F-actin. The actin filament consists of two helically wound strands made up of monomeric subunits. The helically wound strands cross over every 360–370 Å along the actin filament and the number of monomer units per turn is between 13 and 14 (*5*). The length of the actin filaments in muscle measured from their attachment to the Z-band to the free end of the filaments can vary from 0.3 μ in the squid mantle to 6.0 μ in slow fibers from crab leg or eyestalk levator muscle (*7*). In vertebrate muscles the length of the actin filament is around 1 μ (*59*).

Unlike the myosin filament, the thin actin-containing filament is known to have other proteins associated with it in the myofibril. By using fluorescent antibody staining, both tropomyosin (*12,13*) and troponin (*13,14*) have been shown to be associated with the thin filaments of the striated myofibril. Antitroponin staining in electron microscopy showed the troponin to be localized periodically every 400 Å along the length of the thin filament (*14*). This repeat periodicity is the same as the naturally occurring period observed in the I-band. This periodicity can also be seen in aggregates of crude F-actin filaments, which contain tropomyosin and troponin as impurities, when these aggregates are sectioned along the longitudinal axis of the filaments (*15*). Aggregates of highly purified F-actin filaments do not show the periodicity. Therefore, although the crossover repeat for the actin filament is 360–370 Å, the association of tropomyosin and troponin with the actin leads to the appearance of a 400-Å repeat period. As discussed in connection with the proposed scheme for the action of the myosin cross bridge (Section II.C), this interval may correspond to the turning-on and turning-off points in a cycle of interaction between the end of the bridge and the actin filament. Recently, troponin was shown to be made up of two components which may be related to inhibition and activation of the actin–myosin interaction (*60*). This fits nicely into the scheme previously presented (Section II.C).

If the HMM fragment of the myosin molecule (Section II.A) is reacted with the actin filament, the HMM molecules attach to the actin filament in such a way that they appear as arrowheads all pointing in one direction along the filament (*19*). This indicates that the actin filaments are polarized in one direction. Applying the same procedure to actin filaments that were still attached to the Z-band, Huxley (*19*) showed that the arrowheads on filaments on one side of the Z-band pointed in the opposite direction to those on filaments on the other side of the Z-band. This indicated that the actin filaments are polarized in opposite directions on each side of the Z-band. This is consistent with the opposite orientation of myosin molecules in the thick filaments on each side of the center of the A-band. The direction of the arrowheads in relation to the Z-band was also consistent with the angling of the bridges observed in sections of insect flight muscle in rigor (54,55). If one considers that the myosin cross bridges reach out and attach to the thin filaments at the point farthest away from the origin of the bridge, it would be expected that arrowhead structures would form with the arrowheads pointing toward the Z-band. However, the arrowheads were observed to point away from the Z-band. This indicates that the free end of the myosin molecule contributing to the bridge (the HMM end) must point toward the middle of the A-band. Since the LMM end in the core of the myosin filament also points toward the middle of the A-band, the myosin molecule must be shaped into a hairpin loop when interaction of the bridge with the actin filament occurs in situ. How could this relationship be established? From fluorescent anti-HMM staining of the A-band (18,22), it was found that at sarcomere lengths at which there is a zone of no overlap of filaments the staining of HMM sites in the no overlap region of the A-band shows up as a doublet with the unstained region in the middle corresponding to the pseudo-H-zone (the region where no myosin cross bridges are present). As the thin filaments approach the region of the pseudo-H-zone, the previously unstained region stains with anti-HMM. When complete overlap of filaments is achieved, all the HMM staining is abolished. This was interpreted as representing bending of the cross bridges ahead of the advancing end of the thin filament so that as the thin filaments approach the speudo-H-zone cross bridges are bent into this region and when there is complete overlap these bridges then become attached to the thin filaments. However, since the staining took place in glycerol, the muscle was in rigor and one could argue that the unreacted bridges in front of the end of the thin filaments were passively bent into the pseudo-H-zone when rigor was established. At present, we cannot choose between these alternatives. However, if the cross bridge is indeed involved in a micellar aggregate which includes the composite actin filament (actin + tropomyosin + troponin) as envisioned in Section II.C, one would expect that the incorporation of the cross bridge into the micellar aggregate would occur at the advancing end of the thin filament. Here the proper relationship between the actin filament and the part of the bridge involved in the translation producing interaction could be established and this relationship could be maintained throughout interaction of the bridge with successive actin monomers.

IV. FILAMENT LATTICE

In the I-band the thin filaments are not as precisely organized into a regular lattice as they generally are in the A-band where they interdigitate with the thick filaments. The interaction between the thin and thick filaments in the overlap region of the A-band probably accounts for the increased order. The ratio of thin to thick filaments in vertebrate muscles is 2 : 1 (7,61). The thin and thick filaments in the overlap region are arranged in a hexagonal lattice with the thin filaments occurring in the trigonal position between three thick filaments (16). In vertebrate fast muscles the thin filaments, close to their attachment to the Z-band, are arranged in a square lattice (62). The square lattice arrangement is probably imposed by the organization of the Z-band (see Section VI). In vertebrate slow muscle (62), it is more difficult to determine if there is any regular arrangement of the thin filament attachment to the Z-band since the structure of the Z-band cannot be visualized clearly (see Section VI).

Most invertebrate muscles have a larger thin-to-thick-filament ratio than 2 : 1 (7,61). In insect flight muscles (7,63), the ratio of thin to thick filament is generally 3 : 1, with six thin filaments around each thick one. However, in lepidopterans (41,42) there are from 7 to 10 thin filaments around each thick one. In insect flight muscles with a 3 : 1 ratio of thin to thick filaments, the filaments are arranged in a highly ordered hexagonal lattice with one thin filament positioned between two thick filaments. The hexagonal arrangement of thin filaments persists straight into the Z-band (64), where thin filaments from adjacent sarcomeres overlap (see Section VI). It is not clear what significance can be attributed to the maintenance of the hexagonal lattice of thin filaments both in the overlap region and at the Z-band in insect flight muscle, while in vertebrate muscle the hexagonal lattice of thin filaments in the overlap region converts to a square lattice at the Z-band. In insect flight muscle it might be argued that if the hexagonal lattice in the overlap region had to accommodate to a square lattice at the Z-band some distortion of the thin–thick filament relationship might occur because of the short I-band, in the functional range of sarcomere lengths. Such a problem would not exist in vertebrate muscle in which the I-band is comparatively much longer in the functional range of sarcomere lengths. The maintenance of a precise lattice may be advantageous for the proper functioning of a fast-acting muscle.

Since interaction between the thin and thick filaments occurs by means of the cross bridges on the thick filaments, it seems reasonable that the arrangement of these bridges on the surface of the thick filament may contribute to defining the lattice arrangement. In insect flight muscle the arrangement of myosin cross bridges conforms to a nonintegral helix, while in vertebrate muscle the arrangement of myosin cross bridges conforms to an integral helix (Section II.A). How this difference can be related to the differences found in the lattice arrangements is not clear. Another important contribution to the organization of the lattice probably comes from the electrostatic repulsion between the filaments. Elliott (65) has considered that the changes occurring in the myofibril during contraction are a result of the electrostatic effects, van der Waals' attraction, and hydration effects and may not necessarily involve chemical bonds between the filaments. In the scheme proposed in Section II.C, such interactions are involved in the micellar aggregate which includes the bridge. These occur in addition to the chemical bonding involved in the translation producing interaction occurring at the point of contact between the end of the bridge and the actin filament. It is conceivable that the distribution of charges both on the surface of the filaments and on the proteins, tropomyosin and troponin, associated with the thin filaments influences the organization of the filament lattice. If these charged proteins can effect a contribution to the differences in lattice arrangements observed in different muscles, it probably occurs through differences in the charges on the protein constituents in different muscles or through differences in the organization of the postulated micellar aggregates which include these proteins. When present, a well-organized M-band may also contribute to a well-organized filament lattice. It is therefore clear that the lattice arrangement is probably determined by a combination of different factors including: (1) cross-bridge arrangement on the surface of the thick filament, (2) spatial arrangement of the proteins associated with the thin filaments and their association with the bridge as previously discussed (Section II.C), (3) electrostatic repulsions between the filaments and modification of these by the proteins associated with the thin filaments, and (4) the organization of the M-band when present. All or some of these contributing factors can be expected to be different in different muscles.

V. M-BAND

The M-band, when present, is made up of bridges attached to the middle of the thick filaments and bridge between neighboring filaments (21,66,67). The M-material is not myosin, actin, or tropomyosin (9,12) but a different protein which can be isolated from the myofibril (68). When the myofibril is homogenized in the presence of Mg-ATP, the thick and thin filaments can be separated into suspension (19). This leads to loss of the M-protein, indicating that it loosely adheres to the thick filaments (69). The M-band can be extracted easily from the myofibril without loss of the integrity of the thick filaments (70–72). The extracted material can also be replaced and the M-band can thus be reconstituted (72).

The detailed structural organization of the M-band has been studied in the lateral muscles of the fish and in chicken breast muscle (22). In these muscles the detailed structure was similar. In cross sections of the muscle fibrils through the M-band, six M-bridges can be seen on each thick filament, one to each of the surrounding six thick filaments (21,22,66,67). In longitudinal sections through the fibril, the M-band is seen to be approximately 860 Å in width and to be made up of lines perpendicular to the axis of the filaments (21,22,73). Depending on the plane of longitudinal section through the M-band, the number of lines varies (22). The variation consists of omission of some of the lines without change in the position of the remaining lines. A maximum of five lines is observed. The three central lines are always most distinct and well-defined, while the two lines bordering the M-band are fuzzier. The different patterns observed consist of: (1) a maximum of five lines, (2) four lines, (3) the three central lines, (4) two of the three central lines with the middle one missing, and (5) two of the three central lines with the middle one present (see Fig. 5 in Reference 22). A detailed analysis of the structural organization of the M-band represented by these observations is being prepared.

Double M-bands have been observed in longitudinal sections of some copopod (7,74) and some crab (7,45,46) muscle fibers. In some invertebrate muscle fibers that have no well-defined M-band, Hoyle (7,45) obtained indications that bridges may occur between the thick filaments all along their length. The possible relationships of these filaments and the attached M-material to the model derived for vertebrate muscle are discussed in Section II.A.

Nothing is known for certain about the function of the M-band. It is highly unlikely that it has any direct involvement in contraction. In general, in muscles with no M-band the thick filaments become misaligned longitudinally, giving an indistinct A–I junction, whereas when an M-band is present a very sharp A–I junction is seen. Such misalignment occurs in both vertebrate (62) and invertebrate (75) muscles without an M-band. They are generally slow muscles, although slow muscles can have an M-band (76). It may be that the presence of an M-band in helping to maintain the relative position of the thick filaments is advantageous to the expression of the contractile properties of fast muscles.

Muscles have been described in which there are small tubules dispersed between the filaments in the region of the A-band where the M-band is generally found (7,41–43). There seem to be no direct connections between the filaments and the tubular system in these muscles, although there is a specific relationship in that the tubules surround only a short length of the middle part of the myosin filament. The function of this tubular system is not known. If the M-band is defined structurally as consisting of connections between the thick filaments, the tubular structures would not properly be considered M-band material.

VI. Z-BAND

The detailed structural organization of the Z-band is different for different muscles (7,45). Some muscles have Z-bands with a well-defined fine structure, whereas others have Z-bands with poorly defined fine structure. In general, Z-bands with poorly defined fine structure are associated with slow muscles (62). Knappeis and Carlsen (77) have described the detailed structure of the Z-band in frog skeletal muscle. At the point of attachment of the thin filaments to the structure of the Z-band, the thin filaments give rise to four Z-filaments which branch out to the other side of the Z-band and there each Z-filament attaches to a thin filament to give the tetragonal thin-filament lattice observed close to the Z-band. Therefore each thin filament is always attached to four Z-filaments. The thin filament is known to consist of two strands of actin monomers helically wound with a crossover repeat of 360–370 Å (see Section III). How a two-stranded structure such as this can lead to the four filaments in the structure of the Z-band is difficult to see. A possible model had been proposed by Kelly (78) in which each of the two strands of the actin filament continue into the Z-band, loop back, and continue as one of the strands of a neighboring actin filament. Such looping of the individual strands of actin in the Z-band can result in the structural observation of a two-stranded thin filament giving rise to four Z-filaments. An additional detail of this structural arrangement was observed by Reedy (79) in cross sections of the Z-bands of rat diaphragm muscle. He found that the Z-filaments were tangentially related to the I-filaments in a cross section through the Z-band, and this configuration showed up as either a clockwise or counterclockwise relationship. Analysis of the patterns using stereotechniques and serial sectioning led to the conclusion that the Z-filament configuration at the end of the thin filament was represented by a left-handed coil. Reedy (79) noted that it is not necessary for the thin filament to have the same sense of twist as the four Z-filaments attached to it do. Depue and Rice (80), using the mica replica technique concluded that the helix of the two-stranded actin filament is right-handed. The significance of this relationship between the thin filament and the Z-filaments is difficult to imagine.

In insect flight muscles the thin filaments extend into the structure of the Z-band (64,81). In the flight muscles of the belostomatid water bugs, Ashurst (64) has concluded that the thin filaments are not continuous through the Z-band from one side to the other but that they end in the Z-band where overlap occurs. The hexagonal lattice of the thin filaments entering from both sides of the Z-band is maintained in the Z-band, but with a relative displacement between them so that the two hexagonal lattices combine to form smaller hexagons in the Z-band. The thin filaments in the Z-band are presumably cemented together by the amorphous material present in the Z-band. In dipteran flight muscles, Auber and Couteaux (81) concluded that the thin filaments split on entering the Z-band and recombine on the other side to form the thin filament of the neighboring sarcomere. Also, in the flight muscles of Diptera and Hymenoptera, connections have been reported to occur between the ends of the thick filaments and the Z-band (81–83). These connections were not observed in the belostomatid flight muscles (64).

By controlled extraction it is possible to remove the Z-band from fibrils without loss of the integrity of the fibril (71,72). Exhaustive extraction causes marked I-band damage. The extracted material can be replaced in fibrils not exhaustively extracted, and the Z-band can thus be reconstituted. Tropomyosin had been suggested as a constituent of the Z-band based on the similarity in the lattice structure seen in crystals of tropomyosin and that observed in the structure of the Z-band (19). By using fluorescent antibody staining, tropomyosin can be shown to be associated with the actin filaments (12,13), but it cannot be detected in the Z-band (12). Stromer et al. (71,72), using different fractions of extracted Z-band material, concluded that the fractions that were most efficient for reconstitution of the Z-band were deficient in tropomyosin. Using fluorescent antibody staining methods Masaki et al. (84) identified the 6 S protein component of α-actinin as a constituent of the Z-band. Since the Z-band is made up of both the filamentous structural elements and the amorphous material between them, there is probably more than one component present. Whether α-actinin is a constituent of the structural elements or the amorphous material is not known.

Very little is known about the function of the Z-band. Since it serves as an attachment point for the thin filaments, its major function is probably to maintain the sarcomere as a structural unit during repeated cycles of contraction and relaxation. The differences in structural organization of the Z-band in different muscles may eventually be related to the ability of the Z-band to survive the stresses imposed on it by the different contractile properties of the sarcomere. For instance, if the thick and thin filaments can become out of longitudinal register, which in itself may be related to the contractile property of the fibrils, the Z-band must be capable of rather severe distortion to accommodate this while still being capable of bearing the load. This capability must be built into its structure. If maintenance of a very precise longitudinal register and lattice arrangement of the thick and thin filaments is necessary for expression of the particular contractile properties of a fibril, then a rigidly structured Z-band also capable of bearing the necessary load is desirable. Therefore the structural organization and components of the Z-band may vary relative to the properties of the contractile portion of the sarcomere, the function of the Z-band being primarily to maintain the structural integrity of each contractile unit and thus the myofibril as a whole.

REFERENCES

1. J. Gergely, *Ann. Rev. Biochem.*, **35**, Part II, 691 (1966).
2. J. W. S. Pringle, in *Progress in Biophysics and Molecular Biology* (J. A. V. Butler and H. E. Huxley, eds.), Vol. 17, Pergamon Press, Oxford, 1967, Chapter 1.
3. S. V. Perry, in *Progress in Biophysics and Molecular Biology* (J. A. V. Butler and H. E. Huxley, eds.), Vol. 17, Pergamon Press, Oxford, 1967, Chapter 8.
4. S. Ebashi and M. Endo, in *Progress in Biophysics and Molecular Biology* (J. A. V. Butler and D. Noble, eds.), Vol. 18, Pergamon Press, Oxford, 1968, Chapter 5.
5. J. Hanson, *Quart. Rev. Biophys.*, **1**, 177 (1968).
6. L. D. Peachy, *Ann. Rev. Physiol.*, **30**, 401 (1968).
7. G. Hoyle, *Ann. Rev. Physiol.*, **31**, 43 (1969).
8. M. Young, *Ann. Rev. Biochem.*, **38**, 913 (1969).
9. F. A. Pepe, *Intern. Rev. Cytol.*, **24**, 193 (1968).
10. A. F. Huxley and R. Niedergerke, *Nature*, **173**, 971 (1954).
11. H. E. Huxley and J. Hanson, *Nature*, **173**, 973 (1954).
12. F. A. Pepe, *J. Cell Biol.*, **28**, 505 (1966).
13. M. Endo, Y. Nonomura, T. Masaki, I. Ohtsuki, and S. Ebashi, *J. Biochem.*, **60**, 605 (1966).
14. I. Ohtsuki, T. Masaki, T. Nonomura, and S. Ebashi, *J. Biochem.*, **61**, 817 (1967).
15. J. Hanson, *Nature*, **213**, 353 (1967).
16. H. E. Huxley, *J. Biophys. Biochem. Cytol.*, **3**, 631 (1957).
17. H. E. Huxley and J. Hanson, *Biochim. Biophys. Acta*, **23**, 229 (1957).
18. F. A. Pepe, *J. Mol. Biol.*, **27**, 227 (1967).
19. H. E. Huxley, *J. Mol. Biol.*, **7**, 281 (1963).
20. H. E. Huxley and W. Brown, *J. Mol. Biol.*, **30**, 383 (1967).
21. F. A. Pepe, *J. Mol. Biol.*, **27**, 203 (1967).
22. F. A. Pepe, in *Progress in Biophysics and Molecular Biology* (J. A. V. Butler and D. Noble, eds.), Vol. 22, Pergamon Press, Oxford, 1971, Chapter 3.
23. J. Gergely, *J. Biol. Chem.*, **200**, 543 (1953).
24. E. Mihalyi and A. G. Szent-Györgyi, *J. Biol. Chem.*, **201**, 211 (1953).
25. A. G. Szent-Györgyi, *Arch. Biochem. Biophys.*, **42**, 305 (1953).
26. S. Lowey and C. Cohen, *J. Mol. Biol.*, **4**, 293 (1962).
27. H. Mueller and S. V. Perry, *Biochem. J.*, **85**, 431 (1962).
28. S. Lowey, L. Goldstein, C. Cohen, and S. M. Luck, *J. Mol. Biol.*, **23**, 287 (1967).
29. D. R. Kominz, E. R. Mitchell, T. Nihei, and C. M. Kay, *Biochemistry*, **4**, 2373 (1965).
30. T. Nihei and C. M. Kay, *Biochim. Biophys. Acta*, **160**, 46 (1968).
31. S. Lowey, H. S. Slayter, A. G. Weeds, and H. Baker, *J. Mol. Biol.*, **42**, 1 (1969).
32. H. E. Huxley, in *Muscle* (W. M. Paul, E. E. Daniel, C. M. Kay, and G. Monckton, eds.), Pergamon Press, Oxford, 1965, p. 3.
33. G. F. Elliott, *Proc. Roy. Soc.* (*London*), **B160**, 476 (1964).

34. H. E. Huxley, *Science*, **164,** 1356 (1969).

35. M. K. Reedy, *J. Mol. Biol.*, **31,** 155 (1968).

36. B. Baccetti, *J. Ultrastructure Res.*, **13,** 245 (1965).

37. B. Baccetti, *Bol. Soc. Ital. Biol. Sper.*, **42,** 1181 (1966).

38. V. P. Gilev, *Biochim. Biophys. Acta*, **112,** 340 (1966).

39. V. P. Gilev, in *Electron Microscopy* (R. Uyeda, ed.), Vol. 2, Maruzen, Tokyo, 1966, p. 689.

40. M. Halvarson and B. A. Afzelius, *J. Ultrastructure Res.*, **26,** 289 (1969).

41. J. F. Reger and D. P. Cooper, *J. Cell Biol.*, **33,** 531 (1967).

42. J. Auber, *Compt. Rend.*, **264,** 621 (1967).

43. J. Auber, *Compt. Rend.*, **264,** 2916 (1967).

44. J. F. Reger, *J. Ultrastructure Res.*, **20,** 72 (1967).

45. G. Hoyle, *Am. Zoologist*, **7,** 435 (1967).

46. G. Hoyle and P. A. McNeill, *J. Exptl. Zool.*, **167,** 487 (1968).

47. J. Kendrick-Jones, C. Cohen, A. G. Szent-Györgyi, and W. Longley, *Science*, **163,** 1196 (1969).

48. D. M. Young, S. Himmelfarb, and W. F. Harrington, *J. Biol. Chem.*, **240,** 2428 (1965).

49. G. F. Elliott, J. Lowy, and B. M. Millman, *J. Mol. Biol.*, **25,** 31 (1967).

50. A. F. Huxley, in *Progress in Biophysics and Biophysical Chemistry* (J. A. V. Butler and B. Katz, eds.), Vol. 7, Pergamon Press, Oxford, 1957, Chapter 6.

51. A. G. Loewy, *J. Theoret. Biol.*, **20,** 164 (1968).

52. S. Ebashi and A. Kodama, *J. Biochem.*, **59,** 425 (1966).

53. D. R. Kominz and K. Maruyama, *J. Biochem.*, **61,** 269 (1967).

54. M. K. Reedy, K. C. Holmes, and R. T. Tregear, *Nature*, **207,** 1276 (1965).

55. M. K. Reedy, *Am. Zoologist*, **7,** 465 (1967).

56. R. E. Davies, *Nature*, **199,** 1068 (1963).

57. J. Hanson and J. Lowy, *Circulation Res.*, **15,** 4 (1964).

58. J. Hanson and J. Lowy, *J. Mol. Biol.*, **6,** 46 (1963).

59. S. G. Page and H. E. Huxley, *J. Cell Biol.*, **19,** 369 (1963).

60. D. J. Hartshorne and H. Mueller, *Biochem. Biophys. Res. Commun.*, **31,** 647 (1968).

61. M. Hagopian and D. Spiro, *J. Cell Biol.*, **36,** 443 (1968).

62. S. G. Page, *J. Cell Biol.*, **26,** 477 (1965).

63. H. E. Huxley and J. Hanson, in *Electron Microscopy* (F. S. Sjostrand and J. Rhodin, eds.), Academic Press, New York, 1957, p. 202.

64. D. E. Ashhurst, *J. Mol. Biol.*, **27,** 385 (1967).

65. G. F. Elliott, *J. Gen. Physiol.*, **50,** 171 (1967).

66. D. Spiro, *Trans. N.Y. Acad. Sci.*, Series 2, **24,** 879 (1962).

67. C. Franzini-Armstrong and K. R. Porter, *J. Cell. Biol.*, **22,** 675 (1964).

68. T. Masaki, O. Takaiti, and S. Ebashi, *J. Biochem.*, **64,** 909 (1968).

69. F. A. Pepe and H. E. Huxley, in *Biochemistry of Muscle Contraction* (J. Gergely, ed.), Little, Brown, Boston, 1964, p. 320.

70. N. V. Samosudova, in *Electron Microscopy* (R. Uyeda, ed.), Vol. 2, Maruzen, Tokyo, 1966, p. 691.

71. M. H. Stromer, D. J. Hartshorne, and R. V. Rice, *J. Cell Biol.*, **35,** C23 (1967).

72. M. H. Stromer, D. J. Hartshorne, H. Mueller, and R. V. Rice, *J. Cell Biol.*, **40,** 167 (1969).

73. D. W. Fawcett and N. S. McNutt, *J. Cell Biol.*, **42,** 1 (1969).

74. Y. Bouligand, *J. Microscopie*, **1,** 377 (1962).

75. P. A. Toselli and F. A. Pepe, *J. Cell Biol.*, **37,** 445 (1968).

76. A. Hess, *Invest. Ophthalmol.*, **6,** 217 (1967).

77. G. G. Knappeis and F. Carlsen, *J. Cell Biol.*, **13**, 323 (1962).
78. D. E. Kelly, *J. Cell Biol.*, **34**, 827 (1967).
79. M. K. Reedy, *Proc. Roy. Soc.* (*London*), **B160**, 458 (1964).
80. R. H. Depue, Jr. and R. V. Rice, *J. Mol. Biol.*, **12**, 302 (1965).
81. J. Auber and R. Couteaux, *J. Microscopie*, **2**, 309 (1963).
82. N. Garamvolgyi, *J. Microscopie*, **2**, 107 (1963).
83. N. Garamvolgyi, *J. Ultrastructure Res.*, **13**, 409 (1965).
84. T. Masaki, M. Endo, and S. Ebashi, *J. Biochem.*, **62**, 630 (1967).

CHAPTER 8

MICROTUBULES

R. E. Stephens

DEPARTMENT OF BIOLOGY, BRANDEIS UNIVERSITY
WALTHAM, MASSACHUSETTS

I. THE OCCURRENCE OF MICROTUBULES

It has been nearly a decade since the first identification and description of the cytoplasmic microtubule. The advent of glutaraldehyde fixation in electron microscopy has made possible the visualization of this heretofore elusive structure. Almost simultaneously, Ledbetter and Porter (1963) observed 250-Å tubular elements in plant cells, while Slautterbach (1963) described their existence in *Hydra*. Since that time the number of works cataloging the distribution of microtubules among various protists, plants, and animals of all phyla has been increasing in the exponential manner so typical of research in vogue. In the interest of brevity, therefore, only those studies having some direct biochemical or physical implication are considered in any detail here.

As a general categorization, microtubules may be placed in two broad groups: those associated with overt motile systems and those that apparently serve a somewhat more passive structural role in the production and maintenance of cell shape.

The first category may be subdivided further into microtubules of direct centriolar origin—cilia, flagella, and contractile axostyles—and those of cytoplasmic origin—axopodial fibers, spindle fibers, cytokinetic rings, and the tubules sometimes found associated with streaming cytoplasm, morphogenetic movement, and cell shape.

The outer microtubule doublets of cilia and flagella and the triplet fibers of centrioles were actually described somewhat earlier than the single cylinder commonly referred to as the microtubule (Fawcett and Porter, 1954; Afzelius, 1959). The fine structure of the flagellum was thoroughly defined in the classic work of Gibbons and Grimstone (1960), while Gibbons (1963, 1965) presented evidence of ATPase "arms" associated with the outer fiber microtubules and devised procedures for the bulk isolation of the latter. The contractile axostyle of certain protozoans is comprised of an orderly array of singlet microtubules connected by "bridges" (Grassé, 1956; Grimstone and Cleveland, 1965) and, although the mode of beat of this internal compound flagellum is similar to that of a normal flagellum, it has not been demonstrated that the connecting elements correspond to ATPase structures.

In motile systems not directly associated with centrioles, less orderly structures are generally observed. One notable exception to this is the axopodia of some radiolaria which consist of a "jelly roll"–like array of single microtubules, showing 12-fold symmetry and bridging structures (Kitching, 1964; Tilney and Porter, 1965a). Mitotic spindle fibers are made up of microtubule bundles (Roth and Daniels, 1962), which occasionally show intratubular bridging (Wilson, 1969). A cytokinetic ring and "cleavage apparatus" composed of microtubules has been described during the division cycle of *Chlamydomonas* (Johnson and Porter, 1968), although cytokinesis more generally involves a cytokinetic ring of filamentous material rather than microtubules (cf. Schroeder, 1968). Nuclear elongation during spermatid development occurs through contrarotation of a double helical array of microtubules interconnected with cross bridges (McIntosh and Porter, 1967). Pigment migration in melanophores has been postulated to occur by means of microtubules found associated with the cell processes (Bikle et al., 1966). Direction of fusing vesicles during plant cell plate formation is apparently mediated through microtubules in the phragmoplast (Hepler and Jackson, 1968).

Microtubules of an apparently passive nature are often associated with cellular asymmetry. Rings of microtubules seem to be involved in producing the discoidal shape of some blood cells (Fawcett and Witebsky, 1964). Bundles of microtubules are found associated with elongating cells of the developing lens rudiment (Byers and Porter, 1964), with elongating myotubes during myogenesis (Warren, 1968), in mesenchymal cell processes during gastrulation (Gibbins et al., 1969; Tilney and Gibbins, 1969) and in neuron elongation (Olmsted et al., 1970). In all of these systems one can question the passivity of the microtubules since energy obviously must be expended in order to elongate or modify cell shape.

II. THE STRUCTURE OF THE MICROTUBULE

André and Thiéry (1963) and Pease (1963) observed that negatively stained microtubules of mammalian sperm are made up of at least 10 protofilaments, 35–40 Å wide and laterally associated to form the cylindrical tubule wall. These protofilaments appeared to be made up of 80- to 88-Å repeating units. Observations of such frayed tubules have led to the assumption that the longitudinal bonding between the repeating units is somewhat stronger than the lateral bonding between the resultant protofilaments.

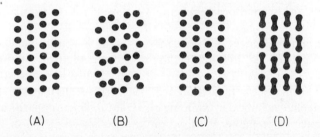

(A)   (B)   (C)   (D)

Fig. 1. Possible microtubule surface lattices. (A) 40 × 50 Å lattice based on light diffraction of negatively stained microtubules (Grimstone and Klug, 1966). (B) Perturbed lattice incorporating an 80-Å repeat distance (Grimstone and Klug, 1966). (C) 40 × 53 Å half-staggered lattice inferred from X-ray diffraction of wet gels (Cohen et al., 1971). (D) The same as (C) but incorporating 80-Å dimers of tubulin.

Ledbetter and Porter (1964) and numerous subsequent workers (Ringo, 1967) determined, through the Markham rotation technique (Markham et al., 1963), that many microtubules in cross section are composed of 13 circumferential, 40-Å, globular subunits. Kiefer et al. (1966) observed that a 33-Å particle constituted the building block of the tubule wall in negatively stained surface-spread mitotic microtubules from sea urchin eggs and tentatively identified it as the 3.5 S mitotic apparatus protein described by Sakai (1966). Similar subunits which measured about 35 Å in diameter were noted by Barnicot (1966) in surface-spread mitotic tubules of tissue culture cells.

Employing optical diffraction analysis of negatively stained protozoan flagella, Grimstone and Klug (1966) determined that the microtubule was composed of a basic 40 × 50 Å surface lattice. This lattice manifested itself in collapsed or disintegrating tubules as an array of longitudinal filaments spaced 50 Å apart and composed of 40-Å beads, in essence confirming the observations of earlier workers (Fig. 1A). When the fibers were intact, 80- and 160-Å longitudinal spacings became apparent and were explained as being attributable to perturbation of the simple 40×50 Å lattice both radially and in the plane of the surface lattice (Fig. 1B). The 160-Å periodicity corresponds roughly with that observed by Gibbons and Grimstone (1960) for the spacing of the ATPase dynein "arms" along the A-subfiber, but it has yet to be shown how dynein binds to the A-tubule. The dynein periodicity may be inherent in the dynein polymer rather than in any site on the tubule itself. Grimstone and Klug (1966) also observed a 480-Å fundamental repeat period and attributed it in part, at least, to interfibrillar material joining the central and outer fibers. It may be significant that a 960-Å repeat period is found for nexin bridges connecting the adjacent A-tubules of the $9+2$ structure (see below); perhaps the 480-Å spacing represents the second-order spacing of this 960-Å fundamental repeat period.

Cohen and associates (1971) recently observed X-ray diffraction patterns from wet gels of sea urchin flagellar outer fiber doublets and also single A-tubules (Stephens, 1970a). They found that the previous patterns obtained from sperm tails (Pautard, 1962) were attributable primarily to the flagellar membrane, while more recent diffraction patterns of flagellar axonemes prepared by digitonin extraction (Forslind et al., 1968) were attributable to cholesterol digitonide. However, when microtubules were delipidated with Triton X-100, the basic 40×53 Å surface lattice became evident (Cohen et al., 1971). The 80-Å spacing prominent in light-diffraction studies was a minor feature of the X-ray diffraction patterns. The most significant features of the wet gel X-ray diffraction data were the significantly off-meridional 40-Å reflection, a near-meridional second layer line, and a strong 52-Å equatorial reflection (Fig. 2A). Such spacings are consistent with a pseudo-centered lattice of subunits with 40×53 Å packing distances (Fig. 1C) and possibly 12- or 13-fold rotational symmetry. Quite significantly, upon drying, the strong 40-Å reflection becomes meridional (Fig. 2B) as the lateral coherence of the surface lattice is destroyed. The resultant fiber then diffracts as a simple linear array of subunits, just as was observed in the negatively stained preparations already discussed (Fig. 1A). Thomas (1970) reports surface lattices such as these within the same microtubule.

Of what significance is the 80-Å spacing sometimes observed in the electron microscope and consistently seen in light diffraction of such micrographs? As discussed in detail below, the structural protein of the microtubule exists as a dimer, that is, two 40-Å globular subunits. Evidently, this 80-Å particle is quite prominent in the mammalian outer fibers of André and Thiéry (1963) and Pease (1963), since 40-Å globules as such were not observed. The sea urchin outer fibers used for the X-ray studies showed no obvious pairing of subunits into 80-Å particles (Stephens, 1968a), although in vitro polymerization studies indicate that the dimer is the effective polymerizing unit (Stephens, 1969). Simple differences in staining properties or asymmetry in the monomer-to-monomer vs. dimer-to-dimer bonding could easily give rise to the 80-Å (dimer) spacing as a prominent feature in some cases (Fig. 1D).

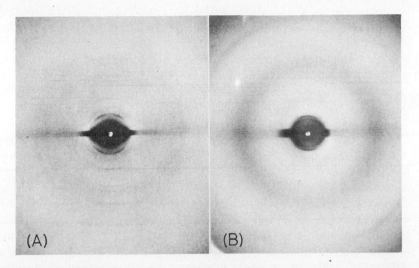

Fig. 2. X-Ray diffraction from microtubules. (A) Diffraction from wet gels of flagellar outer fibers; four layer lines of the 40-Å repeat are seen, but diffuse scatter produces the horizontal streak almost obscuring a 52-Å equatorial reflection. (B) Diffraction from air-dried outer fibers showing a strong 40-Å meridional and diffuse 50-Å equatorial reflection (Cohen et al., 1971).

III. THE ROLE OF DYNEIN AND RELATED ATPases

When cilia or flagella are stipped of their membranes by detergent treatment and then dialyzed against a low-ionic-strength buffer (1 mM tris–HCl, pH 8; 0.1 mM EDTA), nearly all of the associated ATPase goes into solution, leaving outer fiber doublet microtubules in suspension (Gibbons, 1963, 1965). After such treatment the outer fibers often retain their ninefold cylindrical arrangement but lack the characteristic arms originally present in the axoneme. In the cilia of *Tetrahymena pyriformis*, the solubilized ATPase protein, dynein, exists in two forms, a 14 S "monomer" and a 30 S polymer. In the presence of magnesium, the latter polymer can be added back to the outer fiber preparation, resulting in the re-formation of arms (Gibbons, 1963, 1965). Thus the principal ATPase of the cilium is localized in the armlike structures of the A-subfiber microtubule. Recent studies by Allen (1968), using the rotation technique of Markham et al. (1963), indicate that the two arms of *Tetrahymena* cilia are significantly dissimilar.

The 14 S form of dynein is a globular particle $85 \times 90 \times 140$ Å with a molecular weight of 600,000, while the 30 S form appears to be a linear polymer of 14 S particles (Gibbons and Rowe, 1965). The mode of bonding

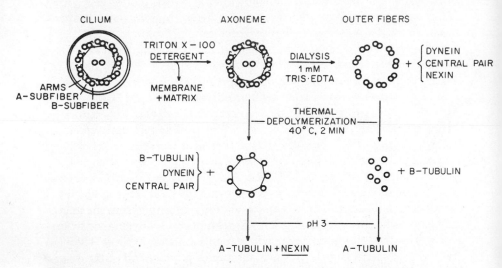

Fig. 3. Summary chart of procedures for the fractionation of cilia or flagella as described in Sections III, V, VI, and XI.

between these subunits remains unclear since the 30 S form cannot be depolymerized to the 14 S "monomer" under conditions in which breakage of covalent bonds cannot be unequivocally eliminated (Gibbons, 1965).

Sperm flagella yield chiefly the 14 S form of dynein, the properties of which closely parallel those of the ciliary counterpart (Gibbons, 1965). This dynein may be extracted either by low-ionic-strength dialysis or by extraction with strong salt solutions.

Gibbons (1966) studied extensively the enzymology of dynein. Either form is quite specific for ATP or deoxy-ATP, splitting other nucleoside triphosphates at about 1/10 the rate of ATP hydrolysis. The two forms of dynein differ somewhat when properties other than specificity are considered. The 14 S dynein has a maximum ATPase activity of 3.5 μM of phosphate per minute per milligram of protein while the 30 S form has a specific activity of 1.3, both being measured at 20°C with magnesium as cofactor. With calcium activation, the 30 S specific activity remains constant, but the 14 S activity is halved. With increasing ionic strength, whether magnesium or calcium activated, the specific activity of 30 S dynein increases markedly while that for 14 S dynein decreases correspondingly. Both forms show pH optima around pH 8.5–9, but the 30 S form has a second optimum near pH 6.

Various studies of isolated flagella or whole sperm indicate enzymic properties rather similar to those of dynein, but such studies have been complicated by the presence of membrane ATPases (cf. Mohri, 1958). Intact axonemes from gill cilia, freed of membranes by detergent treatment, resemble 14 S dynein in nucleotide specificity, pH optimum, and magnesium-to-calcium activation ratio but differ from either form of dynein in their relative insensitivity to ionic strength (Stephens and Levine, 1970). What modification in enzymic properties is brought about by the association of dynein with axonemal microtubules is still uncertain. These studies with axonemes imply that no gross changes in basic properties are evident, in contrast with myosin in which interaction with actin greatly enhances ATPase activity and subsequent interaction with troponin lends calcium sensitivity to the system (Ebashi et al., 1967).

Preparation of dynein by low ionic strength dialysis of gill cilia from *Aequipecten* (*Pecten*) *irradians* yields one-half of its dynein in the 14 S form; the remainder is tightly bound and only removable from the microtubule by either brief trypsinization, to yield more 14 S dynein, or by dissolution of the microtubule, to yield a highly polymeric form. Identical treatment of sperm flagella of this same species produces all of the dynein as the 14 S monomer after only dialysis (Linck, 1970, 1971). Thus, in the two cases studied, cilia yield both "monomeric" and polymeric forms of the ATPase while all flagella reportedly yield only the 14 S form.

A set of dynein arms on the A-subfiber of the microtubule doublet directed toward the B-subfiber of the adjacent doublet offers an inviting basis upon which to propose a sliding filament theory for ciliary or flagellar motion. By analyzing serial sections of the tips of bending and relaxed gill cilia, Satir (1968) has concluded that the outer fibers do not change in length during ciliary beat but slide over one another to produce the characteristic bending and relaxing waves. His observation that the fibers on the inside of the bend extend further out at the tip than do those at the outside of the bend does support constancy in length, but passive bending of the structure would produce an identical tip configuration. To date, no evidence exists that the dynein arms of the A-subfiber interact with the adjacent B-subfiber, that is, no musclelike cross bridges have been discovered.

ATPase activity has been reported to be present in the mitotic apparatus. Mazia et al. (1961) demonstrated enzymic activity in sea urchin mitotic apparatuses isolated in dithiodiglycol. Miki (1963) found a three-fold enhancement of ATPase activity in alcohol–digitonin-isolated mitotic apparatus preparations when compared with that of the cytoplasm and further demonstrated histochemically that ATPase was clearly localized in the spindle region. Weisenberg and Taylor (1968) isolated a 14 S ATPase from sea urchin mitotic apparatuses prepared in hexylene glycol but found that this component was also present in the cytoplasm in a similar amount. The nucleotide specificity, divalent cation activation, and pH optima of these mitotic apparatus-related ATPases are strikingly reminiscent of ciliary or flagellar dynein.

Cross bridges have been reported to occur between microtubules of the mitotic apparatus (Wilson, 1969; McIntosh et al., 1969). By performing careful microtubule counts in serial cross sections throughout the mitotic spindle, McIntosh et al. (1969) have mounted evidence for a sliding filament mechanism for mitosis wherein chromosomal spindle fibers of chromosome pairs move in opposite directions over continuous pole-to-pole fibers. Cross bridges similar to those in the mitotic apparatus have been reported in helical perinuclear microtubules of fowl spermatids (McIntosh and Porter, 1967), the motile force generated by the tubule system apparently being responsible for nuclear elongation. In certain protozoans bridgelike structures are prominent between microtubules in the contractile axostyle, an organelle capable of rapid sinousoidal beat (Grimstone and Cleveland, 1965). Thus far, there is no evidence that any such bridges possess ATPase activity, however.

The relationship between mitotic ATPases or nonciliary microtubule cross bridges and ciliary dynein must await isolation and characterization of the constituent proteins. The possible interaction of dynein or dynein-like proteins and the microtubule is an issue approachable only through detailed biochemical studies of these ATPases both in the intact motile structure and in isolated form, with particular emphasis on the probable participation of nucleotides in the motile process.

IV. GUANINE NUCLEOTIDES IN MICROTUBULE SYSTEMS

Various nucleotides, aside from those of adenine which are prerequisite for motility, have been reported to be components of sperm or sperm flagella (cf. Stephens et al., 1967). Recent work has shown that one of these is an integral part of the microtubular structural protein, both in cilia or flagella and in neurotubules.

By using sea urchin sperm outer fiber doublet microtubules isolated by selective solubilization (Gibbons, 1965) and 6 S outer fiber protein from acetone powders of *Tetrahymena* cilia (Renaud et al., 1966), it has been demonstrated that guanine derivatives are associated with the outer fiber protein in a ratio of 1 mole of guanine compound per 50,000–60,000 g of protein (Stephens et al., 1967) (Fig. 4). The sperm outer fibers contained

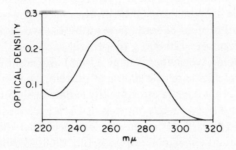

Fig. 4. Neutralized perchloric acid filtrate from sea urchin outer fibers. The spectrum is that of a guanine nucleotide, and paper chromatography identifies the material as a mixture of guanosine di- and triphosphate (Stephens, 1969).

GTP, GDP, and GMP in varying ratios, while the *Tetrahymena* cilia contained chiefly guanine and guanosine, roughly one-half of each being bound tightly to the protein while the remainder was removable with gel filtration. Since whole *Tetrahymena* cilia were found to contain guanine nucleotides and little free base or nucleoside, it appears as if the phosphoribosyl moiety is susceptible to degradation during isolation. This is in contrast to muscle actin in which the bound ADP is not normally susceptible to phosphorylation or dephosphorylation when in the polymerized form. The presence of free base and nucleoside in *Tetrahymena* ciliary outer fibers and variable ratios of nucleotides in sea urchin outer fibers also implies that these compounds are bound in some manner to the protein via the purine ring.

Careful reanalysis of guanine nucleotide binding has somewhat modified the originally observed ratio of bound nucleotides in flagellar outer fibers. Employing perchloric acid precipitation of protein from exhaustively dialyzed outer fibers, Stephens (1969) showed that 1 mole of guanine nucleotide per 100,000–120,000 g of protein was present in tightly bound form, being found primarily as the tri- and diphosphates. A similar amount of these nucleotides was found to be loosely bound. This result brings the ratio of tightly bound flagellar nucleotide in line with that observed for the ciliary counterpart (Stephens et al., 1967) and for the neurotubule structural protein (Weisenberg et al., 1968).

Yanagisawa and co-workers (1968) have, in general, confirmed these results in outer fibers from several other species of sea urchin, although the amount of guanine nucleotide is more in accord with the earlier binding ratio of 1 mole per 60,000 g of protein. In addition, these workers have shown that the terminal phosphate of ATP is transferred to the bound guanine nucleotide of the outer fibers during sperm motility, presumably through an interaction with dynein, implying that the bound nucleotide of the microtubule has some role in the motile process.

In the structural protein of brain neurotubules, two moles of guanosine di- and triphosphate per 120,000 g of protein are found, one of which is freely exchangeable (Weisenberg et al., 1968). In addition, this protein binds 1 mole of colchicine per 120,000 g of protein. The central pair microtubule from sea urchin flagella as originally isolated (Shelanski and Taylor, 1967) contained no bound nucleotide, but when prepared in the presence of magnesium and excess GTP, at least 0.50 mole of nucleotide per 120,000 g of protein was found (Shelanski and Taylor, 1968). Similar to the neurotubule protein, this microtubule subunit bound 1 mole of colchicine per 120,000 g of protein. These workers have also confirmed the presence of GTP and GDP in outer fibers in a ratio of 1 mole per 60,000 g of protein, but unlike the central pair and neurotubule protein, the outer fibers did not bind colchicine.

The presence of bound nucleotides, guanine or otherwise, in microtubular systems other than those discussed above has yet to be conclusively demonstrated. However, the identical nucleotide-binding properties of such diverse systems as ciliary and flagellar outer fiber or central pair microtubules and brain neurotubules favor the universality of this property.

V. PHYSICOCHEMICAL PROPERTIES OF TUBULIN

The first unequivocal characterization of a microtubule protein was carried out on *T. pyriformis* ciliary outer fiber doublets by Renaud et al. (1966, 1968), employing both acetone powders of whole cilia and preparations of outer fibers isolated by differential solubilization (Gibbons, 1965). The protein was found to migrate as a single band on acrylamide gel electrophoresis, possess an actinlike amino acid composition, and exist as a 6.0 S dimer of 103,000 molecular weight at low ionic strength and as a 2.4 S monomer with a molecular weight averaging 55,000 in guanidine hydrochloride. The dimer was found to bind 2 moles of mixed guanine derivatives, half of which were tightly bound (Stephens et al., 1967).

Employing brief dialysis in the usual Gibbons (1965) fractionation procedure, Shelanski and Taylor (1967) isolated a 6 S colchicine-binding protein from sea urchin flagella and identified it as the central pair protein.

In a later report, these workers compare this central pair protein with the outer doublet protein (Shelanski and Taylor, 1968). Both proteins are 6 S particles with a molecular weight of 120,000 at neutrality and both form monomers with an average molecular weight of 60,000 in guanidine hydrochloride. The two proteins were indistinguishable on the basis of gel electrophoresis and almost identical in regard to amino acid composition. Only the central pair protein bound colchicine. Unlike the outer fibers which bound 1 mole of guanine nucleotide per 60,000 g, the central pair contained bound nucleotide only when prepared in the presence of excess GTP and magnesium.

Outer fiber doublet protein from another species of sea urchin was characterized by Stephens (1968a). Urea or guanidine hydrochloride solubilization produced a 3 S monomer with an average molecular weight of 59,000, while 50 mM Salyrgan produced a 130,000-molecular-weight 5.4 S dimer and an apparent 9 S tetramer. The last-named particle showed an α-helix content of 25–30% when measured by optical rotatory dispersion. Amino acid composition indicated an actinlike protein similar to, if not identical with, those already described.

Sea urchin outer fibers in the detergent Sarkosyl (Stephens, 1969) sediment as a 2.8 S particle with a molecular weight of 66,000. The outer fiber protein from isolated gill cilia sediments at this same rate and has a molecular weight of 65,000 (Stephens and Linck, 1969). However, analysis of the 2.8 S subunits for Sarkosyl indicates that at least 10 moles of the detergent are tightly bound to the protein and thus increase the apparent molecular weight by 4000 or more.

Using outer fibers from yet another sea urchin species, Mohri (1968) reported the amino acid compositions and compared these values with those of actin, flagellin, myosin, tropomyosin, and various ciliary extracts reported by previous workers. These outer fiber proteins were found to resemble actin most closely, but were considered to be distinctive, hence were generically named "tubulin." Neither this report nor a previous one dealing with the isolation and the flow birefringence of outer fibers (Mohri et al., 1967) included any physicochemical data.

The 6 S colchicine-binding protein from brain homogenates (Weisenberg et al., 1968), apparently derived from neurotubules, has properties virtually identical to the various proteins derived from ciliary or flagellar sources. Molecular weights of 60,000 and 120,000 were found for the monomer and dimer, respectively. As discussed above, the protein binds 1 mole of colchicine and 2 moles of guanine nucleotide per dimer.

The physicochemical data and the amino acid composition for tubulin from ciliary, flagellar, and neurotubular material are summarized in Tables I and II. The molecular weight data in one case has been adjusted for a difference in $\bar{v}$. The amino acid compositions have been normalized to express residues of amino acid per 100,000 g of protein. In contrast to some other areas of protein chemistry, the agreement on the molecular weight, amino acid composition, nucleotide content, and colchicine binding among these proteins of various sources and by different laboratories is rather phenomenal. It is also generally agreed that the 60,000-molecular-weight monomer corresponds to the 40-Å globular subunit seen in negative staining and X-ray diffraction, while the 80-Å repeat quite likely represents the 120,000-molecular-weight dimer (cf. Shelanski and Taylor, 1968 or Fig. 1D).

Less complete data are available for tubulin from other microtubule sources. This is mainly attributable to the inability to obtain adequate

TABLE I

Physicochemical Parameters of Tubulin

Source	Solvent	$s^0_{20,w}$	Molecular weight	Ref.
Tetrahymena outer fibers	1 m*M* tris	6.0	110,000[a]	Renaud et al., 1968
	5 *M* G-HCl	2.4	58,000[a]	Renaud et al., 1968
Strongylocentrotus purpuratus central pair	0.1 *M* KCl	6.2	62,000 and 122,000	Shelanski and Taylor, 1968
	6 *M* G-HCl	—	61,000	Shelanski and Taylor, 1968
Strongylocentrotus purpuratus outer fibers	0.1 *M* KCl	6.0	60,000 and 118,000	Shelanski and Taylor, 1968
	6 *M* G-HCl	—	61,000	Shelanski and Taylor, 1968
Strongylocentrotus droebachiensis outer fibers	8 *M* Urea	3.1	62,000	Stephens, 1968a
	5 *M* G-HCl	3.2	56,000	Stephens, 1968a
	Salyrgan	5.4	130,000	Stephens, 1968a
	Sarkosyl	2.8	66,000[b]	Stephens, 1969
Pecten irradians outer fibers	Sarkosyl	2.8	65,000[b]	Stephens and Linck, 1969
Pig brain neurotubules	0.1 *M* KCl	5.8	119,000	Weisenberg et al., 1968
	5 *M* G-HCl	—	57,000	Weisenberg et al., 1968

[a] Recalculated for $\bar{v} = 0.73$.
[b] Not corrected for bound Sarkosyl.

TABLE II

Comparative Amino Acid Composition of Various Tubulins[a]

	OF[b]	CP[c]	OF[c]	OF[d]	OF[e]	OF[f]	NT[g]	NT[h]	NT[i]
Lysine	51	54	51	48	55	51	34	56	40
Histidine	22	25	25	22	20	21	21	21	24
Arginine	41	45	42	47	45	49	38	48	43
Aspartic acid	94	85	85	90	89	91	89	88	92
Threonine	46	55	54	53	50	61	54	55	47
Serine	54	47	47	47	45	51	51	47	59
Glutamic acid	117	124	123	113	104	111	122	116	113
Proline	39	46	41	43	42	41	45	45	46
Glycine	80	70	70	62	63	60	68	71	73
Alanine	56	68	71	64	61	61	65	70	63
Cysteine	13	13	13	14	25	6	16	15	20
Valine	53	56	58	55	51	56	54	58	50
Methionine	26	22	26	26	32	24	20	20	25
Isoleucine	49	40	41	41	43	40	39	44	39
Leucine	66	70	67	68	67	66	66	76	63
Tyrosine	29	27	29	29	25	28	33	26	30
Phenylalanine	39	34	39	36	34	33	41	35	36
Tryptophan	7	7	7	7	15	7	6	—	—

[a] Values are expressed in moles per 100,000 g.
[b] Outer fiber protein from *Tetrahymena* (Renaud et al., 1968).
[c] Central pair and outer fiber from *S. purpuratus* (Shelanski and Taylor, 1968).
[d] Outer fiber protein from *S. droebachiensis* (Stephens, 1968a).
[e] Outer fiber protein from *Pseudocentrotus depressus* (Mohri, 1968).
[f] Outer fiber protein from *Pecten irradians* (Stephens and Linck, 1969).
[g] Neurotubule protein from pig brain (Weisenberg et al., 1968).
[h] Neurotubule protein from mouse neuroblastoma (Olmsted et al., 1970).
[i] Neurotubule protein from squid (Davison and Huneeus, 1970).

quantities of pure protein for biochemical characterization. Borisy and Taylor (1967a, 1967b) isolated 6 S colchicine-binding proteins from sources as varied as tissue culture cells, isolated sea urchin mitotic apparatuses, and brain tissue; in all probability these are microtubule subunits of the same nature as those already described.

A protein obtained from the isolated sea urchin mitotic apparatus and identified as the subunit of the microtubule (Sakai, 1966; Kiefer et al., 1966) has somewhat different properties from those of tubulin as described above. This protein is obtained as a 3.5 S particle with a molecular weight of 68,000. This is in turn reducible by means of sulfite to a 2.5 S monomer with a molecular weight of 34,000; oxidation restores the particle to the dimer form. The only similar behavior reported for tubulin involved a 30,000-molecular-weight component detected through meniscus depletion ultracentrifugation after treatment of the outer fiber and central pair protein with 6 M guanidine hydrochloride and mercaptoethanol for 2 weeks (Shelanski and Taylor, 1968). On the basis of such significant molecular weight differences, it is difficult to see how this mitotic apparatus subunit and the tubulins are related.

The major component of hexylene glycol–isolated mitotic apparatuses is a 22 S protein with a molecular weight of 880,000 and composed of eight monomers of 110,000 molecular weight (Kane, 1967; Stephens, 1967). Breakdown products of this octamer can explain most of the "protein components" found in isolated mitotic preparations prior to those of Sakai (1966). The 22 S particle and tubulin are clearly dissimilar on the basis of minimal molecular weight and amino acid composition. The 22 S protein is likely a matrix component associated with the isolated mitotic unit. This has been demonstrated independently with ferritin-labeled anti-22 S antibody by Bibring and Baxandall (1969); no labeling of the mitotic microtubules was observed, nor did acid-extracted tubulin (Bibring and Baxandall, 1968) cross-react with the 22 S antiserum.

Recently Burns and Kane (1970) demonstrated that both the 3.5 S and the 22 S mitotic apparatus proteins were ultimately derived from yolk granules. Furthermore, nearly complete removal of yolk granules through centrifugation prior to fertilization produced half-eggs that divided quite normally. Thus neither protein is related to the microtubule structural protein nor is essential to mitotic function.

VI. HOMOLOGY AMONG MICROTUBULES

Biologically, the apparent centriolar origin of ciliary, flagellar, and mitotic microtubules suggests homology among the constituent proteins of these morphologically similar (if not identical) structures. Chemically, at the level of size, shape, or composition (as discussed in the preceding section), the proteins of the outer fibers and central pair of either cilia or flagella (whether of protozoan, echinoderm, or molluscan origin) and mammalian neurotubule proteins are nearly indistinguishable.

However, Behnke and Forer (1967) have presented evidence for four classes of microtubules based on their differential resistance to enzymic digestion and to heat or cold treatment. In order of decreasing general stability of these agents, these workers have distinguished the A-subfiber, B-subfiber, central pair, and cytoplasmic microtubules as distinct entities.

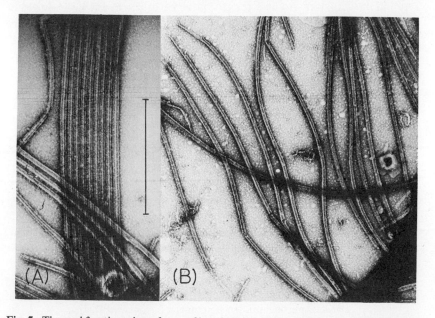

Fig. 5. Thermal fractionation of outer fiber doublets at low ionic strength. (A) Doublet preparation in 1 mM tris–HCl, pH 8.0. (B) The same as in (A) after treatment at 40°C for 2 min. Scale marker = 1 μ (Stephens, 1970a).

Indirect chemical evidence regarding solubility properties of the isolated microtubules also supports such categorization. Salt treatment or slight pH change solubilizes the mitotic apparatus microtubules (Kane, 1963) as does simply standing for several hours in isolation medium (Kane and Forer, 1965). Gibbons (1965) noted that long-term dialysis caused some preferential breakdown of the B-subfiber in *Tetrahymena* cilia. Brief dialysis dissolves the central pair of flagella (Shelanski and Taylor, 1967) but high pH or denaturing solvents are needed to solubilize the outer fiber doublets (Renaud et al., 1968; Shelanski and Taylor, 1968; Stephens, 1968a).

Treatment of outer fibers with dilute detergent preferentially dissolved the B-subfiber (Stephens, 1969), but inadequate separation has prohibited further chemical analysis of the two components. Differential solubility has also been observed between the central pair microtubules (Jacob et al., 1968).

The differential thermal behavior of the A- and B-subfibers observed by Behnke and Forer (1967) gave impetus to a study in which the two sub-fibers were successfully separated through thermal depolymerization (Stephens, 1970a). When sea urchin flagellar outer fibers were heated for 2 min at 40°C in a low-ionic-strength buffer (1 mM tris–HCl, pH 8.0), the B-subfiber, an incomplete microtubule sharing a portion of the A-tubule

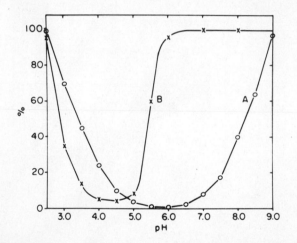

Fig. 6. Solubility of A- and B-tubulin at various pH values. Samples were brought to pH 3 and back-titrated to the appropriate pH, at which point the amount of protein remaining in solution was determined (Stephens, 1970a).

wall, was selectively depolymerized (Fig. 5). Further subjection of the A-tubule to this treatment ultimately resulted in its depolymerization, but the time course differed by nearly a factor of 10 from that of the B-subfiber, allowing a very clear separation. Because of their origin, these two components have been named A- and B-tubulin.

A- and B-tubulin are almost indistinguishable on alkaline urea-containing polyacrylamide gels. A-Tubulin has an R_f of 0.42, while B-tubulin has a value of 0.45 in 5% gels. However, under acid (pH 4.5) conditions, A-tubulin migrates, while B-tubulin does not penetrate the gel. Both components are soluble below pH 3 (cf. Bibring and Baxandall, 1968) but show different properties as the pH is raised. A-Tubulin is minimally soluble at pH 6, while B-tubulin is least soluble at pH 4.5 (Fig. 6).

TABLE III

Comparative Amino Acid Composition of Performic-
Acid-Oxidized A- and B-Tubulin[a]

	A	B	A−B
Lysine	30.8	25.6	+5
Histidine	12.7	12.6	—
Arginine	28.8	25.6	+3
(Ammonia)	(98.3)	(99.6)	—
Cysteic acid	7.8 (8)[b]	8.9 (7)[b]	−1
Aspartic acid	57.5	55.2	—
Methionine sulfone	12.5	13.0	—
Threonine	28.7[c]	29.3[c]	—
Serine	22.6[c]	25.3[c]	—
Glutamic acid	69.6	69.9	—
Proline	26.1	24.5	+1
Glycine	37.7	43.0	−5
Alanine	39.5	40.1	—
Valine	33.4	35.9	−2
Isoleucine	25.3	25.1	—
Leucine	40.0	38.2	+2
Tyrosine[d]	17.0	17.0	—
Phenylalanine	21.0	21.5	—
Tryptophan[d]	4.2	4.3	—

[a] Values are expressed in moles of amino acid per 60,000 g (from
   Stephens, 1970a).
[b] Cysteine by DTNB titration.
[c] Extrapolated to zero hydrolysis time.
[d] Spectrophotometric determination.

Amino acid analysis of the tubulins has indicated definite chemical differences (Table III). A-Tubulin has at least six more lysine and arginine residues than B-tubulin, while the latter has an equivalent number of additional glycine and valine residues. Minor differences are seen in proline and leucine. A-Tubulin has eight cysteine residues and no disulfide bonds, while B-tubulin has seven free cysteine residues and one disulfide bond. The electrophoretic and solubility properties are rather easily rationalized on the basis of these differences in basic amino acid content.

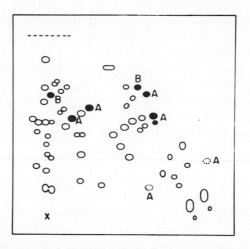

Fig. 7. Composite tryptic peptide map of A- and B-tubulin. Chromatography in ascending direction with chloroform, methanol, and ammonia (2 : 2 : 1) and electrophoresis with pyridine, acetic acid, and water (1 : 10 : 489), cathode on the right. Unique peptides are solid and marked with source; dotted peptides are faint but always present (Stephens, 1970a).

Such differences in lysine and arginine suggest that tryptic peptide mapping should serve as an effective means for comparing A- and B-tubulin. A composite map, representing spots found in common on six sets of determinations is illustrated in Fig. 7. Peptides unique to each tubulin have been appropriately marked. As would be predicted from amino acid analysis, A-tubulin yields 56 to 58 peptides (59 to 60 expected) while B-tubulin yields 53 (52 to 53 expected). The two proteins have about 50 peptides in common, indicating a high degree of similarity, roughly comparable to the α- and β-chains of hemoglobin or other nonallelic gene products. It might be proposed that the two moderately large nonpolar peptides unique to B-tubulin are the sites of the basic amino acid substitutions that give rise to A-tubulin. However, peptide mapping and amino acid composition data are inconsistent with the idea of the tubulins being related through a terminal addition of some basic peptide to a common polypeptide chain to form A-tubulin.

When central pair tubulin was prepared according to Shelanski and Taylor (1967), and a tryptic peptide map was run, the most intense spots coincided fairly well with those of A-tubulin (Stephens, unpublished). This result might have been predicted from the amino acid composition differences between it and the predominently A-tubulin-containing outer fiber (see Table II), but the opposite might have been predicted on the basis of solubility or stability properties. Such a comparison is complicated by many secondary peptide spots in the central pair maps, indicative of the presence of several molecular species.

Employing dialysis and salt extraction to fractionate *Chlamydomonas* flagella, Jacobs and McVittie (1970) have likewise demonstrated that the A and B subfibers are each composed of a different protein subunit, corresponding in electrophoretic mobility to A and B tubulin described above. In addition, these workers presented evidence that both kinds of protein subunit appear to be present in the central pair.

Isoelectric focusing of sulfonated A and B tubulin from sea urchin flagella reveal isoelectric points of approximately 6.2 and 5.9 for these respective derivatives. Upon close examination, either electrofocused band appears double, the equally dense sub-bands differing in both cases by about 0.05 pH unit. This may imply that each type of tubulin dimer, A and B, is in turn composed of two different polypeptide chains, neither of which is held in common (Stephens, unpublished).

In contrast to the studies of Stephens (1970a) and of Jacobs and McVittie (1970), wherein each subfiber was composed of a distinct subunit, Witman (1970) reports that A and B subfibers of *Chlamydomonas* are composed of several subunits, some of which are held in common. Such microheterogeneity is in one sense a rather attractive hypothesis in that it would provide the possibility of unique bonding sites at various locations along and around the microtubule, something that is not easily accomplished when the tubule is uniformly composed of identical subunits.

Perhaps complicating simple morphology even further is the finding by Linck (1970) that exhaustive dialysis of *Aequipecten* gill *ciliary* axonemes showed solubilization of all B-tubules and one central pair (forming a 9 + 1 configuration of singlets) while identical treatment of *Aequipecten* sperm *flagellar* axonemes resulted in the production of doublet outer fibers only (forming a sheet of 9 parallel doublets). Such striking solubility differences in morphologically identical organelles of the same species need not imply further subunit differences between cilia and flagella. Secondary proteins may somehow influence stability, or the presence of guanine nucleo*tide* in flagella versus guanine nucleo*side* in cilia (Stephens et al., 1967; Stephens and Linck, 1969) may also be a factor in the behavior of the microtubules to which these groups are bound.

Future work on the comparative peptide mapping of pure tubulin fractions should show conclusively how closely related chemically the morphologically similar though functionally different microtubules really are.

Immunological studies with an antibody prepared against both outer fiber doublet tubulins from sea urchin flagella indicate a common antigenic site in tubulins prepared from various sources, at least within the Echinoidea (Fulton et al., in press). Outer fiber and central pair tubulin from both sea urchin blastula cilia and sperm tail flagella from several species cross-react with antibody against flagellar outer fibers; crude mitotic apparatus tubulin shows similar cross reactivity. Bibring and Baxandall (1971) have prepared an antiserum against purified sea urchin flagellar outer fibers and have demonstrated cross reactivity against pure mitotic apparatus tubulin by means of immunoelectrophoresis, eliminating any doubt of immune response to unavoidable minor impurities in outer fiber preparations.

This evidence of a common antigenic site, coupled with the overall gross similarity in peptide composition, implies that the tubulins are nearly identical, with rather localized substitutions occurring in regions of the polypeptide chain involved in either intratubular bonding of the tubulin itself (stability differences), intertubular bonding of different tubulins (doublet or triplet formation), or specific bonding to other proteins (dynein arms or nexin bridges).

VII. HOMOLOGY WITH ACTIN?

Obviously related to a desire for biological unity, considerable speculation has been put forth relating the protein of the microtubule to muscle actin. Roslansky (Mazia, 1955) carried out an amino acid analysis of a dissolved mitotic apparatus and found a striking resemblance to rabbit actin. Inoué (1964) pointed out that the thermodynamics of mitotic spindle formation (and presumably microtubule polymerization) were essentially the same as that for the polymerization of TMV protein and also G-ADP actin. When amino acid composition and physicochemical data for pure tubulin became available, the apparent homology became more strongly grounded. Renaud et al. (1966, 1968) observed that the similarity in amino acid composition between *Tetrahymena* outer fiber tubulin and *Pecten* actin was no more distant than that among actins of different species. In addition, tubulin and actin have identical electrophoretic mobilities. At that time, tubulin and actin were thought to have the same molecular weight (60,000) and each contained 1 mole of bound nucleotide per mole (Stephens et al., 1967), although the nucleotides were not the same.

The issues of the similarity of molecular weight and nucleotide binding however, were open to some debate. For many years, 60,000 was a well-accepted value for the molecular weight of G-actin, and consistent values of 55,000–60,000 were found for central pair and outer fiber tubulin (Table I). Regarding actin, Johnson and associates (1967) estimated a minimal subunit weight of 47,600 based on 3-methylhistidine content, while Rees and Young (1967) and Tsuboi (1968) found that actin bound 1 mole of ADP per 45,000–47,000 g of protein. As already mentioned, reexamination of the guanine-binding ratio in tubulin has revealed that 1 mole of nucleotide per 120,000 g of protein is tightly bound while another mole is loosely bound (Weisenberg et al., 1968; Stephens, 1969). Also, unlike actin, which contains 1 mole of 3-methylhistidine, tubulin contains none (Weisenberg and Taylor, 1968; Stephens, unpublished). Thus, accepting this recent data, actin binds 1 mole of ADP per monomer molecular weight of 47,000, while tubulin binds 1 mole of guanine nucleotide per dimer molecular weight of 120,000.

The amino acid composition data represented the point of strongest similarity, but all previous comparisons between actin and tubulin were made on widely divergent species, thus making difficult any estimation of homology. Employing tubulin from outer fibers of gill cilia and actin from the striated adductor muscle of *Pecten irradians*, Stephens and Linck (1969) directly compared homospecific proteins. The comparative amino acid composition data are given in Table IV. Even though the existence of A- and B-tubulin was only speculation at that time and the comparison was made, therefore, on mixed tubulins, the data indicate that tubulins from sea urchins and *Pecten* resemble each other far more closely (16 residues per mole difference) than do tubulin and actin commonly derived from *Pecten* (84 residues difference). The differences between A- and B-tubulin (Table III) do not change this conclusion.

Comparative peptide mapping of *Pecten* actin was also carried out to detect independently any similarity to tubulin in gross primary structure. A tryptic peptide map of *Pecten* actin is shown in Fig. 8A, while that for A-tubulin from *Pecten* is shown in Fig. 8B. For added comparison, chymotryptic maps of *Pecten* actin and tubulin are illustrated in Fig. 8C and D, respectively (Stephens, 1970b). The tubulins (whether sea urchin or mollusc) show at least 50 tryptic peptides, as expected from their lysine-arginine content and molecular weight, while actin has less than 40 tryptic peptides, as its composition and lower molecular weight would indicate. The lesser specificity of chymotrypsin does not allow such an argument to

TABLE IV

Comparative Amino Acid Composition of Tubulin
and Actin[a,b]

	Tubulin		Actin	
	Urchin	Pecten	Pecten	Rabbit
Lysine	**28.8**	*30.9*	*26.0*	26.0
Histidine	12.8	*12.4*	*10.7*	10.3
Arginine	28.2	*29.4*	*24.3*	25.1
Aspartic acid	53.4	*54.6*	*44.2*	**48.3**
Threonine	**31.4**	36.7	36.2	**40.6**
Serine	**28.0**	*30.4*	*34.6*	34.3
Glutamic acid	67.6	*66.9*	*53.2*	**56.1**
Proline	25.4	*24.6*	*28.0*	**26.3**
Glycine	37.2	*35.8*	*39.8*	39.5
Alanine	38.1	*36.6*	*43.2*	42.1
Cysteine	**8.4**	*3.5*	*4.9*	**6.6**
Valine	33.0	*33.6*	*29.2*	**26.2**
Methionine	**15.8**	*14.2*	*20.8*	**17.4**
Isoleucine	24.3	*24.2*	*37.0*	38.7
Leucine	40.3	39.9	39.8	**35.8**
Tyrosine	17.1	*16.8*	*19.8*	20.3
Phenylalanine	21.6	20.0	19.5	**16.3**
Tryptophan	**4.3**	*5.7*	*4.0*	**5.1**
Residues per mole difference	16	84	28	

[a] From Stephens and Linck (1969).
[b] Values are in moles per 60,000 g. Values for amino acids differing by more than 4% (tubulin and actin in same species) are italic; those for amino acids differing by more than 4% (homologous proteins from different species) are boldface.

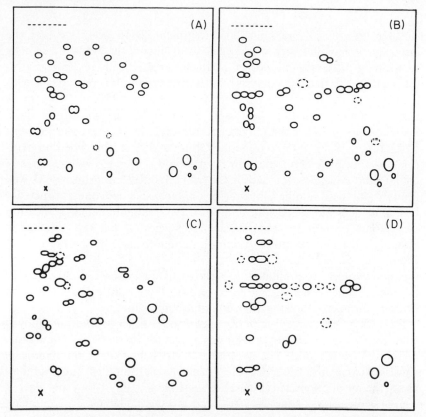

Fig. 8. Comparative tryptic and chymotryptic peptide maps of actin and A-tubulin from *P. irradians* muscle and cilia. (A) Tryptic digest of actin. (B) Tryptic digest of A-tubulin. (C) Chymotryptic digest of actin. (D) Chymotryptic digest of A-tubulin. Conditions were the same as in Fig. 7 (Stephens, 1970b).

be made. In either case, however, it should be obvious from the maps that the major portions of the peptide chains in these two proteins are clearly different. Because of the complexity of the maps it is not possible to unequivocally rule out coincidence of some peptides, so that it is still possible that small local regions of homology do exist (for example, at the nucleotide binding sites), but this appears unlikely.

VIII. FORMATION OF MICROTUBULES IN VIVO

Studies of microtubule formation were begun on a rather sophisticated level nearly a decade before their actual discovery in fixed material. Inoué (1952a, 1952b) noted that the mitotic spindle of various living marine eggs was sensitive to cold or colchicine treatment. Both agents caused the birefringent spindle fibers to disappear; warming of the cold-treated cells or washing the colchicine out of colchicine-treated cells restored the birefringence. In a detailed analysis of variation in spindle birefringence with temperature, Inoué (1964) concluded that the endothermic process of spindle tubule formation possessed thermodynamic parameters quite analogous to those for the polymerization of TMV coat protein (Stevens and Lauffer, 1965) or G-ADP-actin (Grant, 1965), at least when birefringence as a measure of polymer concentration was used to evaluate directly equilibrium constants for use in the van't Hoff relationship. D_2O was found to increase spindle birefringence and cause a concomitant increase in spindle microtubules (Inoué and Sato, 1967). Simultaneous application of D_2O and variation in temperature has suggested that D_2O affects the "pool" of protein available for spindle tubule formation (Carolan et al., 1966), but perhaps brings to light some of the difficulties in applying thermodynamics to simplified models of living systems. Regardless of thermodynamic purity or accuracy, however, there is little doubt that such cytoplasmic microtubules are in a dynamic equilibrium.

Following the early observations of Inoué, other workers employed cold, colchicine, D_2O, and pressure as agents for investigating tubule formation and disruption. The axopodia of the heliozoan *Actinosphaerium* consist of two concentric "jelly rolls" of single microtubules. These tubules, as are the mitotic spindle tubules, are sensitive to cold (Tilney and Porter, 1965b), colchicine (Tilney, 1968), and pressure (Tilney et al., 1966), apparently breaking down into filamentous material and, during recovery, re-forming from this same material. In a study of the re-formation of the axopodial tubules after cold treatment, Tilney and Byers (1969) showed that the form of the array of microtubules could be explained best in terms of specific linkages between the tubules. Behnke (1967), investigating repolymerization of cold-depolymerized microtubules of blood platelets, found that upon warming microtubules form through lateral association of protofilaments, since newly formed tubules appeared initially as "C" or "S" shapes in cross section. By analogy to virus sheath growth, one might expect to find a helical addition of subunits to the end of a cylindrical tubule during growth and thus see only cylindrical cross sections. However, as already described, isolated tubules fray into protofilaments rather than split into doughnuts; the following section describes in vitro tubule formation from protofilament ribbons.

The temperature and D_2O sensitivity of cytoplasmic microtubules has been interpreted as being related to changes in structured water and its concurrent influence on hydrophobic bonding (cf. Inoué and Sato, 1967). The colchicine effect has been postulated to be a competitive binding phenomenon, blocking bonding sites between polymerizing monomers (cf. Borisy and Taylor, 1967a). Although both of these proposals are reasonable and quite consistent with in vivo data, the actual physico-chemical process of polymerization has yet to be demonstrated in a system of cytoplasmic tubulin polymerizing in vitro.

Unlike cytoplasmic or axopodial microtubules, the $9+2$ microtubule system of cilia and flagella, *once formed*, is unresponsive to cold, colchicine, or pressure. It is well-known that isolated outer fibers are incapable of binding colchicine (Shelanski and Taylor, 1968). However, developing flagella are sensitive to colchicine, at least in *Tetrahymena* or *Chlamydomonas* (Rosenbaum and Carlson, 1969), while thermal depolymerization of B-tubulin from outer fibers (Stephens, 1970a) results in the formation of a colchicine-binding 6 S dimer. Apparently, the subunits of ciliary or flagellar microtubules *when assembled* have their colchicine-binding site buried within the structure; *before assembly* the 6 S dimer can evidently bind colchicine, and polymerization is effectively blocked. Regarding the lack of cold or pressure effect on the assembled structure, one must simply conclude that the bonding in ciliary or flagellar structures is considerably stronger than in cytoplasmic microtubules. These agents do affect the developing systems, but what else in the cell is being affected is unclear. Once again, only in vitro evidence can provide a final answer.

The classic observation that cilia and flagella "arise" from basal bodies has gained unequivocal support from electron microscope observation of flagella formation in the water mold *Allomyces* (Renaud and Swift, 1964) and in the ameba-flagellate *Naegleria* (Dingle and Fulton, 1966). In the former case a preexisting basal body serves as a growth point upon which invaginating vesicles of the cell membrane coalesce, forming an intra-cellular primary vesicle into which a flagellar axoneme develops. In the case of *Naegleria*, the basal body apparently arises de novo during the ameba-to-flagellate transformation, after which an axoneme grows out-ward from within an outpocketing of the cell membrane. As would be predicted from this and other morphological evidence, the flagellum or cilium probably grows through the addition of new material at the tip of the advancing axoneme.

Control of flagellar growth, and hence microtubule length, has been the subject of two very intriguing reports. Rosenbaum et al. (1969) demon-strated that removal of one of the two flagella of *Chlamydomonas* resulted in the immediate shortening of the remaining intact flagellum and in the simultaneous outgrowth of the amputated one. When both approached a common length they elongated together. Through inhibition of protein synthesis, Coyne and Rosenbaum (1970) then showed that the proteins of the resorbing flagellum were reutilized for elongation and that the equilibrium length for flagellar outgrowth after amputation is therefore a function of the original degree of amputation. Perhaps the real beauty in these observations lies in the question of how one flagellum knows what the other is doing!

Although cytoplasmic and axopodial microtubules may arise randomly throughout the cytoplasm, ciliary and flagellar axonemes arise only from centrioles or basal bodies, and only in these axonemal structures are double or triple microtubules ever seen. One might speculate that the centriole serves as a "crystalization center" upon which the 9 + 2 arrangement forms, and that the outer fiber triplet of the centriole or basal body is necessary for the growth of the axonemal doublet. In order to disprove this notion, it will be necessary to grow outer doublets from dimeric A- and B-tubulin alone, without the presence of already ordered form.

IX. REASSOCIATION OF TUBULIN IN VITRO

Central to the question of microtubule function is the mechanism of their assembly. In mitotic and axopodial systems, the assembly-disassembly process appears to be related to motility, while the three-dimensional assemblage of proteins in ciliary and flagellar axonemes must, as in muscle, relate in some manner to their biochemical function in motility.

Early attempts to reassociate tubulin met with only limited success. Renaud et al. (1966) reported tactoids of 40-Å protofilaments arising upon magnesium addition to 6 S tubulin derived from acetone powders of *Tetrahymena* cilia, but electron microscopy indicated little hierarchal structure. Similarly, 6 S tubulin from sea urchin sperm flagella acetone powders and from Salyrgan-solubilized outer fibers formed fiber bundles of 40-Å filaments in high yield but these in no way resembled microtubules, even when fragmented tubules were used as "seeds" (Stephens, 1969). It is important to point out that in none of these 6 S dimers was guanosine di- or triphosphate bound to the protein.

Employing dilute (0.5%) solutions of the detergent Sarkosyl (sodium lauroyl sarcosinate, Stephens (1968b, 1969) obtained a 3 S monomer of tubulin, already described, from sea urchin outer fibers (Fig. 9A and 9B). When the outer fibers were detergent-depolymerized in the presence of GTP, the resulting monomer contained 0.5–0.6 moles of guanine nucleotide per 60,000 g of protein. When this nucleotide-containing monomer was freed of any higher aggregates by high-speed ultracentrifugation, no fibrous or tubular aggregation could be induced by removal of detergent, either by dialysis or dilution, or by salt addition to the detergent-free solution. Only amorphous precipitate formed in the latter case.

However, when aggregated protein of the order of 1000-Å particles or smaller was left in solution, and when sonicated tubules of this same size were added to the 3 S solution, considerable ordered aggregation resulted upon simple 1 : 10 dilution of the Sarkosyl solution. Ribbons of subunits 150–250 Å wide and many microns in length resulted (Fig. 9C). Subsequent addition of 0.05–0.1 M NaCl caused considerable lateral and longitudinal growth of these ribbons, but no obvious tubules were formed (Fig. 9D). Further, it was found that magnesium in low concentration (0.1 mM) could grossly change the mode of aggregation if present when salt was added to the diluted ribbon-containing tubulin solution. Under such circumstances seemingly endless aggregates of singlet tubules resulted (Fig. 9E), often appearing as macroscopic "cotton balls" in the vessel used for polymerization. Upon close examination the negatively stained images (Fig. 9F) are essentially identical to those of single tubules. The formation of ribbons, higher-order fibrous aggregates, and apparent tubules from outer fiber tubulin in Sarkosyl solutions has been confirmed by Mohri and co-workers (1968) using other species of sea urchins.

Only singlet microtubules have been reconstituted. Even when fragmented but still doublet microtubules were used as seed material, no reassociated doublet tubules were observed, nor could greater than 50% of the protein be recovered in reassociated form as tubules or otherwise. In retrospect, this observation is easily consistent with the possibility that only A-tubulin polymerizes from Sarkosyl solutions. Indeed, Sarkosyl solutions of strength comparable to that obtained after dilution of Sarkosyl–tubulin solutions were originally used in preliminary fractionation experiments to remove the B-subfibers from outer fiber doublets (Stephens, 1969).

Later resolution of doublet microtubules into A and B subfiber components (Stephens, 1970a) has clearly confirmed some of these ideas. Purified A tubules dissolved in GTP-Sarkosyl and then diluted 1:20 with 50 mM NaCl and 10^{-4} M MgCl$_2$ results in the formation of singlet microtubules in 60–80% yield. The B-tubulin, however, will form no obvious aggregates when diluted similarly out of Sarkosyl solution. This 6 S B-tubulin from thermal fractionation will form massive fiber bundles in the presence of salt and magnesium and in the absence of Sarkosyl. When A-tubules are present as "seed" material, 6 S B-tubulin will not add on to the tubules to form doublets when salt and magnesium are added; rather, protofilament segments add parallel to the A-tubule at somewhat random radial locations about the microtubule, with no apparent "closing" of adjacent segments to form B-subfibers (Stephens, unpublished). Perhaps a basal body is critical for proper placement of the B-subfiber.

The above observations were repeatable only if the tubulin retained bound GTP. The amount of reassociable tubulin appears to be directly

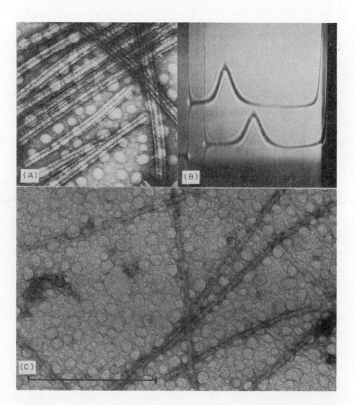

Fig. 9. Reassociation of tubulin from Sarkosyl solutions. (A) Flagellar outer fibers. (B) Outer fibers dissolved in 0.5% Sarkosyl and 1 mM GTP, with (lower trace) and without (upper trace) 0.1 M KCl. (C) Preparation as in (B) diluted 1:10 with distilled water. (D) Same as (B) but made 0.1 M in KCl. (E) Same as (B) but made 0.1 mM in magnesium and 0.1 M in KCl. (F) High magnification of tubules illustrating protofilament ribbons issuing from the tips. Scale marker in each is 0.5 μ (Stephens, 1969).

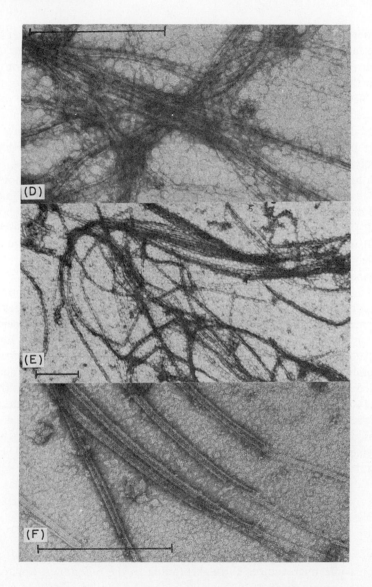

Fig. 9. (*cont.*).

proportional to the GTP content, but the exact stoichiometry has yet to be determined. It is also not clear whether or not GTP must be split to GDP upon polymerization. This question is clouded by the observation that both the di- and triphosphate are present in native isolated microtubule doublets (Stephens et al., 1967; Yanagisawa et al., 1968) and in the reconstituted fibers (Stephens, 1969). It is further complicated by the inherent instability of GTP and by the small amounts of inorganic phosphate that must be detected. ATP can be substituted for GTP during the initial depolymerization in Sarkosyl; identical polymer formation results, often with higher yields. The repolymerized tubulin contains up to 2 moles of bound ATP per 60,000 g of protein as opposed to 0.5–0.6 moles of guanosine di- and triphosphate for tubulin polymerization in the presence of GTP (Stephens, 1969). During polymerization in the presence of ATP, GTP is lost, for the reassociated fibers contain no detectable guanine nucleotide. No morphological differences were evident between fibers polymerized with these two nucleotides.

Nucleotide triphosphate and divalent cations appear to be essential for the polymerization; some form of nucleation is also apparently necessary. Polymerization of tubulin from Sarkosyl is unaffected by cold or colchicine. One might conclude that in this somewhat artificial system the 3 S monomer alone cannot polymerize as such; higher-order polymers (ultracentrifugation suggests at least 9 S or higher) must be present to provide nucleation. Sarkosyl effectively prevents polymerization until diluted out; thereafter the equilibrium may be strongly in favor of the polymer, and thus unaffected by cold. The monomer does not bind colchicine, nor do the reassociated fibers, hence one would not expect this alkaloid to block polymerization competitively. Magnesium appears to be involved in the lateral aggregation of protofibrils into tubules (Stephens, 1969).

X. AGGREGATION OF TUBULIN IN THE PRESENCE OF ALKALOIDS

As mentioned above, colchicine reversibly disrupts mitotic and other cytoplasmic microtubules through binding to the 6 S dimer, apparently blocking polymerization. Other alkaloids have similar effects on a gross level but, in addition to causing dissolution of the microtubule structure, can induce formation of aggregates of differing molecular arrangement.

Bensch and Malawista (1969) have demonstrated that vinblastine and vincristine in less-than-millimolar amounts induce microtubule "crystals" in mammalian fibroblasts and leukocytes. These "crystals" are hexagonal in cross section, measuring 280×240 Å for the major and minor axes of the hexagon. In longitudinal section the crystals are made up of a rhomboidal array of dots, spaced at 200×280 Å. Other sections at varying orientation suggest a helical arrangement of subunits with a pitch of 68°. It is not obvious how this geometry is related to that of the native tubule. Malawista and Sato (1969), using polarization microscopy, report that uniaxial crystals can be induced by these same alkaloids in marine oocytes.

Weisenberg and Timasheff (1970) have shown that vinblastine and magnesium cause aggregation of 6 S colchicine-binding tubulin or porcine brain, apparently forming several fairly well-defined polymers of the basic 6 S subunit; the stoichiometry of polymerization is as yet undetermined. A number of workers found almost simultaneously that vinblastine or vincristine induce in vitro "crystal" formation from high-speed super- natants of various cell homogenates. These "crystals" yield tubulin, as judged by the various physicochemical criteria discussed previously (Bensch et al., 1969; Marantz et al., 1969; Olmsted et al., 1969).

These observations that *Vinca* alkaloids induce the crystalline precipita- tion of microtubule protein to a morphological form quite unlike the original tubule lend support to the concept that antimitotic alkaloids function by competition for protein-to-protein bonding regions (Borisy and Taylor, 1967b) or modify the association to favor a nontubular form (Weisenberg and Timasheff, 1970). Colchicine or its relatives may thus bind to produce an "unpolymerizable" dimer, while the *Vinca* alkaloids, binding to a different site (Weisenberg and Timasheff, 1970), modify the dimer in such a way that nontubular "crystals" form at the expense of the microtubule through decreasing the effective amount of available poly- merizable dimer.

XI. SYMMETRY IN MICROTUBULAR ARRAYS

The ninefold symmetry of ciliary and flagellar axonemes is a ubiquitous form in nature. The possible intratubular interaction with dynein has already been discussed. Since dynein cross bridges have not been detected, it is unlikely that dynein itself is responsible for retaining the overall con- formation. When dynein and central pair proteins are removed by dialysis of detergent-extracted axonemes at low ionic strength, the ninefold cylin- drical structure is retained by the outer fibers of cilia or as an opened cylinder of nine doublets in flagella (Gibbons, 1965). Fine bridges from A-subfiber to adjacent A-subfiber are detectable and appear to be the linkage material.

If intact flagellar axonemes are subjected to the thermal depolymeriza- tion process outlined above, B-tubulin is lost through melting while dynein is removed as a result of the low ionic strength. Remaining are A-tubules, linked together by periodic fibrous material (Fig. 10) spaced at intervals of about 960 Å. Mild acidification (Bibring and Baxandall, 1968; Stephens, 1970b) solubilizes the A-tubulin, leaving insoluble the periodic material responsible for the supramolecular structure. On alkaline acrylamide gels this protein, *nexin* (from *nexus*—a tie binding the members of a group), moves as a single, sharp band, distinct from the tubulins. Once identified, nexin can readily be detected in gel patterns of whole axonemes (Fig. 11). Nexin probably represents less than 2% of the total protein of the purified

Fig. 10. Thermal depolymerization of whole flagellar axonemes. Only A-subfibers connected via periodic nexin bridges remain. Scale marker is 1 μ (Stephens, 1970c).

axoneme; as a result of such limited quantities, its physicochemical properties have yet to be determined, except for a molecular weight of 165,000 from SDS-polyacrylamide gel electrophoresis (Stephens, 1970c).

Lacking more data, it is probably premature to speculate on the nature of the linkages in axopodia (Tilney and Byers, 1969) or in axostyles (Grimstone and Cleveland, 1965), but nexin bridges, by virtue of their size could easily correspond to at least some of the bridges seen in these very uniform arrays.

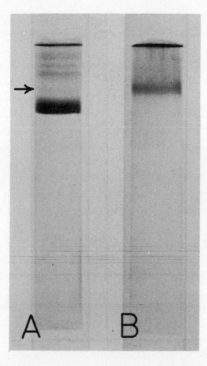

Fig. 11. Acrylamide gel electrophoresis of whole flagella (A) and the nexin fraction (B) remaining after removal of the A-tubulin by mild acid treatment of preparations such as in Fig. 10. Arrow indicates faint band of nexin identifiable in the original unfractionated preparation (Stephens, 1970c).

XII. SUMMARY

Microtubules are composed of tubulin, a 40-Å globular protein with a molecular weight of 55,000–60,000. X-ray diffraction and electron microscope studies show that the subunits of the microtubule are arranged in a 40×53 Å surface lattice with 12- or 13-fold rotational symmetry.

The outer fiber doublet microtubules of flagella yield two distinct tubulins, differing chiefly in basic amino acid content. Tryptic peptide mapping indicates that these two tubulins are very closely related, having over 80% of their peptides in common.

Though closely resembling muscle actin in amino acid composition, tubulin and actin from the same species differ significantly from each other in this aspect, resembling homologous proteins from other species far more closely. Like actin, tubulin binds 1 mole of nucleotide per monomer but the nucleotide is guanosine di- or triphosphate. With a substantially lower molecular weight of 47,000, actin yields few if any tryptic peptides that indicate a relationship to tubulin.

Tubulin from a variety of sources exists as a 6 S dimer under physiological conditions; it is presumably this form that actively polymerizes. The monomer alone does not polymerize unless higher aggregates are present to act as nucleation. Singlet microtubules can be salt-polymerized from dilute solutions of Sarkosyl-dissolved outer fiber doublets, the process requiring GTP and magnesium. Without nucleotide, only protofibrils re-form.

The dimer binds colchicine and related antimitotic alkaloids. Colchicine evidently prevents tubule formation through competitive binding, while *Vinca* alkaloids promote the formation of polymorphic forms of tubulin at the expense of microtubules.

In flagellar doublet A-subfibers, dynein arms bind to the tubule at 170-Å intervals, while nexin bridges connect adjacent A-subfibers every 960 Å. Comparative enzymological studies indicate that interaction of dynein with tubulin, if any, does not appreciably affect the ATPase properties of dynein alone. No dynein cross bridges to adjacent subfibers, analogous to myosin-actin cross bridges in muscle, have yet been reported in cilia or flagella, although evidence for constancy in subfiber length is compatible with a sliding filament mechanism.

Most of the foregoing discussion is based on studies of ciliary, flagellar, or neurotubular microtubule systems. Few chemical data exist on cytoplasmic or mitotic microtubules and none is available on axopodial or contractile axostyle systems.

Much critical information is lacking in regard to the function of the microtubule. Does the tubule have a direct role in motility or does it merely serve some mechanical support function? In terms of proteins, does the ATPase dynein interact chemically with tubulin? What role might guanine nucleotides play in such an interaction? Or are the guanine nucleotides simply involved in the maintenance of tubulin conformation? Is GTP dephosphorylated during the polymerization of tubulin? Can substitutions in the polypeptide chain of tubulin be correlated with its bonding properties in tubule formation and stability? Do perturbations in tubule structure correlate with bonding sites for dynein arms, nexin bridges, or auxiliary tubules? Or are there several types of tubulin contained within the ciliary or flagellar A-subfiber in order to bond to these different entities? If a multitude of proteins are involved, does the basal body act as a "crystallizing center" for the precise geometrical array of the axoneme? How are cytoplasmic, spindle, or axopodial microtubules related to those from ciliary, flagellar, or neurotubular sources?

Even though these questions remain speculative and unanswered at this writing, it is hoped that this somewhat selective compendium of references and discussion will kindle further biochemical interest in this nearly ubiquitous cell organelle.

REFERENCES

Afzelius, B. (1959). *J. Biophys. Biochem. Cytol.*, **5**, 269.

Allen, R. D. (1968). *J. Cell Biol.*, **37**, 825.

André, J., and J. -P. Thiéry (1963). *J. Microscopie*, **2**, 71.

Barnicot, N. A. (1966). *J. Cell. Sci.*, **1**, 217.

Behnke, O. (1967). *J. Cell. Biol.*, **34**, 697.

Behnke, O., and A. Forer (1967). *J. Cell Sci.*, **2**, 169.

Bensch, K. G., and S. E. Malawista (1969). *J. Cell. Biol.*, **40**, 95.

Bensch, K., R. Marantz, H. Wisniewski, and M. Shelanski (1969). *Science*, **165**, 495.

Bibring, T., and J. Baxandall (1968). *Science*, **161**, 377.

Bibring, T., and J. Baxandall (1969). *J. Cell Biol.*, **41**, 577.

Bibring, T., and J. Baxandall (1971). *J. Cell Biol.*, **48**, 324.

Bikle, D., L. G. Tilney, and K. R. Porter (1966). *Protoplasma*, **61**, 322.

Borisy, G. G., and E. W. Taylor (1967a). *J. Cell Biol.*, **34**, 525.

Borisy, G. G., and E. W. Taylor (1967b). *J. Cell Biol.*, **34**, 535.

Burns, R. G., and R. E. Kane (1970). *J. Cell Biol.*, **47**, 27a.

Byers, B., and K. R. Porter (1964). *Proc. Natl. Acad. Sci. U.S.*, **52**, 1091.

Carolan, R. M., H. Sato, and S. Inoué (1966). *Biol. Bull.*, **131**, 385.

Cohen, C., S. E. Harrison, and R. E. Stephens (1971). *J. Mol. Biol.* In press.

Coyne, B., and J. L. Rosenbaum (1970). *J. Cell Biol.*, **47**, 777.

Davison, P. F., and F. C. Huneeus (1970). *J. Mol. Biol.*, **52**, 429.

Dingle, A. D., and C. Fulton (1966). *J. Cell Biol.*, **31**, 43.

Ebashi, S., F. Ebashi, and A. Kodama (1967). *J. Biochem.*, **62**, 137.

Fawcett, D. W., and K. R. Porter (1954). *J. Morphol.*, **94**, 221.

Fawcett, D. W., and F. Witebsky (1964). *Z. Zellforsch.*, **62**, 785.

Forslind, B., G. Swanbeck, and H. Mohri (1968). *Exptl. Cell Res.*, **53**, 678.

Fulton, C. M., R. E. Kane, and R. E. Stephens. In press.

Gibbons, I. R. (1963). *Proc. Natl. Acad. Sci. U.S.*, **50**, 1002.

Gibbons, I. R. (1965). *Arch. Biol., Liège*, **76**, 317.

Gibbons, I. R. (1966). *J. Biol. Chem.*, **241**, 5590.

Gibbons, I. R., and A. V. Grimstone (1960). *J. Biophys. Biochem. Cytol.*, **7**, 697.

Gibbons, I. R., and A. J. Rowe (1965). *Science*, **149**, 424.

Gibbins, J. R., L. G. Tilney, and K. R. Porter (1969). *J. Cell. Biol.*, **41**, 201.

Grant, R. J. (1965). Ph.D. Dissertation, Columbia University, New York.

Grassé, P. (1956). *Arch. Biol.*, **67**, 595.

Grimstone, A. V., and L. R. Cleveland (1965). *J. Cell Biol.*, **24**, 387.

Grimstone, A. V., and A. Klug (1966). *J. Cell Sci.*, **1**, 351.

Hepler, P. K., and W. T. Jackson (1968). *J. Cell Biol.*, **38**, 437.

Inoué, S. (1952a). *Biol. Bull.*, **103**, 316.

Inoué, S. (1952b). *Exptl. Cell Res. Suppl.*, **2**, 305.

Inoué, S. (1964). In *Primitive Motile Systems in Cell Biology* (R. Allen and N. Kamiya, eds.), Academic Press, New York, p. 549.

Inoué, S., and H. Sato (1967). *J. Gen. Physiol.*, **50**, 259.

Jacobs, M., J. M. Hopkins and J. T. Randall (1968). *J. Cell Biol.*, **39**, 66a.

Jacobs, M., and A. McVittie (1970). *Exptl Cell Res.*, **63**, 53.

Johnson, P., C. I. Harris, and S. V. Perry (1967). *Biochem. J.*, **105**, 361.

Johnson, U. G., and K. R. Porter (1968). *J. Cell Biol.*, **38**, 403.

Kane, R. E. (1963). *J. Cell Biol.*, **12**, 47.

Kane, R. E. (1967). *J. Cell Biol.*, **32**, 243.

Kane, R. E., and A. Forer (1965). *J. Cell Biol.*, **25**, 31.

Kiefer, B., H. Sakai, A. J. Solari, and D. Mazia (1966). *J. Mol. Biol.*, **20**, 75.

Kitching, J. A., (1964). In *Primitive Motile Systems in Cell Biology* (R. Allen and N. Kamiya, eds.), Academic Press, New York, p. 445.

Ledbetter, M. C., and K. R. Porter (1963). *J. Cell Biol.*, **19**, 239.

Ledbetter, M. C., and K. R. Porter (1964). *Science*, **144**, 872.

Linck, R. W. (1970). *Biol. Bull.*, **139**, 429.

Linck, R. W. (1971). Ph.D. Thesis, Brandeis University.

McIntosh, J. R., and K. R. Porter (1967). *J. Cell Biol.*, **35**, 153.

McIntosh, J. R., P. K. Hepler, and D. G. Van Wie (1969). *Nature*, **224**, 659.

Malawista, S. E., and H. Sato (1969). *J. Cell Biol.*, **42**, 596.

Marantz, R., M. Ventilla, and M. Shelanski (1969). *Science*, **165**, 498.

Markham, R., S. Frey, and G. J. Hills (1963). *Virology*, **20**, 88.

Mazia, D. (1955). *Symp. Soc. Exptl. Biol.*, **IX**, 355.

Mazia, D., R. R. Chaffee, and R. M. Iverson (1961). *Proc. Natl. Acad. Sci. U.S.*, **47**, 788.

Miki, T. (1963). *Exptl. Cell Res.*, **29**, 92.

Mohri, H. (1958). *J. Fac. Sci. Tokyo*, **8**, 307.

Mohri, H. (1968). *Nature*, **217**, 1053.

Mohri, H., S. Murakami, and K. Maruyama (1967). *J. Biochem.*, **61**, 518.

Mohri, H., S. Hasegawa, and K. Maruyama (1968). *Dobutsugaku Zasshi*, **77**, 399.

Olmsted, J. B., K. Carlson, R. Klebe, F. Ruddle, and J. Rosenbaum (1970). *Proc. Natl. Acad. Sci. U.S.*, **65**, 129.

Pautard, F. G. (1962). In *Spermatozoan Motility* (D. W. Bishop, ed.), AAAS, Washington, D.C., p. 189.

Pease, D. C. (1963). *J. Cell Biol.*, **18**, 313.

Rees, M. K., and M. Young (1967). *J. Biol. Chem.*, **242**, 4449.

Renaud, F. L., and H. Swift (1964). *J. Cell Biol.*, **23**, 339.

Renaud, F. L., A. J. Rowe, and I. R. Gibbons (1966). *J. Cell Biol.*, **31**, 92A.

Renaud, F. L., A. J. Rowe, and I. R. Gibbons (1968). *J. Cell Biol.*, **36**, 79.

Ringo, D. L. (1967). *J. Ultrastruct. Res.*, **17**, 266.

Rosenbaum, J.L., and K. Carlson (1969). *J. Cell Biol.*, **40**, 415.

Rosenbaum, J. L., J. E. Moulder, and D. L. Ringo (1969). *J. Cell Biol.*, **41**, 600.

Roth, L. E., and E. W. Daniels (1962). *J. Cell Biol.*, **12**, 57.

Sakai, H. (1966). *Biochim. Biophys. Acta*, **112**, 132.

Satir, P. (1968). *J. Cell Biol.*, **39**, 77.

Schroeder, T. E. (1968). *Exptl. Cell Res.*, **53**, 272.

Shelanski, M. L., and E. W. Taylor (1967). *J. Cell Biol.*, **34**, 549.

Shelanski, M., and E. W. Taylor (1968). *J. Cell Biol.*, **38**, 304.

Slautterbach, D., (1963). *J. Cell Biol.*, **18**, 367.

Stephens, R. E., (1967). *J. Cell Biol.*, **32**, 255.

Stephens, R. E., (1968a). *J. Mol. Biol.*, **32**, 277.

Stephens, R. E., (1968b). *Symp. Soc. Exptl. Biol.*, **XXII**, 43.

Stephens, R. E., (1968c). *J. Mol. Biol.*, **33**, 517.

Stephens, R. E., (1969). *Quart. Rev. Biophys.*, **I**, 377.

Stephens, R. E. (1970a). *J. Mol. Biol.*, **47**, 353.

Stephens, R. E. (1970b). *Science*, **168,** 845.

Stephens, R. E. (1970c). *Biol. Bull.*, **139,** 438.

Stephens, R. E., and E. E. Levine (1970). *J. Cell Biol.*, **46,** 416.

Stephens, R. E., and R. W. Linck (1969). *J. Mol. Biol.*, **40,** 497.

Stephens, R. E., F. L. Renaud, and I. R. Gibbons (1967). *Science*, **156,** 1606.

Stevens, C. L., and M. A. Lauffer (1965). *Biochemistry*, **4,** 31.

Thomas, M. B. (1970). *Biol. Bull.*, **138,** 219.

Tilney, L. G. (1968). *J. Cell Sci.*, **3,** 549.

Tilney, L. G., and B. Byers (1969). *J. Cell Biol.*, **43,** 148.

Tilney, L. G., and J. R. Gibbins (1969). *J. Cell Biol.*, **41,** 227.

Tilney, L. G., and K. R. Porter (1965a). *Protoplasma*, **60,** 317.

Tilney, L. G., and K. R. Porter (1965b). *J. Cell Biol.*, **34,** 327.

Tilney, L. G., Y. Hiramoto, and D. Marsland (1966). *J. Cell Biol.*, **29,** 77.

Tsuboi, K. K. (1968). *Biochim. Biophys. Acta*, **160,** 420.

Warren, R. H. (1968). *J. Cell Biol.*, **39,** 544.

Warren, R. H. (1968). *J. Cell Biol.*, **39,** 544.

Weisenberg, R. C., and E. W. Taylor (1968). *Exptl. Cell Res.*, **53,** 372.

Weisenberg, R. C., and S. N. Timasheff (1970). *Biochemistry*, **9,** 4094.

Weisenberg, R. C., G. G. Borisy, and E. W. Taylor (1968). *Biochemistry*, **7,** 4466.

Wilson, H. J. (1969). *J. Cell Biol.*, **40,** 854.

Witman, G. (1970). *J. Cell Biol.*, **47,** 229a.

Yanagisawa, T., S. Hasegawa, and H. Mohri (1968). *Exptl. Cell Res.*, **52,** 86.

AUTHOR INDEX

Numbers in parentheses are reference numbers and indicate that an author's work is referred to although his name is not cited in the text. Numbers in italics give the page on which the complete reference is listed.

A

Adelman, M., 304(83), *322*
Adelstein, R. S., 232(70), *257*
Afzelius, B. A., 331, 356, *352, 389*
Aktipis, S., 67(16), 72(16), *102*
Albert, M., 4(23), *51*
Allen, R. D., 360, *389*
Ambler, R. P., 232(71), *257*
Amiconi, G., 82(33), 100(33), *103*
Anderer, F. A., 150(3), 152(3, 15), *198*
Anderson, T., 22(53), *52*
André, J., 357, 359, *389*
Andree, P. J., 39(98), 45(98), 48, *53*
Ansevin, A. T., 150(5), 152, 163, 188(5), *198*
Antonini, E., 82(33), 100(33), *103*
Appel, P., 137, 240(90), *148, 257*
Asai, H., 277(31), 283(38a), 302, 311(87), *321, 322*
Asakura, S., 262(5), 266(8, 10, 14), 267(10, 11, 18), 269(5, 8, 15), 273(5), 282(37), 286(44), 289(58), 290, 291(58), 292(8, 10, 14, 44), 294(64), 295(8), 296(11), 297(5, 70), 298(5, 72), 300, 313(37), 315(70), *320, 321, 322*
Ashhurst, D. E., 239(89), 346(64), 349(64), 350(64), *257, 352*
Auber, J., 331, 346(42), 348(42, 43), 349(81), 350(81), *352, 353*

B

Baccetti, B., 330(36), *352*
Bailey, K., 254(126), *258*
Bailin, G., 227(52), 230(52), 238(52), *257*
Baker, H., 207(36), 209(36), 211(36), 213–226(36), 251(36), *256*
Bálint, M., 253(119), *258*
Baltimore, D., 238(87), *257*
Bancroft, C., 16(113), *53*
Banerjee, K., 150(4), 152, 153(19), 163(4),

169, 170, 175, 176, 177, 181, 183, 188(19), 189, *198, 199*
Bannister, W. H. 5(29), *51*
Bárány, K., 227(48, 52), 230(52), 232(74), 233(75), 238(52), 239(88), 286(42), 292(61), 298(73), *256, 257*
Barber, A. A., 48, *53*
Barnicot, N. A., 357, *389*
Bartell, L. S., 141(62), *148*
Bartels, W. J., 29(75), 47(107), *52, 53*
Bates, G., 82(33), 100(33), *103*
Bates, R. G., 141(61), *148*
Baxandall, J., 368, 370, 386, 374, *389*
Bayer, E., 4(20), 5(20), 45, *51*
Baylar, R. B., 22(54), *52*
Berns, D. S., 106(11, 12), 109(15, 16), 110(20), 111(11, 23), 112(23), 113(11, 20, 24, 26), 114(11), 115(20, 23, 26), 115(27), 116(24), 118(24, 38), 119(11, 20, 24), 120(20), 121(20, 31), 122(32), 123(32, 33), 124(20), 125(32, 34), 126(32, 34, 35, 36), 127(20, 36, 37, 38), 129, 130(35), 131(35, 36), 132(36), 133(20, 33, 34), 134(33, 34), 135(16, 33), 135(48), 136(12, 26, 49), 137(49), 138(12, 26, 49), 139(12, 49), 140(20, 27), 142(49), 143(16, 48), 144(16, 33), 145(16, 48), 146(11, 15, 16, 33, 35), 146(73), 147, *147, 148*
Beuverg, E. C., 42(92), *52*
Beychock, S., 43(94), *53*
Bibring, T., 368, 370, 386, 374, *389*
Bikle, D., 356, *389*
Biró, N. A., 253(119), 292(61), *258, 321*
Blanchard, M. H., 231(65), 232(65), *257*
Blasius, W., 2, *50*
Blaton, V., 4(23), 45(97), 46(97), *51*
Blazsó, M., 253(119), *258*
Bloomfield, V., 33(78, 79), *52*
Blout, E. R., 196, *199*
Blow, D. M., 238(86), *257*

P

Paabo, M., 141, *148*
Padan, E., 133, 146, *148*
Page, S. G. 251(114), 344(59), 346(62), 348(62), 349(62), *258*, *352*
Paglini, S., 143(67), 153, 167, 175, 181, 182, 183, 195, *148*, *198*, *199*
Palmer, G., 72(22), *103*
Pardee, A. B., 214(24), *256*
Pautard, F. G., 358, *390*
Peachy, L. D., 323(6), *351*
Pease, D. C., 357, 359, *390*
Pecci, J., 115(29), 127(29), *148*
Pedersen, K. O., 10(39), 13, 18, *51*
Peeters, B., 4(23), *51*
Pepe, F. A., 249(104), 251(104), 323(3, 9), 325(9, 12, 18), 326(21, 22), 327(21, 22), 328(21, 22), 329(9, 12, 18, 21, 22), 330(21, 22), 331(12, 18, 21, 22), 333(9, 12, 18, 21, 22), 334(12, 18, 21), 335(18, 22), 336(12, 18, 22), 337(22), 342(9, 12, 18, 21, 22), 343, 345(18, 22), 347(9, 12, 21, 69), 348(21, 22, 75), 350(12), *258*, *351*, *352*
Perry, S. V., 202(4), 218(37), 232(4, 68), 253(124), 326(27), 375, *255*, *256*, *257*, *258*, *351*, *390*
Perutz, M. F., 56(3, 4), 68(3, 4), *102*
Peticolas, W. L., 82(31), *103*
Phillipi, E., 4(15), 10(15), *50*
Philpott, D. E., 214(30), 243(30), *256*
Pickett, S. M., 42(120), *53*
Pilbrow, J. R., 45(100), *53*
Pilson, M. E. Q., 3(11), *50*
Polson, A. G., 22(55), *52*
Porath, J., 106(6), 110(6), *147*
Porter, K. R., 347(67), 348(67), 355, 356, 357, 362, 378, *352*, *389*, *390*
Porter, R. R., 213(23), 238(23), *256*
Premsela, H. F., 23, *52*
Prevost, S., 285(386), *321*
Pringle, S. W. S., 323(2), 336(2), 342, *351*
Printz, M. P., 16(114), *53*
Prior, G., 313(95), *322*
Prosser, C. L., 46(104), *53*
Putney, F., 86(40, 41), *103*

Q

Quaife, M. L., 22(54), *52*

R

Raffery, M. A., 112(70), *148*
Randall, J. T., 370, *390*
Redfield, A. C., 2(1), 36, 40(85, 86, 87, 88, 89), 42(81, 88), *50*, *52*
Redmond, J. P., 42(119), *53*
Redmond, J. R., 2(3), 47(105), *50*, *53*
Reedy, M. K., 251(109), 329(35), 330(35), 336(35), 339(54, 55), 345(54, 55), 349 *258*, *352*, *353*
Rees, M. K., 286(46), 375, *321*, *390*
Rees, M. W., 232(71), *257*
Reger, J. F., 331, 332(44), 346(41), 348(41), *352*
Renaud, F. L., 362, 363, 364, 366, 367, 369, 374, 375, 379, 380, 384, *390*, *391*
Reuss, E., 3(10), 5(10), *50*
Rice, S., 208(11), 210(13), 212(13, 20), 347(71, 72), 349, 350(71, 72), *255*, *256*, *352*, *353*
Rich, A., 238(84), *257*
Richards, E. G., 227(54), 240(90), *257*
Riggs, A. F., 42(120), 113, 119, 123, 124(25), 125, 126(25), 129, *53*, *147*
Rill, R. L., 81, *103*
Ringo, D. L., 357, 379, *390*
Riordan, J. F., 81(27), *103*
Risby, D., 239(139), *259*
Rizzino, A. A., 233(80), *257*
Robinson, R. A., 141(61), *148*
Rome, E., 252(115), *258*
Rosemeyer, M. A., 88(47), *103*
Rosenbaum, J. L., 356, 367, 385, 379, *389*, *390*
Rosenbluth, R., 281(32), 282(32), *321*
Roskos, R. R., 141(62), *148*
Roth, L. E., 356, *390*
Rowe, A. J., 360, 363, 364, 366, 367, 369, 374, 382, *389*, *390*
Ruddle, F., 356, 367, 385, *390*
Rüdiger, W., 106, 113(9), *147*

S

Sakai, H., 357, 368, *390*
Sakakibara, I., 286(49), 311(87), *321*, *322*
Salem, L., 142(64), *148*
Samosudova, N. V., 347(70), *352*
Sarkar, S., 238(137), *259*
Satir, P., 361, *390*

SUBJECT INDEX

A

Actin, 261–322
  amino acid composition, 287
  carboxymethylated, renaturation of, 296–297
  chemically modified, polymers of, 309–310
  effect of environments, 291–297
    organic solvents, 295–296
    pressure, 295
    salts and pH, 291–292
    temperature, 292–295
  F-actin ultrastructure, 275–277
  G-actin, plasmodium type, 304
    structure, 286–289
  G-F transformation and splitting of ATP, 262–264
  G- and F-actins, dynamic balance between, 264–266
  interaction with other muscle proteins, 310–317
    myosin, 232–235, 310–315
    tropomyosin, 315–317
  microtubules and, 374–377
  monomer structure, 286–291
    interaction with divalent cations, 291
    interaction with nucleotides, 289–290
  myosin binding of, 232–235
  physiological functions of, 317–318
  plasmodium type, 304–310
    polymers, 304–309
  polymer
    ATP splitting, 302–304
    helical form, 277–279
    polymorphism, 297–298
    sonic vibration effects, 298–302
    structure, 275–279
  polymerization, 262–269
    as condensation phenomenon, 266–269
    cooperative nature, 269
    divalent cations in, 285
    kinetics, 273–275
    nucleotides as regulators, 282–285
      exchangeability, 285–286
  regulation, 279–286
  theoretical aspects, 269–273
  without nucleotide participation, 279–282
Algae, phycocyanins in, 106, 109
Apohemocyanin, 35
  formation, 4
A-protein of tobacco mosaic virus, see under Tobacco mosaic virus protein
Araneida, hemocyanin occurrence in, 3
Arthropoda, hemocyanin of, 2, 12–14, 48–49
  molecular architecture, 23–29
  oxygen binding by, 42
ATPases, effect on microtubules, 359–362

B

Brachiopods, hemerythrin in, 100

C

Callinectes sapidus, hemocyanin from, 4
Ceruloplasmin, comparison with hemocyanins, 39
Chlorocruorin, 46
Copper, binding by hemocyanins, 4–6, 10
Copper proteins, comparison with hemocyanins, 39
Crab, hemocyanin from, 3

D

Deaminase, hemocyanin comparison with, 9
Decapoda, hemocyanin occurrence in, 3
Deoxyhemerythrin
  Mössbauer spectra, 77
  structure, 79
Deuterated proteins, 135–147
Dynein, role in microtubules, 359–362

G

Glycoproteins, hemocyanins as, 8–10
Goldfingia gouldii, hemerythrin studies on, 57–101
Guanine nucleotides, in microtubule systems, 362–364